Springer Series in Statistics

Advisors:

Rob J. Hyndman, Anne B. Koehler,
J. Keith Ord and Ralph D. Snyder

Forecasting with Exponential Smoothing

The State Space Approach

Springer

Professor Rob Hyndman
Department of Econometrics & Business Statistics
Monash University
Clayton VIC 3800
Australia
Rob.Hyndman@buseco.monash.edu.au

Professor Anne Koehler
Department of Decision Sciences & Management Information Systems
Miami University
Oxford, Ohio 45056
USA
koehleab@muohio.edu

Professor Keith Ord
McDonough School of Business
Georgetown University
Washington DC 20057
USA
ordk@georgetown.edu

Associate Professor Ralph Snyder
Department of Econometrics & Business Statistics
Monash University
Clayton VIC 3800
Australia
Ralph.Snyder@buseco.monash.edu.au

ISBN 978-3-540-71916-8 ISBN 978-3-540-71918-2 (eBook)

Library of Congress Control Number: 2008924784

Cover design: Deblik, Berlin, Germany

9 8 7 6 5 4 3 2 1

springer.com

Preface

Exponential smoothing methods have been around since the 1950s, and are still the most popular forecasting methods used in business and industry. Initially, a big attraction was the limited requirements for computer storage. More importantly today, the equations in exponential smoothing methods for estimating the parameters and generating the forecasts are very intuitive and easy to understand. As a result, these methods have been widely implemented in business applications.

However, a shortcoming of exponential smoothing has been the lack of a statistical framework that produces both prediction intervals and point forecasts. The innovations state space approach provides this framework while retaining the intuitive nature of exponential smoothing in its measurement and state equations. It provides prediction intervals, maximum likelihood estimation, procedures for model selection, and much more.

As a result of this framework, the area of exponential smoothing has undergone a substantial revolution in the past ten years. The new innovations state space framework for exponential smoothing has been discussed in numerous journal articles, but until now there has been no systematic explanation and development of the ideas. Furthermore, the notation used in the journal articles tends to change from paper to paper. Consequently, researchers and practitioners struggle to use the new models in applications. In writing this book, we have attempted to compile all of the material related to innovations state space models and exponential smoothing into one coherent presentation. In the process, we have also extended results, filled in gaps and developed totally new material. Our goal has been to provide a comprehensive exposition of the innovations state space framework for forecasting time series with exponential smoothing.

Outline of the Book

We have written this book for people wanting to apply exponential smoothing methods in their own area of interest, as well as for researchers wanting to take the ideas in new directions. In attempting to cater for this broad audience, the book has been structured into four parts, providing increasing levels of detail and complexity.

Part I: Introduction (Chaps. 1 and 2)

If you only want a snack, then read Part I. It provides an overview of our approach to forecasting and an introduction to the state space models that underlie exponential smoothing. You will then be able to appreciate how to implement exponential smoothing in the statistical framework of innovations state space models.

Chapter 1 includes some general information on forecasting time series and provides an historical context. In Chap. 2, we establish the linkage between standard exponential smoothing methods and the innovations state space models. Then, we describe all parts of the forecasting process using innovations state space models in the following order: initialization, estimation, forecasting, evaluation of forecasts, model selection, and an automatic procedure for the entire process which includes finding prediction intervals.

Part II: Essentials (Chaps. 3–7)

Readers wanting a more substantial meal should go on to read Chaps. 3–7. They fill out many of the details and provide links to the most important papers in the literature. Anyone finishing the first seven chapters will be ready to begin using the models for themselves in applied work.

We examine linear models more closely in Chap. 3, before adding the complexity of nonlinear and heteroscedastic models in Chap. 4. These two chapters also introduce the concepts of stationarity, stability, and forecastability. Because the linear models are a subset of the general innovations state space model, the material on estimation (Chap. 5), prediction (Chap. 6), and model selection (Chap. 7) relates to the general model, with considerations of linear models and other special subgroups where informative.

Part III: Further Topics (Chaps. 8–17)

If you want the full banquet, then you should go on to read the rest of the book. Chapters 8–17 provide more advanced considerations of the details of the models, their mathematical properties, and extensions of the models. These chapters are intended for people wanting to understand the modeling framework in some depth, including other researchers in the field.

We consider the normalization of seasonal components in Chap. 8, and the addition of regressors to the model in Chap. 9. In Chap. 10, we address the important issue of parameter space specification, along with the concept of the minimal dimension of a model. The relationship with other standard time series models is investigated. In particular, Chap. 11 looks at ARIMA models, and Chap. 13 examines conventional state space models, which have multiple sources of randomness. An information filter for estimating the parameters in a state space model with a random seed vector is detailed in Chap. 12. The advantages of the information filter over the Kalman filter, which was originally developed for stationary data, are explained. The remaining four chapters address special issues and models for specific types of time series as follows: time series with multiple seasonal patterns in Chap. 14, time series with strictly positive values in Chap. 15, count data in Chap. 16, and vectors of time series in Chap. 17.

Part IV: Applications (Chaps. 18–20)
The final part of the book provides the after-dinner cocktails and contains applications to inventory control, economics and finance.

These applications are intended to illustrate the potentially wide reach and usefulness of the innovations state space models. Procedures for addressing the important inventory problems of nonstationary demand and the use of sales data when true demand is unknown are covered in Chap. 18 for a reorder inventory system. In Chap. 19, the natural implementation of conditional heteroscedasticity in the innovations state space models framework (i.e., a GARCH-type model) is shown and applied to examples of financial time series. In Chap. 20, the Beveridge-Nelson decomposition of a univariate time series into transitory and permanent components is presented in the linear innovations state space framework. The advantages of this formulation over other approaches to the Beveridge-Nelson decomposition are explained.

Website

The website `http://www.exponentialsmoothing.net` provides supplementary material for this book, including data sets, computer code, additional exercises, and links to other resources.

Forecasting Software

Time series forecasting is not a spectator sport, and any serious forecaster needs access to adequate computing power. Most of the analyses presented in this book can readily be performed using the **forecast** package for

R (Hyndman 2007), which is available on CRAN (`http://cran.r-project.org/`). All of the data in the book are available in the **expsmooth** package for **R**. In addition, we provide **R** code at `http://www.exponentialsmoothing.net` for producing most of the examples in the book.

Acknowledgements

No writing project of this size is undertaken without assistance from many people. We gratefully acknowledge the contributions of several colleagues who co-authored individual chapters in the book. Their collective expertise has greatly added to the depth and breadth of coverage of the book. They are:

Muhammad Akram	Chap. 15;
Heather Anderson	Chap. 20;
Ashton de Silva	Chap. 17;
Phillip Gould	Chap. 14;
Chin Nam Low	Chap. 20;
Farshid Vahid-Araghi	Chap. 14.

We are particularly grateful to Cathy Morgan for her careful copyediting work throughout. We also received valuable suggestions from Andrey Kostenko and programming assistance from Adrian Beaumont. Their attention to detail has been greatly appreciated.

Each of us owes gratitude to our universities for providing excellent environments in which to work, and the research facilities necessary to write this book. Monash University was especially helpful in providing the opportunity for all four authors to spend some time together at crucial points in the process.

We would also like to thank Lilith Braun from Springer for keeping us on-track and for making the finished product possible.

Finally, we are each thankful to our families and friends for their support, even when we were neglectful and distracted. We do appreciate it.

Melbourne, Australia,	*Rob J. Hyndman*
Oxford, Ohio, USA,	*Anne B. Koehler*
Washington DC, USA,	*J. Keith Ord*
Melbourne, Australia,	*Ralph D. Snyder*

February 2008

Contents

Part I

Introduction

1

Basic Concepts

1.1 Time Series Patterns

Time series arise in many different contexts including minute-by-minute stock prices, hourly temperatures at a weather station, daily numbers of arrivals at a medical clinic, weekly sales of a product, monthly unemployment figures for a region, quarterly imports of a country, and annual turnover of a company. That is, time series arise whenever something is observed over time. While a time series may be observed either continuously or at discrete times, the focus of this book is on discrete time series that are observed at regular intervals over time.

A graph of a time series often exhibits patterns, such as an upward or downward movement (trend) or a pattern that repeats (seasonal variation), that might be used to forecast future values. Graphs of four time series that display such features are presented in Fig. 1.1.

- Figure 1.1a shows 125 monthly US government bond yields (percent per annum) from January 1994 to May 2004. This time series appears to have a changing level with a downward drift that one would be reluctant to forecast as continuing into the future, and it seems to have no discernable seasonal pattern.
- Figure 1.1b displays 55 observations of annual US net electricity generation (billion kwh) for 1949 through 2003. This time series contains a definite upward trend that changes somewhat over time.
- Figure 1.1c presents 113 quarterly observations of passenger motor vehicle production in the UK (thousands of cars) for the first quarter of 1977 through the first quarter of 2005. For this time series there is a constant variation around a changing level. As with Fig. 1.1a, there is no trend that one would want to forecast as continuing into the future. However, there is a possibility of a seasonal pattern.
- Figure 1.1d shows 240 monthly observations of the number of short term overseas visitors to Australia from May 1985 to April 2005. There is a

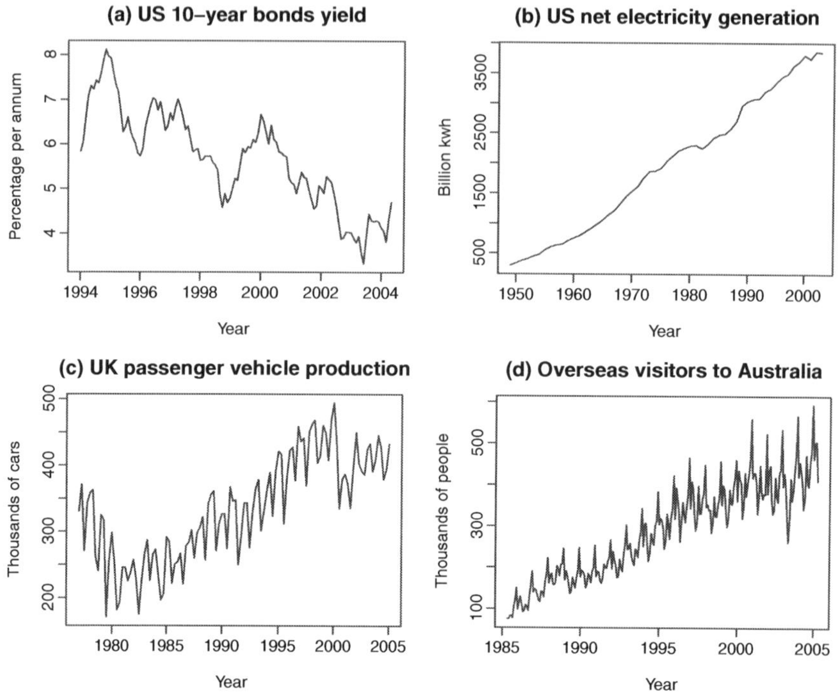

Fig. 1.1. Four time series showing patterns typical of business and economic data.

definite seasonal pattern in this time series, and the variation increases as the level of the time series increases. It is not possible to tell visually whether the increase is due to an increase in the seasonal fluctuations or is caused by some other factors. While there is an upward drift, it might not be a good idea to forecast it as continuing into the future.

From these few examples it is clear that there is frequently a need for forecasting that takes into account trend, seasonality, and other features of the data. Specifically, we are interested in the situation where we observe a time series $y_1, \ldots, y_n$, and we wish to forecast a future observation at time $n + h$. In order to exploit the patterns like those in Fig. 1.1, many different forecasting methods and models have been proposed.

1.2 Forecasting Methods and Models

A forecasting *method* is an algorithm that provides a point forecast: a single value that is a prediction of the value at a future time period. On the other hand, a statistical *model* provides a stochastic data generating process that may be used to produce an entire probability distribution for a future time period $n + h$. A point forecast can then be obtained easily by taking the mean (or median) of the probability distribution. A model also allows the computation of prediction (forecast) intervals with a given level of confidence.

We use the notation $\hat{y}_{n+h|n}$ to denote a point forecast of y_{n+h} using the information available at time n. This notation does not need to distinguish between point forecasts that arise from forecasting methods and those that are derived from statistical models, because the statistical models will lead directly to point forecasting methods.

1.3 History of Exponential Smoothing

Historically, exponential smoothing describes a class of forecasting *methods*. In fact, some of the most successful forecasting methods are based on the concept of exponential smoothing. There are a variety of methods that fall into the exponential smoothing family, each having the property that forecasts are weighted combinations of past observations, with recent observations given relatively more weight than older observations. The name "exponential smoothing" reflects the fact that the weights decrease exponentially as the observations get older.

The idea seems to have originated with Robert G. Brown in about 1944 while he was working for the US Navy as an Operations Research analyst. He used the idea in a mechanical computing device for tracking the velocity and angle used in firing at submarines (Gardner 2006). In the 1950s he extended this method from continuous to discrete time series, and included terms to handle trend and seasonality. One of his first applications was forecasting demand for spare parts in the US Navy inventory system. This latter work was presented at a meeting of the Operations Research Society of America in 1956 and formed the basis of his first book on inventory control (Brown 1959). The ideas were further developed in Brown's second book (1963).

Independently, Charles Holt was also working on an exponential smoothing method for the US Office of Naval Research (ONR). Holt's method differed from Brown's with respect to the smoothing of the trend and seasonal components. His original work was reproduced in an ONR memorandum (Holt 1957), which has been very widely cited, but was unpublished until recently when it appeared in the *International Journal of Forecasting* in 2004. Holt's work on additive and multiplicative seasonal exponential smoothing became well known through a paper by his student Peter Winters (1960) which provided empirical tests for Holt's methods. As a result, the seasonal versions of Holt's methods are usually called Holt-Winters' methods (and sometimes just Winters' methods, which is rather unfair to Holt).

Another of Holt's collaborators was John Muth, who later became famous in economics for formulating the concept of rational expectations. In exponential smoothing he is known for introducing two statistical models

(Muth 1960) for which the optimal forecasts are equivalent to those obtained from simple exponential smoothing.

Muth's models were the first in a long series of statistical models that are related to forecasting using exponential smoothing. The success of the exponential smoothing methods for forecasting, and for controlling inventory, has resulted in many researchers looking for models that produce the same point forecasts as these methods. Many of these models, including those of Muth, are state space models for which the minimum mean squared error forecasts are the forecasts from simple exponential smoothing.

1.4 State Space Models

State space models allow considerable flexibility in the specification of the parametric structure. In this book, we will use the *innovations* formulation of the model (e.g., Anderson and Moore 1979; Aoki 1987; Hannan and Deistler 1988). Let y_t denote the observation at time t, and let $\boldsymbol{x}_t$ be a "state vector" containing unobserved components that describe the level, trend and seasonality of the series. Then a linear innovations state space model can be written as

$$y_t = \boldsymbol{w}'\boldsymbol{x}_{t-1} + \varepsilon_t, \tag{1.1a}$$
$$\boldsymbol{x}_t = \boldsymbol{F}\boldsymbol{x}_{t-1} + \boldsymbol{g}\varepsilon_t, \tag{1.1b}$$

where $\{\varepsilon_t\}$ is a white noise series and $\boldsymbol{F}$, $\boldsymbol{g}$ and $\boldsymbol{w}$ are coefficients. Equation (1.1a) is known as the *measurement* (or observation) equation; it describes the relationship between the unobserved states $\boldsymbol{x}_{t-1}$ and the observation y_t. Equation (1.1b) is known as the *transition* (or state) equation; it describes the evolution of the states over time. The use of identical errors (or innovations) in these two equations makes it an "innovations" state space model. Several exponential smoothing methods are equivalent to point forecasts of special cases of model (1.1); examples are given in Sect. 2.5.

The philosophy of state space models fits well with the approach of exponential smoothing because the level, trend and seasonal components are stated explicitly in the models. In contrast, one cannot see these components as easily in autoregressive integrated moving average (ARIMA) models (Box et al. 1994).

Nonlinear state space models are also possible. One form that we use in Chap. 2 is

$$y_t = w(\boldsymbol{x}_{t-1}) + r(\boldsymbol{x}_{t-1})\varepsilon_t, \tag{1.2a}$$
$$\boldsymbol{x}_t = f(\boldsymbol{x}_{t-1}) + g(\boldsymbol{x}_{t-1})\varepsilon_t. \tag{1.2b}$$

An alternative, and more common, specification is to assume that the errors in the two equations are mutually independent. That is, $\boldsymbol{g}\varepsilon_t$ in (1.1b) is replaced by z_t, where z_t consists of independent white noise series that are also independent of ε_t, the error in the measurement equation. The assumption that z_t and ε_t are independent provides enough constraints to ensure that the remaining parameters are *estimable* (termed *just identified* in the econometrics literature).

There are an infinite number of ways in which the parameter space could be constrained to achieve estimability. The purpose of this book is to present the theory and applications of the *innovations* formulation, wherein all of the error sources are perfectly correlated. In some papers, these are known as *single source of error* (SSOE) models (e.g., Ord et al. 1997). By contrast, we refer to the more common form of the state space model as having *multiple sources of error* (MSOE).

At first it may seem that innovations state space models are more restrictive than MSOE models, but this is not the case. In fact, the reverse is true. Any linear MSOE model can be written in innovations form, and any linear innovations model can be written in MSOE form. However, the innovations models can have a larger parameter space. The innovations models have several other advantages over the models with multiple sources of error, as will be seen in Chap. 13.

Moreover, MSOE state space models, like ARIMA models, are linear models that require both the components and the error terms to be additive. While nonlinear versions of both MSOE and ARIMA models exist, these are much more difficult to work with. In contrast, it is relatively easy to use a nonlinear innovations state space model for describing and forecasting time series data and we will use them frequently in this book.

MSOE models that are similar to the types of models considered in this book include dynamic linear models (Harrison and Stevens 1976; Duncan and Horn 1972; West and Harrison 1997) and structural models (Harvey 1989).

Modern work on state space models began with Kalman (1960) and Kalman and Bucy (1961), following which a considerable body of literature developed in engineering (e.g., Jazwinski 1970; Anderson and Moore 1979). Early work in the statistical area included the Markovian representation developed by Akaike (1973, 1974). Hannan and Deistler (1988) provided a unifying presentation of the work by engineers and statistical time series analysts for stationary time series. In economics, Aoki and Havenner (1991) looked at multivariate state space models and suggested procedures for both stationary and nonstationary data. For a review of the books in the area, see Durbin and Koopman (2001, p. 5).

2

Getting Started

Although exponential smoothing methods have been around since the 1950s, a modeling framework incorporating stochastic models, likelihood calculations, prediction intervals, and procedures for model selection was not developed until relatively recently, with the work of Ord et al. (1997) and Hyndman et al. (2002). In these (and other) papers, a class of state space models has been developed that underlies all of the exponential smoothing methods.

In this chapter, we provide an introduction to the ideas underlying exponential smoothing and the associated state space models. Many of the details will be skipped over in this chapter, but will be covered in later chapters.

Figure 2.1 shows the four time series from Fig. 1.1, along with point forecasts and 80% prediction intervals. These were all produced using exponential smoothing state space models. In each case, the particular models and all model parameters were chosen automatically with no intervention by the user. This demonstrates one very useful feature of state space models for exponential smoothing—they are easy to use in a completely automated way. In these cases, the models were able to handle data exhibiting a range of features, including very little trend, strong trend, no seasonality, a seasonal pattern that stays constant, and a seasonal pattern with increasing variation as the level of the series increases.

2.1 Time Series Decomposition

It is common in business and economics to think of a time series as a combination of various components such as the trend (T), cycle (C), seasonal (S), and irregular or error (E) components. These can be defined as follows:

Trend (T):	The long-term direction of the series
Seasonal (S):	A pattern that repeats with a known periodicity (e.g., 12 months per year, or 7 days per week)

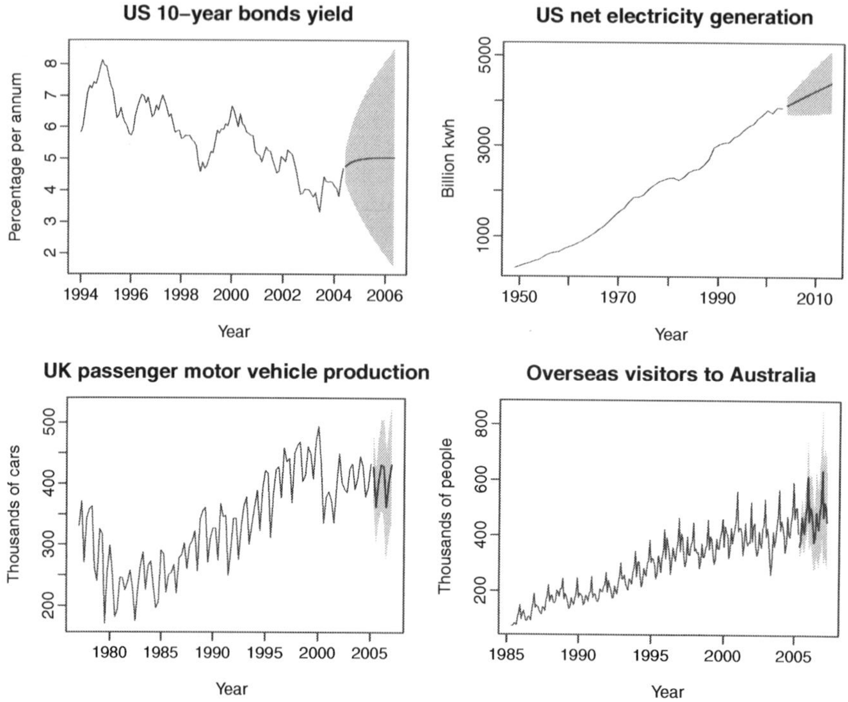

Fig. 2.1. Four time series showing point forecasts and 80% prediction intervals obtained using exponential smoothing state space models.

Cycle (C):	A pattern that repeats with some regularity but with unknown and changing periodicity (e.g., a business cycle)
Irregular or error (E):	The unpredictable component of the series

In this monograph, we focus primarily upon the three components T, S and E. Any cyclic element will be subsumed within the trend component unless indicated otherwise.

These three components can be combined in a number of different ways. A purely additive model can be expressed as

$$y = T + S + E,$$

where the three components are added together to form the observed series. A purely multiplicative model is written as

$$y = T \times S \times E,$$

where the data are formed as the product of the three components. A *seasonally adjusted* series is then formed by extracting the seasonal component from the data, leaving only the trend and error components. In the additive model, the seasonally adjusted series is $y - S$, while in the multiplicative model, the seasonally adjusted series is y/S. The reader should refer to Makridakis et al. (1998, Chap. 4) for a detailed discussion of seasonal adjustment and time series decomposition.

Other combinations, apart from simple addition and multiplication, are also possible. For example,

$$y = (T + S) \times E$$

treats the irregular component as multiplicative but the other components as additive.[1]

2.2 Classification of Exponential Smoothing Methods

In exponential smoothing, we always start with the trend component, which is itself a combination of a level term (ℓ) and a growth term (b). The level and growth can be combined in a number of ways, giving five future trend types. Let T_h denote the forecast trend over the next h time periods, and let ϕ denote a damping parameter ($0 < \phi < 1$). Then the five trend types or growth patterns are as follows:

$$\begin{aligned} \text{None:} \quad & T_h = \ell \\ \text{Additive:} \quad & T_h = \ell + bh \\ \text{Additive damped:} \quad & T_h = \ell + (\phi + \phi^2 + \cdots + \phi^h)b \\ \text{Multiplicative:} \quad & T_h = \ell b^h \\ \text{Multiplicative damped:} \quad & T_h = \ell b^{(\phi + \phi^2 + \cdots + \phi^h)} \end{aligned}$$

A damped trend method is appropriate when there is a trend in the time series, but one believes that the growth rate at the end of the historical data is unlikely to continue more than a short time into the future. The equations for damped trend do what the name indicates: dampen the trend as the length of the forecast horizon increases. This often improves the forecast accuracy, particularly at long lead times.

Having chosen a trend component, we may introduce a seasonal component, either additively or multiplicatively. Finally, we include an error, either additively or multiplicatively. Historically, the nature of the error component has often been ignored, because the distinction between additive and multiplicative errors makes no difference to point forecasts.

If the error component is ignored, then we have the fifteen exponential smoothing methods given in the following table. This classification of methods originated with Pegels' (1969) taxonomy. This was later extended by

[1] See Hyndman (2004) for further discussion of the possible combinations of these components.

Trend component	Seasonal component		
	N (None)	A (Additive)	M (Multiplicative)
N (None)	N,N	N,A	N,M
A (Additive)	A,N	A,A	A,M
A_d (Additive damped)	A_d,N	A_d,A	A_d,M
M (Multiplicative)	M,N	M,A	M,M
M_d (Multiplicative damped)	M_d,N	M_d,A	M_d,M

Gardner (1985), modified by Hyndman et al. (2002), and extended again by Taylor (2003a), giving the fifteen methods in the above table.

Some of these methods are better known under other names. For example, cell (N,N) describes the simple exponential smoothing (or SES) method, cell (A,N) describes Holt's linear method, and cell (A_d,N) describes the damped trend method. Holt-Winters' additive method is given by cell (A,A), and Holt-Winters' multiplicative method is given by cell (A,M). The other cells correspond to less commonly used but analogous methods.

For each of the 15 methods in the above table, there are two possible state space models, one corresponding to a model with additive errors and the other to a model with multiplicative errors. If the same parameter values are used, these two models give equivalent point forecasts although different prediction intervals. Thus, there are 30 potential models described in this classification.

We are careful to distinguish exponential smoothing *methods* from the underlying state space *models*. An exponential smoothing method is an algorithm for producing point forecasts only. The underlying stochastic state space model gives the same point forecasts, but also provides a framework for computing prediction intervals and other properties. The models are described in Sect. 2.5, but first we introduce the much older point-forecasting equations.

2.3 Point Forecasts for the Best-Known Methods

In this section, a simple introduction is provided to some of the best-known exponential smoothing methods—simple exponential smoothing (N,N), Holt's linear method (A,N), the damped trend method (A_d,N) and Holt-Winters' seasonal method (A,A and A,M). We denote the observed time series by $y_1, y_2, \dots, y_n$. A forecast of y_{t+h} based on all the data up to time t is denoted by $\hat{y}_{t+h|t}$. For one-step forecasts, we use the simpler notation $\hat{y}_{t+1} \equiv \hat{y}_{t+1|t}$. Usually, forecasts require some parameters to be estimated; but for the sake of simplicity it will be assumed for now that the values of all relevant parameters are known.

2.3.1 Simple Exponential Smoothing (N,N Method)

Suppose we have observed data up to and including time $t-1$, and we wish to forecast the next value of our time series, y_t. Our forecast is denoted by $\hat{y}_t$. When the observation y_t becomes available, the forecast error is found to be $y_t - \hat{y}_t$. The method of simple exponential smoothing,[2] due to Brown's work in the mid-1950s and published in Brown (1959), takes the forecast for the previous period and adjusts it using the forecast error. That is, the forecast for the next period is

$$\hat{y}_{t+1} = \hat{y}_t + \alpha(y_t - \hat{y}_t), \tag{2.1}$$

where α is a constant between 0 and 1.

It can be seen that the new forecast is simply the old forecast plus an adjustment for the error that occurred in the last forecast. When α has a value close to 1, the new forecast will include a substantial adjustment for the error in the previous forecast. Conversely, when α is close to 0, the new forecast will include very little adjustment.

Another way of writing (2.1) is

$$\hat{y}_{t+1} = \alpha y_t + (1-\alpha)\hat{y}_t. \tag{2.2}$$

The forecast $\hat{y}_{t+1}$ is based on weighting the most recent observation y_t with a weight value α, and weighting the most recent forecast $\hat{y}_t$ with a weight of $1-\alpha$. Thus, it can be interpreted as a weighted average of the most recent forecast and the most recent observation.

The implications of exponential smoothing can be seen more easily if (2.2) is expanded by replacing $\hat{y}_t$ with its components, as follows:

$$\begin{aligned}\hat{y}_{t+1} &= \alpha y_t + (1-\alpha)[\alpha y_{t-1} + (1-\alpha)\hat{y}_{t-1}] \\ &= \alpha y_t + \alpha(1-\alpha)y_{t-1} + (1-\alpha)^2\hat{y}_{t-1}.\end{aligned}$$

If this substitution process is repeated by replacing $\hat{y}_{t-1}$ with its components, $\hat{y}_{t-2}$ with its components, and so on, the result is

$$\begin{aligned}\hat{y}_{t+1} &= \alpha y_t + \alpha(1-\alpha)y_{t-1} + \alpha(1-\alpha)^2 y_{t-2} + \alpha(1-\alpha)^3 y_{t-3} \\ &\quad + \alpha(1-\alpha)^4 y_{t-4} + \cdots + \alpha(1-\alpha)^{t-1} y_1 + (1-\alpha)^t \hat{y}_1.\end{aligned} \tag{2.3}$$

So $\hat{y}_{t+1}$ represents a weighted moving average of all past observations with the weights decreasing exponentially; hence the name "exponential smoothing." We note that the weight of $\hat{y}_1$ may be quite large when α is small and the time series is relatively short. The choice of starting value then becomes particularly important and is known as the "initialization problem," which we discuss in detail in Sect. 2.6.

---

[2] This method is also sometimes known as "*single* exponential smoothing."

For longer range forecasts, it is assumed that the forecast function is "flat." That is,

$$\hat{y}_{t+h|t} = \hat{y}_{t+1}, \qquad h = 2, 3, \ldots .$$

A flat forecast function is used because simple exponential smoothing works best for data that have no trend, seasonality, or other underlying patterns.

Another way of writing this is to let $\ell_t = \hat{y}_{t+1}$. Then $\hat{y}_{t+h|t} = \ell_t$ and $\ell_t = \alpha y_t + (1-\alpha)\ell_{t-1}$. The value of ℓ_t is a measure of the "level" of the series at time t. While this may seem a cumbersome way to express the method, it provides a basis for generalizing exponential smoothing to allow for trend and seasonality.

In order to calculate the forecasts using SES, we need to specify the initial value $\ell_0 = \hat{y}_1$ and the parameter value α. Traditionally (particularly in the pre-computer age), $\hat{y}_1$ was set to be equal to the first observation and α was specified to be a small number, often 0.2. However, there are now much better ways of selecting these parameters, which we describe in Sect. 2.6.

2.3.2 Holt's Linear Method (A,N Method)

Holt (1957)[3] extended simple exponential smoothing to linear exponential smoothing to allow forecasting of data with trends. The forecast for Holt's linear exponential smoothing method is found using two smoothing constants, α and β^* (with values between 0 and 1), and three equations:

$$\text{Level:} \qquad \ell_t = \alpha y_t + (1-\alpha)(\ell_{t-1} + b_{t-1}), \tag{2.4a}$$

$$\text{Growth:} \qquad b_t = \beta^*(\ell_t - \ell_{t-1}) + (1-\beta^*)b_{t-1}, \tag{2.4b}$$

$$\text{Forecast:} \qquad \hat{y}_{t+h|t} = \ell_t + b_t h. \tag{2.4c}$$

Here ℓ_t denotes an estimate of the level of the series at time t and b_t denotes an estimate of the slope (or growth) of the series at time t. Note that b_t is a weighted average of the previous growth b_{t-1} and an estimate of growth based on the difference between successive levels. The reason we use β^* rather than β will become apparent when we introduce the state space models in Sect. 2.5.

In the special case where $\alpha = \beta^*$, Holt's method is equivalent to "Brown's double exponential smoothing" (Brown 1959). Brown used a discounting argument to arrive at his forecasting equations, so $1-\alpha$ represents the common discount factor applied to both the level and trend components.

In Sect. 2.6 we describe how the procedure is initialized and how the parameters are estimated.

---

[3] Reprinted as Holt (2004).

One interesting special case of this method occurs when $\beta^* = 0$. Then

$$\begin{aligned} \text{Level:} \quad & \ell_t = \alpha y_t + (1-\alpha)(\ell_{t-1} + b), \\ \text{Forecast:} \quad & \hat{y}_{t+h|t} = \ell_t + bh. \end{aligned}$$

This method is known as "SES with drift," which is closely related to the "Theta method" of forecasting due to Assimakopoulos and Nikolopoulos (2000). The connection between these methods was demonstrated by Hyndman and Billah (2003).

2.3.3 Damped Trend Method (A_d,A Method)

Gardner and McKenzie (1985) proposed a modification of Holt's linear method to allow the "damping" of trends. The equations for this method are:[4]

$$\begin{aligned} \text{Level:} \quad & \ell_t = \alpha y_t + (1-\alpha)(\ell_{t-1} + \phi b_{t-1}), & (2.5a) \\ \text{Growth:} \quad & b_t = \beta^*(\ell_t - \ell_{t-1}) + (1-\beta^*)\phi b_{t-1}, & (2.5b) \\ \text{Forecast:} \quad & \hat{y}_{t+h|t} = \ell_t + (\phi + \phi^2 + \cdots + \phi^h) b_t. & (2.5c) \end{aligned}$$

Thus, the growth for the one-step forecast of y_{t+1} is ϕb_t, and the growth is dampened by a factor of ϕ for each additional future time period. If $\phi = 1$, this method gives the same forecasts as Holt's linear method. For $0 < \phi < 1$, as $h \to \infty$ the forecasts approach an asymptote given by $\ell_t + \phi b_t/(1-\phi)$. We usually restrict $\phi > 0$ to avoid a negative coefficient being applied to b_{t-1} in (2.5b), and $\phi \leq 1$ to avoid b_t increasing exponentially.

2.3.4 Holt-Winters' Trend and Seasonality Method

If the data have no trend or seasonal patterns, then simple exponential smoothing is appropriate. If the data exhibit a linear trend, then Holt's linear method (or the damped method) is appropriate. But if the data are seasonal, these methods on their own cannot handle the problem well.

Holt (1957) proposed a method for seasonal data. His method was studied by Winters (1960), and so now it is usually known as "Holt-Winters' method" (see Sect. 1.3).

Holt-Winters' method is based on three smoothing equations—one for the level, one for trend, and one for seasonality. It is similar to Holt's linear method, with one additional equation for dealing with seasonality. In fact, there are two different Holt-Winters' methods, depending on whether seasonality is modeled in an additive or multiplicative way.

---

[4] We use the same parameterization as Gardner and McKenzie (1985), which is slightly different from the parameterization proposed by Hyndman et al. (2002). This makes no difference to the value of the forecasts.

Multiplicative Seasonality (A,M Method)

The basic equations for Holt-Winters' multiplicative method are as follows:

$$\begin{aligned}
\text{Level:} \quad & \ell_t = \alpha \frac{y_t}{s_{t-m}} + (1-\alpha)(\ell_{t-1} + b_{t-1}) && (2.6a)\\
\text{Growth:} \quad & b_t = \beta^*(\ell_t - \ell_{t-1}) + (1-\beta^*)b_{t-1} && (2.6b)\\
\text{Seasonal:} \quad & s_t = \gamma y_t/(\ell_{t-1} + b_{t-1}) + (1-\gamma)s_{t-m} && (2.6c)\\
\text{Forecast:} \quad & \hat{y}_{t+h|t} = (\ell_t + b_t h)s_{t-m+h_m^+}, && (2.6d)
\end{aligned}$$

where m is the length of seasonality (e.g., number of months or quarters in a year), ℓ_t represents the level of the series, b_t denotes the growth, s_t is the seasonal component, $\hat{y}_{t+h|t}$ is the forecast for h periods ahead, and $h_m^+ = [(h-1) \bmod m] + 1$. The parameters ($\alpha$, β^* and γ) are usually restricted to lie between 0 and 1. The reader should refer to Sect. 2.6.2 for more details on restricting the values of the parameters. As with all exponential smoothing methods, we need initial values of the components and estimates of the parameter values. This is discussed in Sect. 2.6.

Equation (2.6c) is slightly different from the usual Holt-Winters' equations such as those in Makridakis et al. (1998) or Bowerman et al. (2005). These authors replace (2.6c) with

$$s_t = \gamma y_t/\ell_t + (1-\gamma)s_{t-m}.$$

The modification given in (2.6c) was proposed by Ord et al. (1997) to make the state space formulation simpler. It is equivalent to Archibald's (1990) variation of Holt-Winters' method. The modification makes a small but usually negligible difference to the forecasts.

Additive Seasonality (A,A Method)

The seasonal component in Holt-Winters' method may also be treated additively, although in practice this seems to be less commonly used. The basic equations for Holt-Winters' additive method are as follows:

$$\begin{aligned}
\text{Level:} \quad & \ell_t = \alpha(y_t - s_{t-m}) + (1-\alpha)(\ell_{t-1} + b_{t-1}) && (2.7a)\\
\text{Growth:} \quad & b_t = \beta^*(\ell_t - \ell_{t-1}) + (1-\beta^*)b_{t-1} && (2.7b)\\
\text{Seasonal:} \quad & s_t = \gamma(y_t - \ell_{t-1} - b_{t-1}) + (1-\gamma)s_{t-m} && (2.7c)\\
\text{Forecast:} \quad & \hat{y}_{t+h|t} = \ell_t + b_t h + s_{t-m+h_m^+}. && (2.7d)
\end{aligned}$$

The second of these equations is identical to (2.6b). The only differences in the other equations are that the seasonal indices are now added and subtracted instead of taking products and ratios.

As with the multiplicative model, the usual equation given in textbooks for the seasonal term is slightly different from (2.7c). Most books use

$$s_t = \gamma^*(y_t - \ell_t) + (1-\gamma^*)s_{t-m}.$$

If ℓ_t is substituted using (2.7a), we obtain

$$s_t = \gamma^*(1-\alpha)(y_t - \ell_{t-1} - b_{t-1}) + [1-\gamma^*(1-\alpha)]s_{t-m}.$$

Thus, we obtain identical forecasts using this approach by replacing γ in (2.7c) with $\gamma^*(1-\alpha)$.

2.4 Point Forecasts for All Methods

Table 2.1 gives recursive formulae for computing point forecasts h periods ahead for all of the exponential smoothing methods. In each case, ℓ_t denotes the series level at time t, b_t denotes the slope at time t, s_t denotes the seasonal component of the series at time t, and m denotes the number of seasons in a year; α, β^*, γ and ϕ are constants, and $\phi_h = \phi + \phi^2 + \cdots + \phi^h$.

Some interesting special cases can be obtained by setting the smoothing parameters to extreme values. For example, if $\alpha = 0$, the level is constant over time; if $\beta^* = 0$, the slope is constant over time; and if $\gamma = 0$, the seasonal pattern is constant over time. At the other extreme, naïve forecasts (i.e., $\hat{y}_{t+h|t} = y_t$ for all h) are obtained using the (N,N) method with $\alpha = 1$. Finally, the additive and multiplicative trend methods are special cases of their damped counterparts obtained by letting $\phi = 1$.

2.5 State Space Models

We now introduce the state space models that underlie exponential smoothing methods. For each method, there are two models—a model with additive errors and a model with multiplicative errors. The point forecasts for the two models are identical (provided the same parameter values are used), but their prediction intervals will differ.

To distinguish the models with additive and multiplicative errors, we add an extra letter to the front of the method notation. The triplet (E,T,S) refers to the three components: error, trend and seasonality. So the model ETS(A,A,N) has additive errors, additive trend and no seasonality—in other words, this is Holt's linear method with additive errors. Similarly, ETS(M,M_d,M) refers to a model with multiplicative errors, a damped multiplicative trend and multiplicative seasonality. The notation ETS(·,·,·) helps in remembering the order in which the components are specified. ETS can also be considered an abbreviation of *E*xponen*T*ial *S*moothing.

Once a model is specified, we can study the probability distribution of future values of the series and find, for example, the conditional mean

Table 2.1. Formulae for recursive calculations and point forecasts.

Trend	Seasonal N	Seasonal A	Seasonal M
N	$\ell_t = \alpha y_t + (1-\alpha)\ell_{t-1}$ $\hat{y}_{t+h\|t} = \ell_t$	$\ell_t = \alpha(y_t - s_{t-m}) + (1-\alpha)\ell_{t-1}$ $s_t = \gamma(y_t - \ell_{t-1}) + (1-\gamma)s_{t-m}$ $\hat{y}_{t+h\|t} = \ell_t + s_{t-m+h_m^+}$	$\ell_t = \alpha(y_t / s_{t-m}) + (1-\alpha)\ell_{t-1}$ $s_t = \gamma(y_t/\ell_{t-1}) + (1-\gamma)s_{t-m}$ $\hat{y}_{t+h\|t} = \ell_t s_{t-m+h_m^+}$
A	$\ell_t = \alpha y_t + (1-\alpha)(\ell_{t-1} + b_{t-1})$ $b_t = \beta^*(\ell_t - \ell_{t-1}) + (1-\beta^*)b_{t-1}$ $\hat{y}_{t+h\|t} = \ell_t + hb_t$	$\ell_t = \alpha(y_t - s_{t-m}) + (1-\alpha)(\ell_{t-1} + b_{t-1})$ $b_t = \beta^*(\ell_t - \ell_{t-1}) + (1-\beta^*)b_{t-1}$ $s_t = \gamma(y_t - \ell_{t-1} - b_{t-1}) + (1-\gamma)s_{t-m}$ $\hat{y}_{t+h\|t} = \ell_t + hb_t + s_{t-m+h_m^+}$	$\ell_t = \alpha(y_t / s_{t-m}) + (1-\alpha)(\ell_{t-1} + b_{t-1})$ $b_t = \beta^*(\ell_t - \ell_{t-1}) + (1-\beta^*)b_{t-1}$ $s_t = \gamma(y_t/(\ell_{t-1} + b_{t-1})) + (1-\gamma)s_{t-m}$ $\hat{y}_{t+h\|t} = (\ell_t + hb_t)s_{t-m+h_m^+}$
A_d	$\ell_t = \alpha y_t + (1-\alpha)(\ell_{t-1} + \phi b_{t-1})$ $b_t = \beta^*(\ell_t - \ell_{t-1}) + (1-\beta^*)\phi b_{t-1}$ $\hat{y}_{t+h\|t} = \ell_t + \phi_h b_t$	$\ell_t = \alpha(y_t - s_{t-m}) + (1-\alpha)(\ell_{t-1} + \phi b_{t-1})$ $b_t = \beta^*(\ell_t - \ell_{t-1}) + (1-\beta^*)\phi b_{t-1}$ $s_t = \gamma(y_t - \ell_{t-1} - \phi b_{t-1}) + (1-\gamma)s_{t-m}$ $\hat{y}_{t+h\|t} = \ell_t + \phi_h b_t + s_{t-m+h_m^+}$	$\ell_t = \alpha(y_t / s_{t-m}) + (1-\alpha)(\ell_{t-1} + \phi b_{t-1})$ $b_t = \beta^*(\ell_t - \ell_{t-1}) + (1-\beta^*)\phi b_{t-1}$ $s_t = \gamma(y_t/(\ell_{t-1} + \phi b_{t-1})) + (1-\gamma)s_{t-m}$ $\hat{y}_{t+h\|t} = (\ell_t + \phi_h b_t)s_{t-m+h_m^+}$
M	$\ell_t = \alpha y_t + (1-\alpha)\ell_{t-1}b_{t-1}$ $b_t = \beta^*(\ell_t/\ell_{t-1}) + (1-\beta^*)b_{t-1}$ $\hat{y}_{t+h\|t} = \ell_t b_t^h$	$\ell_t = \alpha(y_t - s_{t-m}) + (1-\alpha)\ell_{t-1}b_{t-1}$ $b_t = \beta^*(\ell_t/\ell_{t-1}) + (1-\beta^*)b_{t-1}$ $s_t = \gamma(y_t - \ell_{t-1}b_{t-1}) + (1-\gamma)s_{t-m}$ $\hat{y}_{t+h\|t} = \ell_t b_t^h + s_{t-m+h_m^+}$	$\ell_t = \alpha(y_t / s_{t-m}) + (1-\alpha)\ell_{t-1}b_{t-1}$ $b_t = \beta^*(\ell_t/\ell_{t-1}) + (1-\beta^*)b_{t-1}$ $s_t = \gamma(y_t/(\ell_{t-1}b_{t-1})) + (1-\gamma)s_{t-m}$ $\hat{y}_{t+h\|t} = \ell_t b_t^h s_{t-m+h_m^+}$
M_d	$\ell_t = \alpha y_t + (1-\alpha)\ell_{t-1}b_{t-1}^{\phi}$ $b_t = \beta^*(\ell_t/\ell_{t-1}) + (1-\beta^*)b_{t-1}^{\phi}$ $\hat{y}_{t+h\|t} = \ell_t b_t^{\phi_h}$	$\ell_t = \alpha(y_t - s_{t-m}) + (1-\alpha)\ell_{t-1}b_{t-1}^{\phi}$ $b_t = \beta^*(\ell_t/\ell_{t-1}) + (1-\beta^*)b_{t-1}^{\phi}$ $s_t = \gamma(y_t - \ell_{t-1}b_{t-1}^{\phi}) + (1-\gamma)s_{t-m}$ $\hat{y}_{t+h\|t} = \ell_t b_t^{\phi_h} + s_{t-m+h_m^+}$	$\ell_t = \alpha(y_t / s_{t-m}) + (1-\alpha)\ell_{t-1}b_{t-1}^{\phi}$ $b_t = \beta^*(\ell_t/\ell_{t-1}) + (1-\beta^*)b_{t-1}^{\phi}$ $s_t = \gamma(y_t/(\ell_{t-1}b_{t-1}^{\phi})) + (1-\gamma)s_{t-m}$ $\hat{y}_{t+h\|t} = \ell_t b_t^{\phi_h} s_{t-m+h_m^+}$

In each case, ℓ_t denotes the series level at time t, b_t denotes the slope at time t, s_t denotes the seasonal component of the series at time t, and m denotes the number of seasons in a year; α, β^*, γ and ϕ are constants, $\phi_h = \phi + \phi^2 + \cdots + \phi^h$ and $h_m^+ = \lfloor (h-1) \bmod m \rfloor + 1$.

of a future observation given knowledge of the past. We denote this as $\mu_{t+h|t} = \mathrm{E}(y_{t+h} \mid \boldsymbol{x}_t)$, where $\boldsymbol{x}_t$ contains the unobserved components such as ℓ_t, b_t and s_t. For $h = 1$ we use $\mu_{t+1} \equiv \mu_{t+1|t}$ as a shorthand notation. For most models, these conditional means will be identical to the point forecasts given earlier, so that $\mu_{t+h|t} = \hat{y}_{t+h|t}$. However, for other models (those with multiplicative trend or multiplicative seasonality), the conditional mean and the point forecast will differ slightly for $h \geq 2$.

2.5.1 State Space Models for Holt's Linear Method

We now illustrate the ideas using Holt's linear method.

Additive Error Model: ETS(A,A,N)

Let $\mu_t = \hat{y}_t = \ell_{t-1} + b_{t-1}$ denote the one-step forecast of y_t assuming we know the values of all parameters. Also let $\varepsilon_t = y_t - \mu_t$ denote the one-step forecast error at time t. From (2.4c), we find that

$$y_t = \ell_{t-1} + b_{t-1} + \varepsilon_t, \tag{2.8}$$

and using (2.4a) and (2.4b) we can write

$$\ell_t = \ell_{t-1} + b_{t-1} + \alpha\varepsilon_t, \tag{2.9}$$

$$b_t = b_{t-1} + \beta^*(\ell_t - \ell_{t-1} - b_{t-1}) = b_{t-1} + \alpha\beta^*\varepsilon_t. \tag{2.10}$$

We simplify the last expression by setting $\beta = \alpha\beta^*$. The three equations above constitute a state space model underlying Holt's method. We can write it in standard state space notation by defining the state vector as $\boldsymbol{x}_t = (\ell_t, b_t)'$ and expressing (2.8)–(2.10) as

$$y_t = [1 \;\; 1]\, \boldsymbol{x}_{t-1} + \varepsilon_t, \tag{2.11a}$$

$$\boldsymbol{x}_t = \begin{bmatrix} 1 & 1 \\ 0 & 1 \end{bmatrix} \boldsymbol{x}_{t-1} + \begin{bmatrix} \alpha \\ \beta \end{bmatrix} \varepsilon_t. \tag{2.11b}$$

The model is fully specified once we state the distribution of the error term ε_t. Usually we assume that these are independent and identically distributed, following a Gaussian distribution with mean 0 and variance σ^2, which we write as $\varepsilon_t \sim \mathrm{NID}(0, \sigma^2)$.

Multiplicative Error Model: ETS(M,A,N)

A model with multiplicative error can be derived similarly, by first setting $\varepsilon_t = (y_t - \mu_t)/\mu_t$, so that ε_t is a relative error. Then, following a similar approach to that for additive errors, we find

$$\begin{aligned} y_t &= (\ell_{t-1} + b_{t-1})(1 + \varepsilon_t), \\ \ell_t &= (\ell_{t-1} + b_{t-1})(1 + \alpha\varepsilon_t), \\ b_t &= b_{t-1} + \beta(\ell_{t-1} + b_{t-1})\varepsilon_t, \end{aligned}$$

or

$$y_t = [1\ 1]\,\boldsymbol{x}_{t-1}(1+\varepsilon_t),$$
$$\boldsymbol{x}_t = \begin{bmatrix}1 & 1\\ 0 & 1\end{bmatrix}\boldsymbol{x}_{t-1} + [1\ 1]\,\boldsymbol{x}_{t-1}\begin{bmatrix}\alpha\\ \beta\end{bmatrix}\varepsilon_t.$$

Again we assume that $\varepsilon_t \sim \text{NID}(0,\sigma^2)$.

Of course, this is a nonlinear state space model, which is usually considered difficult to handle in estimating and forecasting. However, that is one of the many advantages of the innovations form of state space models—we can still compute forecasts, the likelihood and prediction intervals for this nonlinear model with no more effort than is required for the additive error model.

2.5.2 State Space Models for All Exponential Smoothing Methods

We now give the state space models for all 30 exponential smoothing variations. The general model involves a state vector $\boldsymbol{x}_t = (\ell_t, b_t, s_t, s_{t-1}, \ldots, s_{t-m+1})'$ and state space equations of the form

$$y_t = w(\boldsymbol{x}_{t-1}) + r(\boldsymbol{x}_{t-1})\varepsilon_t, \tag{2.12a}$$
$$\boldsymbol{x}_t = f(\boldsymbol{x}_{t-1}) + g(\boldsymbol{x}_{t-1})\varepsilon_t, \tag{2.12b}$$

where $\{\varepsilon_t\}$ is a Gaussian white noise process with variance σ^2, and $\mu_t = w(\boldsymbol{x}_{t-1})$. The model with additive errors has $r(\boldsymbol{x}_{t-1}) = 1$, so that $y_t = \mu_t + \varepsilon_t$. The model with multiplicative errors has $r(\boldsymbol{x}_{t-1}) = \mu_t$, so that $y_t = \mu_t(1+\varepsilon_t)$. Thus, $\varepsilon_t = (y_t - \mu_t)/\mu_t$ is the relative error for the multiplicative model. The models are not unique. Clearly, any value of $r(\boldsymbol{x}_{t-1})$ will lead to identical point forecasts for y_t.

Each of the methods in Table 2.1 can be written in the form given in (2.12a) and (2.12b). The underlying equations for the additive error models are given in Table 2.2. We use $\beta = \alpha\beta^*$ to simplify the notation. Multiplicative error models are obtained by replacing ε_t with $\mu_t\varepsilon_t$ in the equations of Table 2.2. The resulting multiplicative error equations are given in Table 2.3.

Some of the combinations of trend, seasonality and error can occasionally lead to numerical difficulties; specifically, any model equation that requires division by a state component could involve division by zero. This is a problem for models with additive errors and either multiplicative trend or multiplicative seasonality, as well as the model with multiplicative errors, multiplicative trend and additive seasonality. These models should therefore be used with caution. The properties of these models are discussed in Chap. 15.

The multiplicative error models are useful when the data are strictly positive, but are not numerically stable when the data contain zeros or negative

Table 2.2. State space equations for each additive error model in the classification.

Trend	Seasonal N	Seasonal A	Seasonal M
N	$\mu_t = \ell_{t-1}$ $\ell_t = \ell_{t-1} + \alpha\varepsilon_t$	$\mu_t = \ell_{t-1} + s_{t-m}$ $\ell_t = \ell_{t-1} + \alpha\varepsilon_t$ $s_t = s_{t-m} + \gamma\varepsilon_t$	$\mu_t = \ell_{t-1}s_{t-m}$ $\ell_t = \ell_{t-1} + \alpha\varepsilon_t/s_{t-m}$ $s_t = s_{t-m} + \gamma\varepsilon_t/\ell_{t-1}$
A	$\mu_t = \ell_{t-1} + b_{t-1}$ $\ell_t = \ell_{t-1} + b_{t-1} + \alpha\varepsilon_t$ $b_t = b_{t-1} + \beta\varepsilon_t$	$\mu_t = \ell_{t-1} + b_{t-1} + s_{t-m}$ $\ell_t = \ell_{t-1} + b_{t-1} + \alpha\varepsilon_t$ $b_t = b_{t-1} + \beta\varepsilon_t$ $s_t = s_{t-m} + \gamma\varepsilon_t$	$\mu_t = (\ell_{t-1} + b_{t-1})s_{t-m}$ $\ell_t = \ell_{t-1} + b_{t-1} + \alpha\varepsilon_t/s_{t-m}$ $b_t = b_{t-1} + \beta\varepsilon_t/s_{t-m}$ $s_t = s_{t-m} + \gamma\varepsilon_t/(\ell_{t-1} + b_{t-1})$
A_d	$\mu_t = \ell_{t-1} + \phi b_{t-1}$ $\ell_t = \ell_{t-1} + \phi b_{t-1} + \alpha\varepsilon_t$ $b_t = \phi b_{t-1} + \beta\varepsilon_t$	$\mu_t = \ell_{t-1} + \phi b_{t-1} + s_{t-m}$ $\ell_t = \ell_{t-1} + \phi b_{t-1} + \alpha\varepsilon_t$ $b_t = \phi b_{t-1} + \beta\varepsilon_t$ $s_t = s_{t-m} + \gamma\varepsilon_t$	$\mu_t = (\ell_{t-1} + \phi b_{t-1})s_{t-m}$ $\ell_t = \ell_{t-1} + \phi b_{t-1} + \alpha\varepsilon_t/s_{t-m}$ $b_t = \phi b_{t-1} + \beta\varepsilon_t/s_{t-m}$ $s_t = s_{t-m} + \gamma\varepsilon_t/(\ell_{t-1} + \phi b_{t-1})$
M	$\mu_t = \ell_{t-1}b_{t-1}$ $\ell_t = \ell_{t-1}b_{t-1} + \alpha\varepsilon_t$ $b_t = b_{t-1} + \beta\varepsilon_t/\ell_{t-1}$	$\mu_t = \ell_{t-1}b_{t-1} + s_{t-m}$ $\ell_t = \ell_{t-1}b_{t-1} + \alpha\varepsilon_t$ $b_t = b_{t-1} + \beta\varepsilon_t/\ell_{t-1}$ $s_t = s_{t-m} + \gamma\varepsilon_t$	$\mu_t = \ell_{t-1}b_{t-1}s_{t-m}$ $\ell_t = \ell_{t-1}b_{t-1} + \alpha\varepsilon_t/s_{t-m}$ $b_t = b_{t-1} + \beta\varepsilon_t/(s_{t-m}\ell_{t-1})$ $s_t = s_{t-m} + \gamma\varepsilon_t/(\ell_{t-1}b_{t-1})$
M_d	$\mu_t = \ell_{t-1}b_{t-1}^{\phi}$ $\ell_t = \ell_{t-1}b_{t-1}^{\phi} + \alpha\varepsilon_t$ $b_t = b_{t-1}^{\phi} + \beta\varepsilon_t/\ell_{t-1}$	$\mu_t = \ell_{t-1}b_{t-1}^{\phi} + s_{t-m}$ $\ell_t = \ell_{t-1}b_{t-1}^{\phi} + \alpha\varepsilon_t$ $b_t = b_{t-1}^{\phi} + \beta\varepsilon_t/\ell_{t-1}$ $s_t = s_{t-m} + \gamma\varepsilon_t$	$\mu_t = \ell_{t-1}b_{t-1}^{\phi}s_{t-m}$ $\ell_t = \ell_{t-1}b_{t-1}^{\phi} + \alpha\varepsilon_t/s_{t-m}$ $b_t = b_{t-1}^{\phi} + \beta\varepsilon_t/(s_{t-m}\ell_{t-1})$ $s_t = s_{t-m} + \gamma\varepsilon_t/(\ell_{t-1}b_{t-1}^{\phi})$

Table 2.3. State space equations for each multiplicative error model in the classification.

Trend	Seasonal		
	N	A	M
N	$\mu_t = \ell_{t-1}$ $\ell_t = \ell_{t-1}(1 + \alpha\varepsilon_t)$	$\mu_t = \ell_{t-1} + s_{t-m}$ $\ell_t = \ell_{t-1} + \alpha(\ell_{t-1} + s_{t-m})\varepsilon_t$ $s_t = s_{t-m} + \gamma(\ell_{t-1} + s_{t-m})\varepsilon_t$	$\mu_t = \ell_{t-1}s_{t-m}$ $\ell_t = \ell_{t-1}(1 + \alpha\varepsilon_t)$ $s_t = s_{t-m}(1 + \gamma\varepsilon_t)$
A	$\mu_t = \ell_{t-1} + b_{t-1}$ $\ell_t = (\ell_{t-1} + b_{t-1})(1 + \alpha\varepsilon_t)$ $b_t = b_{t-1} + \beta(\ell_{t-1} + b_{t-1})\varepsilon_t$	$\mu_t = \ell_{t-1} + b_{t-1} + s_{t-m}$ $\ell_t = \ell_{t-1} + b_{t-1} + \alpha(\ell_{t-1} + b_{t-1} + s_{t-m})\varepsilon_t$ $b_t = b_{t-1} + \beta(\ell_{t-1} + b_{t-1} + s_{t-m})\varepsilon_t$ $s_t = s_{t-m} + \gamma(\ell_{t-1} + b_{t-1} + s_{t-m})\varepsilon_t$	$\mu_t = (\ell_{t-1} + b_{t-1})s_{t-m}$ $\ell_t = (\ell_{t-1} + b_{t-1})(1 + \alpha\varepsilon_t)$ $b_t = b_{t-1} + \beta(\ell_{t-1} + b_{t-1})\varepsilon_t$ $s_t = s_{t-m}(1 + \gamma\varepsilon_t)$
A_d	$\mu_t = \ell_{t-1} + \phi b_{t-1}$ $\ell_t = (\ell_{t-1} + \phi b_{t-1})(1 + \alpha\varepsilon_t)$ $b_t = \phi b_{t-1} + \beta(\ell_{t-1} + \phi b_{t-1})\varepsilon_t$	$\mu_t = \ell_{t-1} + \phi b_{t-1} + s_{t-m}$ $\ell_t = \ell_{t-1} + \phi b_{t-1} + \alpha(\ell_{t-1} + \phi b_{t-1} + s_{t-m})\varepsilon_t$ $b_t = \phi b_{t-1} + \beta(\ell_{t-1} + \phi b_{t-1} + s_{t-m})\varepsilon_t$ $s_t = s_{t-m} + \gamma(\ell_{t-1} + \phi b_{t-1} + s_{t-m})\varepsilon_t$	$\mu_t = (\ell_{t-1} + \phi b_{t-1})s_{t-m}$ $\ell_t = (\ell_{t-1} + \phi b_{t-1})(1 + \alpha\varepsilon_t)$ $b_t = \phi b_{t-1} + \beta(\ell_{t-1} + \phi b_{t-1})\varepsilon_t$ $s_t = s_{t-m}(1 + \gamma\varepsilon_t)$
M	$\mu_t = \ell_{t-1}b_{t-1}$ $\ell_t = \ell_{t-1}b_{t-1}(1 + \alpha\varepsilon_t)$ $b_t = b_{t-1}(1 + \beta\varepsilon_t)$	$\mu_t = \ell_{t-1}b_{t-1} + s_{t-m}$ $\ell_t = \ell_{t-1}b_{t-1} + \alpha(\ell_{t-1}b_{t-1} + s_{t-m})\varepsilon_t$ $b_t = b_{t-1} + \beta(\ell_{t-1}b_{t-1} + s_{t-m})\varepsilon_t/\ell_{t-1}$ $s_t = s_{t-m} + \gamma(\ell_{t-1}b_{t-1} + s_{t-m})\varepsilon_t$	$\mu_t = \ell_{t-1}b_{t-1}s_{t-m}$ $\ell_t = \ell_{t-1}b_{t-1}(1 + \alpha\varepsilon_t)$ $b_t = b_{t-1}(1 + \beta\varepsilon_t)$ $s_t = s_{t-m}(1 + \gamma\varepsilon_t)$
M_d	$\mu_t = \ell_{t-1}b^{\phi}_{t-1}$ $\ell_t = \ell_{t-1}b^{\phi}_{t-1}(1 + \alpha\varepsilon_t)$ $b_t = b^{\phi}_{t-1}(1 + \beta\varepsilon_t)$	$\mu_t = \ell_{t-1}b^{\phi}_{t-1} + s_{t-m}$ $\ell_t = \ell_{t-1}b^{\phi}_{t-1} + \alpha(\ell_{t-1}b^{\phi}_{t-1} + s_{t-m})\varepsilon_t$ $b_t = b^{\phi}_{t-1} + \beta(\ell_{t-1}b^{\phi}_{t-1} + s_{t-m})\varepsilon_t/\ell_{t-1}$ $s_t = s_{t-m} + \gamma(\ell_{t-1}b^{\phi}_{t-1} + s_{t-m})\varepsilon_t$	$\mu_t = \ell_{t-1}b^{\phi}_{t-1}s_{t-m}$ $\ell_t = \ell_{t-1}b^{\phi}_{t-1}(1 + \alpha\varepsilon_t)$ $b_t = b^{\phi}_{t-1}(1 + \beta\varepsilon_t)$ $s_t = s_{t-m}(1 + \gamma\varepsilon_t)$

values. So when the time series is not strictly positive, only the six fully additive models may be applied.

The point forecasts given earlier are easily obtained from these models by iterating (2.12) for $t = n+1, n+2, \ldots, n+h$, and setting $\varepsilon_{n+j} = 0$ for $j = 1, \ldots, h$. In most cases (notable exceptions being models with multiplicative seasonality or multiplicative trend for $h \geq 2$), the point forecasts can be shown to be equal to $\mu_{t+h|t} = \mathrm{E}(y_{t+h} \mid \boldsymbol{x}_t)$, the conditional expectation of the corresponding state space model.

The models also provide a means of obtaining prediction intervals. In the case of the linear models, where the prediction distributions are Gaussian, we can derive the conditional variance $v_{t+h|t} = \mathrm{V}(y_{t+h} \mid \boldsymbol{x}_t)$ and obtain prediction intervals accordingly. This approach also works for many of the nonlinear models, as we show in Chap. 6.

A more direct approach that works for all of the models is to simply simulate many future sample paths, conditional on the last estimate of the state vector, $\boldsymbol{x}_t$. Then prediction intervals can be obtained from the percentiles of the simulated sample paths. Point forecasts can also be obtained in this way by taking the average of the simulated values at each future time period. One advantage of this approach is that we generate an estimate of the complete predictive distribution, which is especially useful in applications such as inventory planning, where the expected costs depend on the whole distribution.

2.6 Initialization and Estimation

In order to use these models for forecasting, we need to specify the type of model to be used (model selection), the value of $\boldsymbol{x}_0$ (initialization), and the values of the parameters α, β, γ and ϕ (estimation). In this section, we discuss initialization and estimation, leaving model selection to Sect. 2.8.

2.6.1 Initialization

Traditionally, the initial values $\boldsymbol{x}_0$ are specified using ad hoc values, or via a heuristic scheme. The following heuristic scheme, based on Hyndman et al. (2002), seems to work very well.

- *Initial seasonal component.* For seasonal data, compute a $2 \times m$ moving average through the first few years of data. Denote this by $\{f_t\}$, $t = m/2 + 1, m/2+2, \ldots$. For additive seasonality, detrend the data to obtain $y_t - f_t$; for multiplicative seasonality, detrend the data to obtain y_t/f_t. Compute initial seasonal indices, $s_{-m+1}, \ldots, s_0$, by averaging the detrended data for each season. Normalize these seasonal indices so that they add to zero for additive seasonality, and add to m for multiplicative seasonality.

- *Initial level component.* For seasonal data, compute a linear trend using linear regression on the first ten seasonally adjusted values (using the seasonal indices obtained above) against a time variable $t = 1, \ldots, 10$. For nonseasonal data, compute a linear trend on the first ten observations against a time variable $t = 1, \ldots, 10$. Then set ℓ_0 to be the intercept of the trend.
- *Initial growth component.* For additive trend, set b_0 to be the slope of the trend. For multiplicative trend, set $b_0 = 1 + b/a$, where a denotes the intercept and b denotes the slope of the fitted trend.

These initial states are then refined by estimating them along with the parameters, as described below.

2.6.2 Estimation

It is easy to compute the likelihood of the innovations state space model (2.12), and so obtain maximum likelihood estimates. In Chap. 5, we show that

$$\mathcal{L}^*(\boldsymbol{\theta}, \boldsymbol{x}_0) = n \log \Big(\sum_{t=1}^{n} \varepsilon_t^2 \Big) + 2 \sum_{t=1}^{n} \log |r(\boldsymbol{x}_{t-1})|$$

is equal to twice the negative logarithm of the likelihood function (with constant terms eliminated), conditional on the parameters $\boldsymbol{\theta} = (\alpha, \beta, \gamma, \phi)'$ and the initial states $\boldsymbol{x}_0 = (\ell_0, b_0, s_0, s_{-1}, \ldots, s_{-m+1})'$, where n is the number of observations. This is easily computed by simply using the recursive equations in Table 2.1. Unlike state space models with multiple sources of error, we do not need to use the Kalman filter to compute the likelihood.

The parameters $\boldsymbol{\theta}$ and the initial states $\boldsymbol{x}_0$ can be estimated by minimizing $\mathcal{L}^*$. Alternatively, estimates can be obtained by minimizing the one-step mean squared error (MSE), minimizing the residual variance σ^2, or via some other criterion for measuring forecast error. Whichever criterion is used, we usually begin the optimization with $\boldsymbol{x}_0$ obtained from the heuristic scheme above and $\boldsymbol{\theta} = (0.1, 0.01, 0.01, 0.99)'$.

There have been several suggestions for restricting the parameter space of α, β and γ. The traditional approach is to ensure that the various equations can be interpreted as weighted averages, thus requiring α, $\beta^* = \beta/\alpha$, $\gamma^* = \gamma/(1-\alpha)$ and ϕ to all lie within $(0, 1)$. This suggests that

$$0 < \alpha < 1, \qquad 0 < \beta < \alpha, \qquad 0 < \gamma < 1 - \alpha, \qquad \text{and} \qquad 0 < \phi < 1.$$

However, we shall see in Chap. 10 that these restrictions are usually stricter than necessary (although in a few cases they are not restrictive enough).

We also constrain the initial states $\boldsymbol{x}_0$ so that the seasonal indices add to zero for additive seasonality, and add to m for multiplicative seasonality.

2.7 Assessing Forecast Accuracy

The issue of measuring the accuracy of forecasts from different methods has been the subject of much attention. We summarize some of the approaches here. A more thorough discussion is given by Hyndman and Koehler (2006).

There are three possible ways in which the forecasts can have arisen:

1. The forecasts may be computed from a common base time, and be of varying forecast horizons. That is, we may compute out-of-sample forecasts $\hat{y}_{n+1|n}, \ldots, \hat{y}_{n+h|n}$ based on data from times $t = 1, \ldots, n$. When $h = 1$, we write $\hat{y}_{n+1} \equiv \hat{y}_{n+1|n}$.
2. The forecasts may be from varying base times, and be of a consistent forecast horizon. That is, we may compute forecasts $\hat{y}_{1+h|1}, \ldots, \hat{y}_{m+h|m}$ where each $\hat{y}_{j+h|j}$ is based on data from times $t = 1, \ldots, j$.
3. We may wish to compare the accuracy of methods between many series at a single forecast horizon. That is, we compute a single $\hat{y}_{n+h|n}$ based on data from times $t = 1, \ldots, n$ for each of m different series.

While these are very different situations, measuring forecast accuracy is the same in each case.

The measures defined below are described for one-step-ahead forecasts; the extension to h-steps-ahead is immediate in each case and raises no new questions of principle.

2.7.1 Scale-Dependent Errors

The one-step-ahead forecast error is simply $e_t = y_t - \hat{y}_t$, regardless of how the forecast was produced. Similarly the h-step-ahead forecast error is $e_{t+h|t} = y_{t+h} - \hat{y}_{t+h|t}$. This is on the same scale as the data. Accuracy measures that are based on e_t are therefore scale-dependent.

The two most commonly used scale-dependent measures are based on the absolute error or squared errors:

$$\text{Mean absolute error (MAE)} = \text{mean}(|e_t|),$$
$$\text{Mean squared error (MSE)} = \text{mean}(e_t^2).$$

When comparing forecast methods on a single series, we prefer the MAE as it is easy to understand and compute. However, it cannot be used to make comparisons between series as it makes no sense to compare accuracy on different scales.

2.7.2 Percentage Errors

The percentage error is given by $p_t = 100e_t/y_t$. Percentage errors have the advantage of being scale-independent, and so are frequently used to compare forecast performance between different data sets. The most commonly used measure is:

$$\text{Mean absolute percentage error (MAPE)} = \text{mean}(|p_t|)$$

Measures based on percentage errors have the disadvantage of being infinite or undefined if $y_t = 0$ for any t in the period of interest, and having an extremely skewed distribution when any y_t is close to zero. Another problem with percentage errors that is often overlooked is that they assume a meaningful zero. For example, a percentage error makes no sense when measuring the accuracy of temperature forecasts on the Fahrenheit or Celsius scales.

They also have the disadvantage that they put a heavier penalty on positive errors than on negative errors. This observation led to the use of the so-called "symmetric" MAPE proposed by Makridakis (1993), which was used in the M3 competition (Makridakis and Hibon 2000). It is defined by

$$\text{Symmetric mean absolute percentage error (sMAPE)} = \text{mean}(200|y_t - \hat{y}_t|/(y_t + \hat{y}_t)).$$

However, if y_t is zero, $\hat{y}_t$ is also likely to be close to zero. Thus, the measure still involves division by a number close to zero. Also, the value of sMAPE can be negative, so it is not really a measure of "absolute percentage errors" at all.

2.7.3 Scaled Errors

The MASE was proposed by Hyndman and Koehler (2006) as a generally applicable measure of forecast accuracy. They proposed scaling the errors based on the *in-sample* MAE from the naïve forecast method. Thus, a scaled error is defined as

$$q_t = \frac{e_t}{\frac{1}{n-1}\sum_{i=2}^{n} |y_i - y_{i-1}|},$$

which is independent of the scale of the data. A scaled error is less than one if it arises from a better forecast than the average one-step naïve forecast computed in-sample. Conversely, it is greater than one if the forecast is worse than the average one-step naïve forecast computed in-sample.

The *mean absolute scaled error* is simply

$$\text{MASE} = \text{mean}(|q_t|).$$

The *in-sample* MAE is used in the denominator as it is always available and effectively scales the errors. In contrast, the out-of-sample MAE for the naïve method can be based on very few observations and is therefore more variable. For some data sets, it can even be zero. Consequently, the in-sample MAE is preferable in the denominator.

The MASE can be used to compare forecast methods on a single series, and to compare forecast accuracy between series as it is scale-free. It is the only available method which can be used in all circumstances.

2.8 Model Selection

The forecast accuracy measures described in the previous section can be used to select a model for a given set of data, provided the errors are computed from data in a hold-out set and not from the same data as were used for model estimation. However, there are often too few out-of-sample errors to draw reliable conclusions. Consequently, a penalized method based on in-sample fit is usually better.

One such method is via a penalized likelihood such as *Akaike's information criterion*:

$$\text{AIC} = \mathcal{L}^*(\hat{\boldsymbol{\theta}}, \hat{\boldsymbol{x}}_0) + 2q,$$

where q is the number of parameters in $\boldsymbol{\theta}$ plus the number of free states in $\boldsymbol{x}_0$, and $\hat{\boldsymbol{\theta}}$ and $\hat{\boldsymbol{x}}_0$ denote the estimates of $\boldsymbol{\theta}$ and $\boldsymbol{x}_0$. (In computing the AIC, we also require that the state space model has no redundant states—see Sect. 10.1, p. 149.) We select the model that minimizes the AIC amongst all of the models that are appropriate for the data.

The AIC also provides a method for selecting between the additive and multiplicative error models. Point forecasts from the two models are identical, so that standard forecast accuracy measures such as the MSE or MAPE are unable to select between the error types. The AIC is able to select between the error types because it is based on likelihood rather than one-step forecasts.

Obviously, other model selection criteria (such as the BIC) could also be used in a similar manner. Model selection is explored in more detail in Chap. 7.

2.8.1 Automatic Forecasting

We combine the preceding ideas to obtain a robust and widely applicable automatic forecasting algorithm. The steps involved are summarized below:

1. For each series, apply all models that are appropriate, optimizing the parameters of the model in each case.
2. Select the best of the models according to the AIC.
3. Produce point forecasts using the best model (with optimized parameters) for as many steps ahead as required.
4. Obtain prediction intervals[5] for the best model either using the analytical results, or by simulating future sample paths for $\{y_{n+1}, \ldots, y_{n+h}\}$ and

[5] The calculation of prediction intervals is discussed in Chap. 6.

finding the $\alpha/2$ and $1-\alpha/2$ percentiles of the simulated data at each forecasting horizon. If simulation is used, the sample paths may be generated using the Gaussian distribution for errors (parametric bootstrap) or using the resampled errors (ordinary bootstrap).

This algorithm resulted in the forecasts shown in Fig. 2.1. The models chosen were:

- ETS(A,A$_d$,N) for monthly US 10-year bond yields
  ($\alpha = 0.99$, $\beta = 0.12$, $\phi = 0.80$, $\ell_0 = 5.30$, $b_0 = 0.71$)
- ETS(M,M$_d$,N) for annual US net electricity generation
  ($\alpha = 0.99$, $\beta = 0.01$, $\phi = 0.97$, $\ell_0 = 262.5$, $b_0 = 1.12$)
- ETS(A,N,A) for quarterly UK passenger vehicle production
  ($\alpha = 0.61$, $\gamma = 0.01$, $\ell_0 = 343.4$, $s_{-3} = 24.99$, $s_{-2} = 21.40$, $s_{-1} = -44.96$, $s_0 = -1.42$)
- ETS(M,A,M) for monthly Australian overseas visitors
  ($\alpha = 0.57$, $\beta = 0.01$, $\gamma = 0.19$, $\ell_0 = 86.2$, $b_0 = 2.66$, $s_{-11} = 0.851$, $s_{-10} = 0.844$, $s_{-9} = 0.985$, $s_{-8} = 0.924$, $s_{-7} = 0.822$, $s_{-6} = 1.006$, $s_{-5} = 1.101$, $s_{-4} = 1.369$, $s_{-3} = 0.975$, $s_{-2} = 1.078$, $s_{-1} = 1.087$, $s_0 = 0.958$)

Although there is a lot of computation involved, it can be handled remarkably quickly on modern computers. The forecasts shown in Fig. 2.1 took a few seconds on a standard PC.

Hyndman et al. (2002) applied this automatic forecasting strategy to the M-competition data (Makridakis et al. 1982) and IJF-M3 competition data (Makridakis and Hibon 2000), and demonstrated that the methodology is particularly good at short-term forecasts (up to about six periods ahead), and especially for seasonal short-term series (beating all other methods in the competition for these series).

2.9 Exercises

Exercise 2.1. Consider the innovations state space model (2.12). Equations (2.12a) and (2.12b) are called the *measurement equation* and *transition equation* respectively:

a. For the ETS(A,A$_d$,N) model, write the measurement equation and transition equations with a separate equation for each of the two states (level and growth).
b. For the ETS (A,A$_d$,N) model, write the measurement and transition equations in matrix form, defining $\boldsymbol{x}_t$, $w(\boldsymbol{x}_{t-1})$, $r(\boldsymbol{x}_{t-1})$, $f(\boldsymbol{x}_{t-1})$, and $g(\boldsymbol{x}_{t-1})$. See Sect. 2.5.1 for an example based on the ETS(A,A,N) model.
c. Repeat parts a and b for the ETS(A,A,A) model.
d. Repeat parts a and b for the ETS(M,A$_d$,N) model.
e. Repeat parts a and b for the ETS(M,A$_d$,A) model.

Exercise 2.2. Use the innovations state space model, including the assumptions about ε_t, to derive the specified point forecast,

$$\hat{y}_{t+h|t} = \mu_{t+h|t} = \mathrm{E}(y_{t+h} \mid \boldsymbol{x}_t),$$

and variance of the forecast error,

$$v_{t+h|t} = \mathrm{V}(y_{t+h} \mid \boldsymbol{x}_t),$$

for the following models:

a. For ETS(A,N,N), show $\hat{y}_{t+h|t} = \ell_t$ and $v_{t+h|t} = \sigma^2[1 + (h-1)\alpha^2]$.
b. For ETS(A,A,N), show $\hat{y}_{t+h|t} = \ell_t + hb_t$ and

$$v_{t+h|t} = \sigma^2\left[1 + \sum_{j=1}^{h-1} (\alpha + \beta j)^2\right]$$

c. For ETS(M,N,N), show $\hat{y}_{t+h|t} = \ell_t$, $v_{t+1|t} = \ell_t^2\sigma^2$, and

$$v_{t+2|t} = \ell_t^2\left[(1 + \alpha^2\sigma^2)(1 + \sigma^2) - 1\right].$$

Exercise 2.3. Use **R** to reproduce the results in Sect. 2.8.1 for each of the four time series: US 10-year bond yields, US net electricity, UK passenger vehicle production, and Australian overseas visitors. The data sets are named `bonds`, `usnetelec`, `ukcars` and `visitors` respectively. The `ets()` function found in the `forecast` package can be used to specify the model or to automatically select a model.

Exercise 2.4. Using the results of Exercise 2.3, use **R** to reproduce the results in Fig. 2.1 for point forecasts and prediction intervals for each of the four time series. The `forecast()` function in the `forecast` package can be used to produce the point forecasts and prediction intervals for each model found in Exercise 2.3.

Part II

Essentials

3

Linear Innovations State Space Models

In Chap. 2, state space models were introduced for all 15 exponential smoothing methods. Six of these involved only linear relationships, and so are "linear innovations state space models." In this chapter, we consider linear innovations state space models, including the six linear models of Chap. 2, but also any other models of the same form. The advantage of working with the general framework is that estimation and prediction methods for the general model automatically apply to the six special cases in Chap. 2 and other cases conforming to its structure. There is no need to derive these results on a case by case basis.

The general linear innovations state space model is introduced in Sect. 3.1. Section 3.2 provides a simple algorithm for computing the one-step prediction errors (or innovations); it is this algorithm which makes innovations state space models so appealing. Some of the properties of the models, including stationarity and stability, are discussed in Sect. 3.3. In Sect. 3.4 we discuss some basic innovations state space models that were introduced briefly in Chap. 2. Interesting variations on these models are considered in Sect. 3.5.

3.1 The General Linear Innovations State Space Model

In a state space model, the observed time series variable y_t is supplemented by unobserved auxiliary variables called *states*. We represent these auxiliary variables in a single vector $\boldsymbol{x}_t$, which is called the *state vector*. The state vector is a parsimonious way of summarizing the past behavior of the time series y_t, and then using it to determine the effect of the past on the present and future behavior of the time series.

The general[1] *linear innovations state space model* is

$$y_t = \boldsymbol{w}'\boldsymbol{x}_{t-1} + \varepsilon_t, \tag{3.1a}$$

$$\boldsymbol{x}_t = \boldsymbol{F}\boldsymbol{x}_{t-1} + \boldsymbol{g}\varepsilon_t, \tag{3.1b}$$

where y_t denotes the observed value at time t and $\boldsymbol{x}_t$ is the state vector. This is a special case of the more general model (2.12). In exponential smoothing, the state vector contains information about the level, growth and seasonal patterns. For example, in a model with trend and seasonality, $\boldsymbol{x}_t = (\ell_t, b_t, s_t, s_{t-1}, \ldots, s_{t-m+1})'$.

From a mathematical perspective, the state variables are essentially redundant. In Chap. 11, it will be shown that the state variables contained in the state vector can be substituted out of the equations in which they occur to give a *reduced form* of the model. So why use state variables at all? They help us to define large complex models by first breaking them into smaller, more manageable parts, thus reducing the chance of model specification errors. Further, the components of the state vector enable us to gain a better understanding of the structure of the series, as can be seen from Table 2.1. In addition, this structure enables us to explore the need for each component separately and thereby to carry out a systematic search for the best model.

Equation (3.1a) is called the *measurement equation*. The term $\boldsymbol{w}'\boldsymbol{x}_{t-1}$ describes the effect of the past on y_t. The error term ε_t describes the unpredictable part of y_t. It is usually assumed to be from a Gaussian white noise process with variance σ^2. Because ε_t represents what is new and unpredictable, it is referred to as the *innovation*. The innovations are the only source of randomness for the observed time series, $\{y_t\}$.

Equation (3.1b) is known as the *transition equation*. It is a first-order recurrence relationship that describes how the state vectors evolve over time. $\boldsymbol{F}$ is the *transition matrix*. The term $\boldsymbol{F}\boldsymbol{x}_{t-1}$ shows the effect of the past on the current state $\boldsymbol{x}_t$. The term $\boldsymbol{g}\varepsilon_t$ shows the unpredictable change in $\boldsymbol{x}_t$. The vector $\boldsymbol{g}$ determines the extent of the effect of the innovation on the state. It is referred to as a *persistence vector*. The transition equation is the mechanism for creating the inter-temporal dependencies between the values of a time series.

The k-vectors $\boldsymbol{w}$ and $\boldsymbol{g}$ are fixed, and $\boldsymbol{F}$ is a fixed $k \times k$ matrix. These fixed components usually contain some parameters that need to be estimated.

The seed value $\boldsymbol{x}_0$ for the transition equation may be fixed or random. The process that generates the time series may have begun before period 1, but data for the earlier periods are not available. In this situation, the start-up time of the process is taken to be $-\infty$, and $\boldsymbol{x}_0$ must be random. We say that the *infinite start-up assumption* applies. This assumption is typically valid in the study of economic variables. An economy may have been operating for many centuries but an economic quantity may not have been measured until relatively recent times. Consideration of this case is deferred to Chap. 12.

[1] An even more general form is possible by allowing $\boldsymbol{w}$, $\boldsymbol{F}$ and $\boldsymbol{g}$ to vary with time, but that extension will not be considered here.

Alternatively, the process that generates a time series may have started at the beginning of period 1, and $\boldsymbol{x}_0$ is then fixed. In this case we say that the *finite start-up assumption* applies. For example, if y_t is the demand for an inventory item, the start-up time corresponds to the date at which the product is introduced. The theory presented in this and most subsequent chapters is based on the finite start-up assumption with fixed $\boldsymbol{x}_0$.

Upon further consideration, we see that even when a series has not been observed from the outset, we may choose to condition upon the state variables at time zero. We then employ the finite start-up assumption with fixed $\boldsymbol{x}_0$.

Model (3.1) is often called the *Gaussian innovations state space model* because it is defined in terms of innovations that follow a Gaussian distribution. It may be contrasted with alternative state space models, considered in Chap. 13, which involve different and uncorrelated sources of randomness in (3.1a) and (3.1b), rather than a single source of randomness (the innovations) in each case.

The probability density function for $\boldsymbol{y} = [y_1, \dots, y_n]$ is a function of the innovations and has the relatively simple form

$$\begin{aligned} p(\boldsymbol{y} \mid \boldsymbol{x}_0) &= \prod_{t=1}^{n} p(y_t \mid y_1, \dots, y_{t-1}, \boldsymbol{x}_0) \\ &= \prod_{t=1}^{n} p(y_t \mid \boldsymbol{x}_{t-1}) \\ &= \prod_{t=1}^{n} p(\varepsilon_t). \end{aligned}$$

If we assume that the distribution is Gaussian, this expression becomes:

$$p(\boldsymbol{y} \mid \boldsymbol{x}_0) = (2\pi\sigma^2)^{-n/2} \exp\left(-\tfrac{1}{2}\sum_{t=1}^{n} \varepsilon_t^2/\sigma^2\right). \tag{3.2}$$

This is easily evaluated provided we can compute the innovations $\{\varepsilon_t\}$. A simple expression for this computation is given in the next section.

3.2 Innovations and One-Step-Ahead Forecasts

If the value for $\boldsymbol{x}_0$ is known, the innovation ε_t is a one-step-ahead prediction error. This can be seen by applying (3.1a) and (3.1b) to obtain

$$\mathrm{E}(y_t \mid y_{t-1}, \dots, y_1, \boldsymbol{x}_0) = \mathrm{E}(y_t \mid \boldsymbol{x}_{t-1}) = \boldsymbol{w}'\boldsymbol{x}_{t-1}.$$

Then the prediction of y_t, given the initial value $\boldsymbol{x}_0$ and observations $y_1, \dots, y_{t-1}$, is $\boldsymbol{w}'\boldsymbol{x}_{t-1}$. If we denote the prediction by $\hat{y}_{t|t-1}$, the innovations can be computed recursively from the series values using the relationships

$$\hat{y}_{t|t-1} = \boldsymbol{w}'\boldsymbol{x}_{t-1}, \tag{3.3a}$$

$$\varepsilon_t = y_t - \hat{y}_{t|t-1}, \tag{3.3b}$$

$$\boldsymbol{x}_t = \boldsymbol{F}\boldsymbol{x}_{t-1} + \boldsymbol{g}\varepsilon_t. \tag{3.3c}$$

This transformation will be called *general exponential smoothing*. It was first outlined by Box and Jenkins (Box et al. 1994, pp. 176–180) in a much overlooked section of their book.

The forecasts obtained with this transformation are linear functions of past observations. To see this, first substitute (3.3a) and (3.3b) into (3.3c) to find

$$\boldsymbol{x}_t = \boldsymbol{D}\boldsymbol{x}_{t-1} + \boldsymbol{g}y_t, \tag{3.4}$$

where $\boldsymbol{D} = \boldsymbol{F} - \boldsymbol{g}\boldsymbol{w}'$. Then back-solve the recurrence relationship (3.4) to give

$$\boldsymbol{x}_t = \boldsymbol{D}^t\boldsymbol{x}_0 + \sum_{j=0}^{t-1} \boldsymbol{D}^j\boldsymbol{g}y_{t-j}. \tag{3.5}$$

This result indicates that the current state $\boldsymbol{x}_t$ is a linear function of the seed state $\boldsymbol{x}_0$ and past and present values of the time series. Finally, substitute (3.5), lagged by one period, into (3.3a) to give

$$\hat{y}_{t|t-1} = a_t + \sum_{j=1}^{t-1} c_j y_{t-j}, \tag{3.6}$$

where $a_t = \boldsymbol{w}'\boldsymbol{D}^{t-1}\boldsymbol{x}_0$ and $c_j = \boldsymbol{w}'\boldsymbol{D}^{j-1}\boldsymbol{g}$. Thus, the forecast is a linear function of the past observations and the seed state vector.

Equations (3.1), (3.3), and (3.4) demonstrate the beauty of the innovations approach. We may start from the state space model in (3.1) and generate the one-step-ahead forecasts directly using (3.3). When a new observation becomes available, the state vector is updated using (3.4), and the new one-step-ahead forecast is immediately available. As we shall see in Chap. 13, other approaches achieve the updating and the transition from model to forecast function with less transparency and considerably more effort.

3.3 Model Properties

3.3.1 Stability and Forecastability

When the forecasts of y_t are unaffected by observations in the distant past, we describe the model as *forecastable*. Specifically, a forecastable model has the properties

$$\sum_{j=1}^{\infty} |c_j| < \infty \qquad \text{and} \qquad \lim_{t\to\infty} a_t = a. \tag{3.7}$$

Our definition of forecastability allows the initial state $\boldsymbol{x}_0$ to have an ongoing effect on forecasts, but it prevents observations in the distant past having any effect. In most cases, $a = 0$, but not always; an example with $a \neq 0$ is given in Sect. 3.5.2.

A sufficient, but not necessary, condition for (3.7) to hold is that the eigenvalues of $\boldsymbol{D}$ lie inside the unit circle. In this case, $\boldsymbol{D}^j$ converges to a null matrix as j increases. This is known as the "stability condition" and such models are called *stable*. $\boldsymbol{D}$ is called the *discount matrix*. In a stable model, the coefficients of the observations in (3.6) decay exponentially. The exponential decline in the importance of past observations is a property that is closely associated with exponential smoothing.

It turns out that sometimes a_t converges to a constant and the coefficients $\{c_j\}$ converge to zero even when $\boldsymbol{D}$ has a unit root. In this case, the forecasts of y_t are unaffected by distant observations, while the forecasts of $\boldsymbol{x}_t$ may be affected by distant past observations even for large values of t. Thus, any stable model is also forecastable, but some forecastable models are not stable. Examples of unstable but forecastable models are given in Chap. 10. The stability condition on $\boldsymbol{D}$ is closely related to the invertibility restriction for ARIMA models; this is discussed in more detail in Chap. 11.

3.3.2 Stationarity

The other matrix that controls the model properties is the *transition matrix*, $\boldsymbol{F}$. If we iterate (3.1b), we obtain

$$
\begin{aligned}
\boldsymbol{x}_t &= \boldsymbol{F}\boldsymbol{x}_{t-1} + \boldsymbol{g}\varepsilon_t \\
&= \boldsymbol{F}^2\boldsymbol{x}_{t-2} + \boldsymbol{F}\boldsymbol{g}\varepsilon_{t-1} + \boldsymbol{g}\varepsilon_t \\
&\;\vdots \\
&= \boldsymbol{F}^t\boldsymbol{x}_0 + \sum_{j=0}^{t-1} \boldsymbol{F}^j\boldsymbol{g}\varepsilon_{t-j}.
\end{aligned}
$$

Substituting this result into (3.1a) gives

$$
y_t = d_t + \sum_{j=0}^{t-1} k_j\varepsilon_{t-j}, \tag{3.8}
$$

where $d_t = \boldsymbol{w}'\boldsymbol{F}^{t-1}\boldsymbol{x}_0$, $k_0 = 1$ and $k_j = \boldsymbol{w}'\boldsymbol{F}^{j-1}\boldsymbol{g}$ for $j = 1, 2, \ldots$. Thus, the observation is a linear function of the seed state $\boldsymbol{x}_0$ and past and present errors. Any linear innovations model may be represented in the form (3.8); this is an example of a finite Wold decomposition (Brockwell and Davis 1991, p. 180).

The model is described as *stationary*[2] if

$$\sum_{j=0}^{\infty} |k_j| < \infty \qquad \text{and} \qquad \lim_{t \to \infty} d_t = d. \tag{3.9}$$

In such a model, the coefficients of the errors in (3.8) converge rapidly to zero, and the impact of the seed state vector diminishes over time.

We may then consider the limiting form of the model, corresponding to the infinite start-up assumption. Equation (3.8) becomes

$$y_t = d + \sum_{j=0}^{\infty} k_j \varepsilon_{t-j}.$$

This form is known as the Wold decomposition for a stationary series. It follows directly that $\mathrm{E}(y_t) = d$ and $\mathrm{V}(y_t) = \sigma^2 \sum_{j=0}^{\infty} k_j^2$.

A sufficient, but not necessary, condition for stationarity to hold is for the absolute value of each eigenvalue of $\boldsymbol{F}$ to lie strictly in the unit interval $(0, 1)$. Then $\boldsymbol{F}^j$ converges to a null matrix as j increases. As with the stability property, it turns out that sometimes d_t converges to a constant and the coefficients $\{k_j\}$ converge to zero even when $\boldsymbol{F}$ has a unit root. However, this does not occur with any of the models we consider, and so it will not be discussed further.

Stationarity is a rare property in exponential smoothing state space models. None of the models discussed in Chap. 2 are stationary. The six linear models described in that chapter have at least one unit root for the $\boldsymbol{F}$ matrix. However, it is possible to define stationary models in the exponential smoothing framework; an example of such a model is given in Sect. 3.5.1, where all of the transition equations involve damping.

3.4 Basic Special Cases

The linear innovations state space model effectively contains an infinite number of special cases that can potentially be used to model a time series; that is, to provide a stochastic approximation to the data generating process of a time series. However, in practice we use only a handful of special cases that possess the capacity to represent commonly occurring patterns such as trends, seasonality and business cycles. Many of these special cases were introduced in Chap. 2.

The simplest special cases are based on polynomial approximations of continuous real functions. A polynomial function can be used to approximate any real function in the neighborhood of a specified point (this is known

[2] The terminology "stationary" arises because the distribution of $(y_t, y_{t+1}, \ldots, y_{t+s})$ does not depend on time t when the initial state $\boldsymbol{x}_0$ is random.

as Taylor's theorem in real analysis). To demonstrate the idea, we temporarily take the liberty of representing the data by a continuous path, despite the fact that business and economic data are typically collected at discrete points of time.

The first special case to be considered, the local level model, is a zero-order polynomial approximation. As depicted in Fig. 3.1a, at any point along the data path, the values in the neighborhood of the point are approximated by a short flat line representing what is referred to as a local level. As its height changes over time, it is necessary to approximate the data path by many local levels. Thus, the local level effectively represents the state of a process generating a time series.

The gap between successive levels is treated as a random variable. Moreover, this random variable is assumed to have a Gaussian distribution that has a zero mean to ensure that the level is equally likely to go up or down.

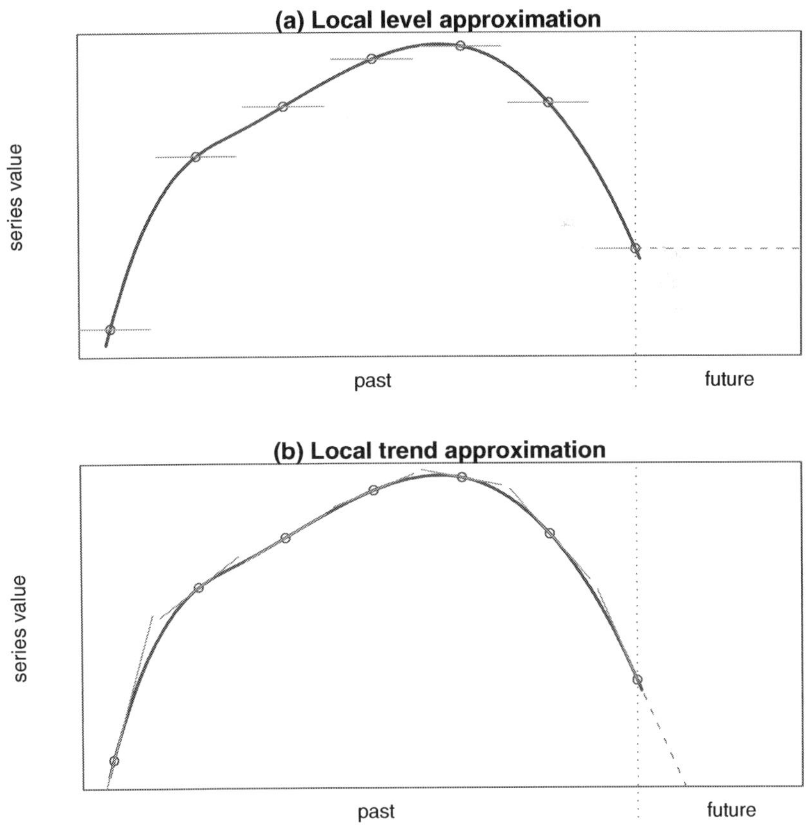

Fig. 3.1. Schematic representation of (**a**) a local level model; and (**b**) a local trend model.

The final local level is projected into the future to give predictions. As the approximation is only effective in a small neighborhood, predictions generated this way are only likely to be reliable in the shorter term.

The second special case involves a first-order polynomial approximation. At each point, the data path is approximated by a straight line. In the deterministic world of analysis, this line would be tangential to the data path at the selected point. In the stochastic world of time series data, it can only be said that the line has a similar height and a similar slope to the data path. Randomness means that the line is not exactly tangential. The approximating line changes over time, as depicted in Fig. 3.1b, to reflect the changing shape of the data path. The state of the process is now summarized by the level and the slope at each point of the path. The stochastic representation is based on the assumption that the gaps between successive slopes are Gaussian random variables with a zero mean. Note that the prediction is obtained by projecting the last linear approximation into the future.

It is possible to move beyond linear functions to higher order polynomials with quadratic or cubic terms. However, these extensions are rarely used in practice. It is commonly thought that the randomness found in real time series typically swamps and hides the effects of curvature.

Another strategy that does often bear fruit is the search for periodic behavior in time series caused by seasonal effects. Ignoring growth for the moment, the level in a particular month may be closer to the level in the corresponding month in the previous year than to the level in the preceding month. This leads to seasonal state space models.

3.4.1 Local Level Model: ETS(A,N,N)

The simplest way to transmit the history of a process is through a single state, ℓ_t, called the level. The resulting state space model is defined by the equations

$$y_t = \ell_{t-1} + \varepsilon_t, \tag{3.10a}$$
$$\ell_t = \ell_{t-1} + \alpha\varepsilon_t, \tag{3.10b}$$

where $\varepsilon_t \sim \text{NID}(0, \sigma^2)$. It conforms to a state space structure with $x_t = \ell_t$, $w = 1$, $F = 1$ and $g = \alpha$. The values that are generated by this stochastic model are randomly scattered about the (local) levels as described in (3.10a). This is illustrated in Fig. 3.2 with a simulated series.

In demand applications, the level ℓ_{t-1} represents the anticipated demand for period t, and ε_t represents the unanticipated demand. Changes to the underlying level may be induced by changes in the customer base such as the arrival of new customers, or by new competitors entering the market. Changes like these transcend a single period and must affect the underlying level. It is assumed that the unanticipated demand includes a persistent and

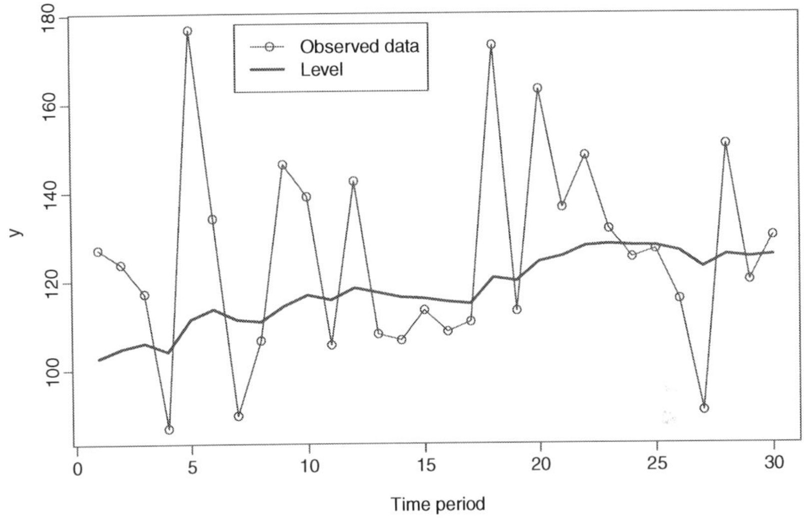

Fig. 3.2. Simulated series from the ETS(A,N,N) model. Here $\alpha = 0.1$ and $\sigma = 5$.

a temporary effect; $\alpha\varepsilon_t$ denotes the persistent effect, feeding through to future periods via the (local) levels governed by (3.10b).

The degree of change of successive levels is governed by the size of the smoothing parameter α. The cases where $\alpha = 0$ and $\alpha = 1$ are of special interest.

Case: $\alpha = 0$ The local levels do not change at all when $\alpha = 0$. Their common level is then referred to as the global level. Successive values of the series y_t are independently and identically distributed. Its moments do not change over time.

Case: $\alpha = 1$ The model reverts to a random walk $y_t = y_{t-1} + \varepsilon_t$. Successive values of the time series y_t are clearly dependent.

The special case of transformation (3.3) for model (3.10) is

$$
\begin{aligned}
\hat{y}_{t|t-1} &= \ell_{t-1}, \\
\varepsilon_t &= y_t - \ell_{t-1}, \\
\ell_t &= \ell_{t-1} + \alpha\varepsilon_t.
\end{aligned}
$$

It corresponds to simple exponential smoothing (Brown 1959), one of the most widely used methods of forecasting in business applications. It is a simple recursive scheme for calculating the innovations from the raw data. Equation (3.4) reduces to

$$\ell_t = (1-\alpha)\ell_{t-1} + \alpha y_t. \tag{3.11}$$

The one-step-ahead predictions obtained from this scheme are linearly dependent on earlier series values. Equation (3.6) indicates that

$$\hat{y}_{t+1|t} = (1-\alpha)^t \ell_0 + \alpha \sum_{j=0}^{t-1} (1-\alpha)^j y_{t-j}. \tag{3.12}$$

This is a linear function of the data and seed level. Ignoring the first term (which is negligible for large values of t and $|1-\alpha| < 1$), the prediction $\hat{y}_{t|t-1}$ is an *exponentially weighted average* of past observations. The coefficients depend on the "discount factor" $1-\alpha$. If $|1-\alpha| < 1$, then the coefficients become smaller as j increases. That is, the stability condition is satisfied if and only if $0 < \alpha < 2$. The coefficients are positive if and only if $0 < (1-\alpha) < 1$, and (3.11) can then be interpreted as a weighted average of the past level ℓ_{t-1} and the current series value y_t. Thus, the prediction can only be interpreted as a weighted average if $0 < \alpha < 1$.

Consequently, there are two possible ranges for α that have been proposed: $0 < \alpha < 2$ on the basis of a stability argument, and $0 < \alpha < 1$ on the basis of an interpretation as a weighted average. The narrower range is widely used in practice.

The impact of various values of α may be discerned from Fig. 3.3. It shows simulated time series from an ETS(A,N,N) model with $\ell_0 = 100$ and $\sigma = 5$ for various values of α. The same random number stream from a Gaussian distribution was used for the three series, so that any perceived differences can be attributed entirely to changes in α. For the case $\alpha = 0.1$, the underlying level is reasonably stable. The plot has a jagged appearance because there is a tendency for the series to switch direction between successive observations. This is a consequence of the fact, shown in Chap. 11, that successive first-differences of the series, Δy_t and Δy_{t-1}, are negatively correlated when α is restricted to the interval $(0,1)$. When $\alpha = 0.5$, the underlying level displays a much greater tendency to change. There is still a tendency for successive observations to move in opposite directions. In the case $\alpha = 1.5$, there is an even greater tendency for the underlying level to change. However, the series is much smoother. This reflects the fact, also established in Chap. 11, that successive first-differences of the series are positively correlated for cases where α lies in the interval $(1,2)$.

3.4.2 Local Trend Model: ETS(A,A,N)

The local level model can be augmented by a growth rate b_t to give

$$y_t = \ell_{t-1} + b_{t-1} + \varepsilon_t, \tag{3.13a}$$
$$\ell_t = \ell_{t-1} + b_{t-1} + \alpha\varepsilon_t, \tag{3.13b}$$
$$b_t = b_{t-1} + \beta\varepsilon_t, \tag{3.13c}$$

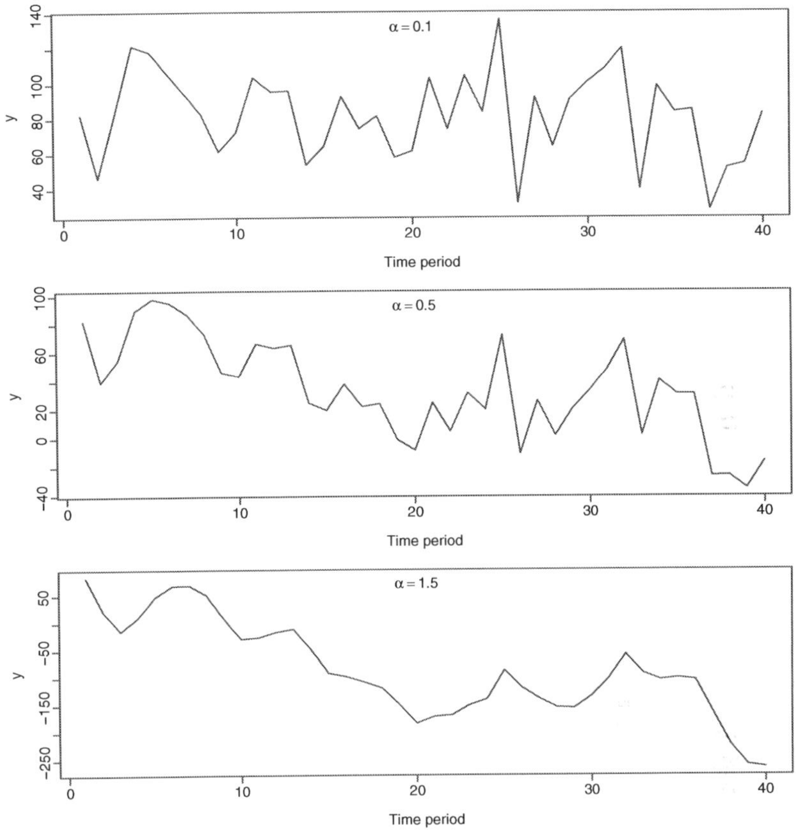

Fig. 3.3. Comparison of simulated time series from a local level model. Here $\sigma = 5$.

where there are now two smoothing parameters α and β. The growth rate (or slope) b_t may be positive, zero or negative. Model (3.13) has a state space structure with

$$\boldsymbol{x}_t = \begin{bmatrix}\ell_t \; b_t\end{bmatrix}', \quad \boldsymbol{w} = \begin{bmatrix}1 \; 1\end{bmatrix}', \quad \boldsymbol{F} = \begin{bmatrix}1 & 1\\ 0 & 1\end{bmatrix} \quad \text{and} \quad \boldsymbol{g} = \begin{bmatrix}\alpha \; \beta\end{bmatrix}'.$$

The size of the smoothing parameters reflects the impact of the innovations on the level and growth rate. Figure 3.4 shows simulated values from the model for different settings of the smoothing parameters. When $\beta = 0$, the growth rate is constant over time. If, in addition, $\alpha = 0$, the level changes at a constant rate over time. That is, there is no random change in the level or growth. This case will be called a *global trend*. The constant growth rate is sometimes interpreted as a long-term growth rate. For other values of the smoothing parameters, the growth rate follows a random walk over time. As

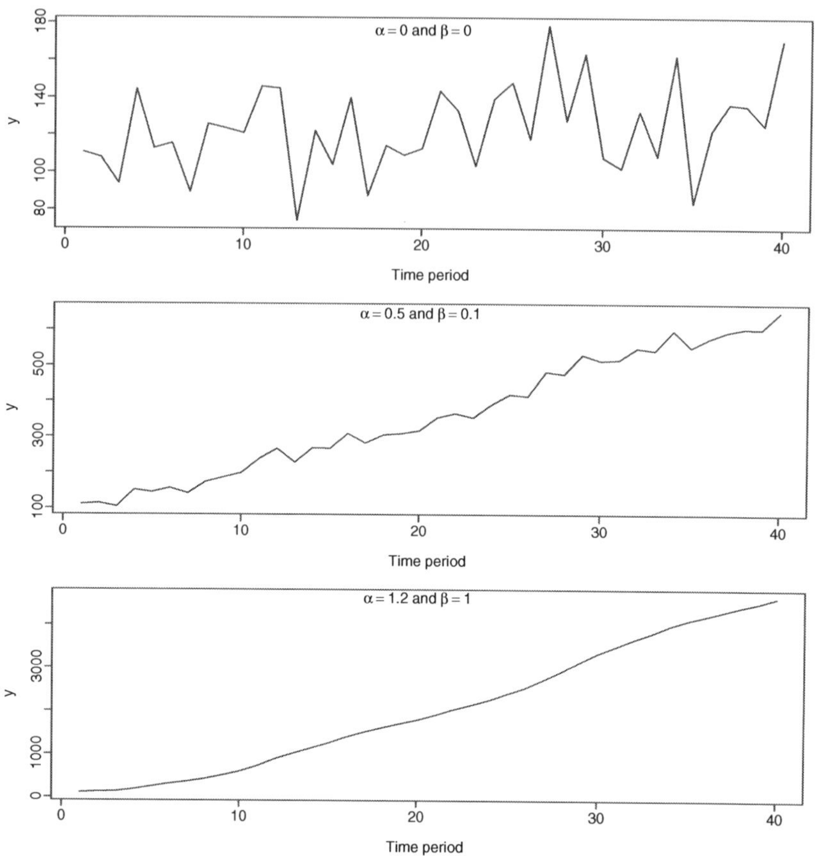

Fig. 3.4. Comparison of simulated time series from a local trend model. Here $\sigma = 5$.

the smoothing parameters increase in size, there is a tendency for the series to become smoother.

For this model, the transformation (3.3) of series values into innovations becomes

$$\begin{aligned}\hat{y}_{t|t-1} &= \ell_{t-1} + b_{t-1},\\ \varepsilon_t &= y_t - \hat{y}_{t|t-1},\\ \ell_t &= \ell_{t-1} + b_{t-1} + \alpha\varepsilon_t,\\ b_t &= b_{t-1} + \beta\varepsilon_t.\end{aligned}$$

This corresponds to Holt's linear exponential smoothing (Holt 1957). An equivalent system of equations is

$$\hat{y}_{t|t-1} = \ell_{t-1} + b_{t-1}, \tag{3.14a}$$

$$\varepsilon_t = y_t - \hat{y}_{t|t-1}, \tag{3.14b}$$
$$\ell_t = \alpha y_t + (1-\alpha)(\ell_{t-1} + b_{t-1}), \tag{3.14c}$$
$$b_t = \beta^*(\ell_t - \ell_{t-1}) + (1-\beta^*)b_{t-1}, \tag{3.14d}$$

where $\beta^* = \beta/\alpha$. The term $\ell_t - \ell_{t-1}$ is often interpreted as the "actual growth" as distinct from the predicted growth b_{t-1}.

Equations (3.14c) and (3.14d) may be interpreted as weighted averages if $0 < \alpha < 1$ and $0 < \beta^* < 1$, or equivalently, if $0 < \alpha < 1$ and $0 < \beta < \alpha$. These restrictions are commonly applied in practice. Alternatively, it can be shown (see Chap. 10) that the model is stable (i.e., the discount matrix $\boldsymbol{D}^j$ converges to $\boldsymbol{0}$ as j increases) when $\alpha > 0$, $\beta > 0$ and $2\alpha + \beta < 4$. This provides a much larger parameter region than is usually allowed.

3.4.3 Local Additive Seasonal Model: ETS(A,A,A)

For time series that exhibit seasonal patterns, the local trend model can be augmented by seasonal effects, denoted by s_t. Often the structure of the seasonal pattern changes over time in response to changes in tastes and technology. For example, electricity demand used to peak in winter, but in some locations it now peaks in summer due to the growing prevalence of air conditioning. Thus, the formulae used to represent the seasonal effects should allow for the possibility of changing seasonal patterns. The ETS(A,A,A) model is

$$y_t = \ell_{t-1} + b_{t-1} + s_{t-m} + \varepsilon_t, \tag{3.15a}$$
$$\ell_t = \ell_{t-1} + b_{t-1} + \alpha\varepsilon_t, \tag{3.15b}$$
$$b_t = b_{t-1} + \beta\varepsilon_t, \tag{3.15c}$$
$$s_t = s_{t-m} + \gamma\varepsilon_t. \tag{3.15d}$$

This model corresponds to the first-order state space model where

$$\boldsymbol{w}' = \begin{bmatrix} 1\,1\,0 \cdots 0\,1 \end{bmatrix},$$

$$\boldsymbol{x}_t = \begin{bmatrix} \ell_t \\ b_t \\ s_t \\ s_{t-1} \\ \vdots \\ s_{t-m+1} \end{bmatrix}, \quad \boldsymbol{F} = \begin{bmatrix} 1\,1\,0\,0 & \cdots & 0\,0 \\ 0\,1\,0\,0 & \cdots & 0\,0 \\ 0\,0\,0\,0 & \cdots & 0\,1 \\ 0\,0\,1\,0 & \cdots & 0\,0 \\ 0\,0\,0\,1 & \cdots & 0\,0 \\ \vdots\,\vdots\,\vdots\,\vdots & \ddots & \vdots\,\vdots \\ 0\,0\,0\,0 & \cdots & 1\,0 \end{bmatrix} \quad \text{and} \quad \boldsymbol{g} = \begin{bmatrix} \alpha \\ \beta \\ \gamma \\ 0 \\ \vdots \\ 0 \end{bmatrix}.$$

Careful inspection of model (3.15) shows that the level and seasonal terms are confounded. If an arbitrary quantity δ is added to the seasonal elements and subtracted from the level, the following equations are obtained

$$\begin{aligned}
y_t &= (\ell_{t-1} - \delta) + b_{t-1} + (s_{t-m} + \delta) + \varepsilon_t, \\
\ell_t - \delta &= \ell_{t-1} - \delta + b_{t-1} + \alpha\varepsilon_t, \\
b_t &= b_{t-1} + \beta\varepsilon_t, \\
(s_t + \delta) &= (s_{t-m} + \delta) + \gamma\varepsilon_t,
\end{aligned}$$

which is equivalent to (3.15). To avoid this problem, it is desirable to constrain the seasonal component so that any sequence $\{s_t, s_{t+1}, \ldots, s_{t+m-1}\}$ sums to zero (or at least has mean zero). The seasonal components are said to be *normalized* when this condition is true. Normalization of seasonal factors involves a subtle modification of the model and will be addressed in Chap. 8. In the meantime, we can readily impose the constraint that the seasonal factors in the initial state $\boldsymbol{x}_0$ must sum to zero. This means that the seasonal components start off being normalized, although there is nothing to constrain them from drifting away from zero over time.

The transformation from series values to prediction errors can be shown to be

$$\begin{aligned}
\hat{y}_{t|t-1} &= \ell_{t-1} + b_{t-1} + s_{t-m}, \\
\varepsilon_t &= y_t - \hat{y}_{t|t-1}, \\
\ell_t &= \ell_{t-1} + b_{t-1} + \alpha\varepsilon_t, \\
b_t &= b_{t-1} + \beta\varepsilon_t, \\
s_t &= s_{t-m} + \gamma\varepsilon_t.
\end{aligned}$$

This corresponds to a commonly used additive version of seasonal exponential smoothing (Winters 1960). An equivalent form of these transition equations is

$$\begin{aligned}
\hat{y}_{t|t-1} &= \ell_{t-1} + b_{t-1} + s_{t-m}, && \text{(3.16a)} \\
\varepsilon_t &= y_t - \hat{y}_{t|t-1}, && \text{(3.16b)} \\
\ell_t &= \alpha(y_t - s_{t-m}) + (1-\alpha)(\ell_{t-1} + b_{t-1}), && \text{(3.16c)} \\
b_t &= \beta^*(\ell_t - \ell_{t-1}) + (1-\beta^*)b_{t-1}, && \text{(3.16d)} \\
s_t &= \gamma^*(y_t - \ell_t) + (1-\gamma^*)s_{t-m}, && \text{(3.16e)}
\end{aligned}$$

where the series value is deseasonalized in the trend equations and detrended in the seasonal equation, $\beta^* = \beta/\alpha$ and $\gamma^* = \gamma/(1-\alpha)$. Equations (3.16c–e) can be interpreted as weighted averages, in which case the natural parametric restrictions are that each of α, β^* and γ lie in the $(0,1)$ interval. Equivalently, $0 < \alpha < 1$, $0 < \beta < \alpha$ and $0 < \gamma < 1 - \alpha$. However, a consideration of the properties of the discount matrix $\boldsymbol{D}$ leads to a different parameter region; this will be discussed in Chap. 10.

3.5 Variations on the Common Models

A number of variations on the basic models of the previous section can be helpful in some applications.

3.5.1 Damped Level Model

One feature of the models in the framework described in Chap. 2 is that the mean and variance are *local* properties. We may define these moments given the initial conditions, but they do not converge to a stable value as t increases without limit. In other words, the models are all nonstationary; the $\boldsymbol{F}$ matrix has at least one unit root in every case. However, it is possible to describe analogous models that are stationary.

Consider the damped local level model

$$\begin{aligned} y_t &= \mu + \phi\ell_{t-1} + \varepsilon_t, \\ \ell_t &= \phi\ell_{t-1} + \alpha\varepsilon_t. \end{aligned}$$

The transition matrix is simply $F = \phi$, which has no roots greater than one provided $|\phi| < 1$. Thus, the model is stationary for $|\phi| < 1$.

The discount matrix is $D = \phi - \alpha$. Thus, the model is stable provided $|\phi - \alpha| < 1$, or equivalently, $\phi - 1 < \alpha < \phi + 1$.

We may eliminate the state variable to arrive at

$$y_t = \mu + \phi^t\ell_0 + \varepsilon_t + \alpha[\phi\varepsilon_{t-1} + \phi^2\varepsilon_{t-2} + \cdots + \phi^{t-1}\varepsilon_1].$$

When $|\phi| < 1$, the mean and variance approach finite limits as $t \to \infty$:

$$\begin{aligned} \mathrm{E}(y_t \mid \ell_0) &= \mu + \phi^t\ell_0 && \to \mu, \\ \mathrm{V}(y_t \mid \ell_0) &= \sigma^2\left[1 + \frac{\alpha^2\phi^2(1-\phi^{2t-2})}{1-\phi^2}\right] && \to \sigma^2\left[1 + \frac{\alpha^2\phi^2}{1-\phi^2}\right]. \end{aligned}$$

Thus, whenever $|\phi| < 1$, the mean reverts to the stable value μ and the variance remains finite. When the series has an infinite past, the limiting values are the unconditional mean and variance. Such stationary series play a major role in the development of Auto Regressive Integrated Moving Average (ARIMA) models, as we shall see in Chap. 11.

There are two reasons why our treatment of mean reversion (or stationarity) is so brief. First, the use of a finite start-up assumption means that stationarity is not needed in order to define the likelihood function. Second, stationary series are relatively uncommon in business and economic applications. Nevertheless, our estimation procedures (Chap. 5) allow mean reverting processes to be fitted if required.

3.5.2 Local Level Model with Drift

A local trend model allows the growth rate to change stochastically over time. If $\beta = 0$, the growth rate is constant and equal to a value that will be denoted by b. The local level model then reduces to

$$\begin{aligned} y_t &= \ell_{t-1} + b + \varepsilon_t, \\ \ell_t &= b + \ell_{t-1} + \alpha\varepsilon_t, \end{aligned}$$

where $\varepsilon_t \sim \text{NID}(0, \sigma^2)$. It is called a "local level model with drift" and has a state space structure with

$$\boldsymbol{x}_t = \begin{bmatrix}\ell_t \; b\end{bmatrix}', \quad \boldsymbol{w} = \begin{bmatrix}1 \; 1\end{bmatrix}', \quad \boldsymbol{F} = \begin{bmatrix}1 & 1 \\ 0 & 1\end{bmatrix} \quad \text{and} \quad \boldsymbol{g} = \begin{bmatrix}\alpha \; 0\end{bmatrix}'.$$

This model can be applicable to economic time series that display an upward (or downward) drift. It is sometimes preferred for longer term forecasting because projections are made with the average growth that has occurred throughout the sample rather than a local growth rate, which essentially represents the growth rate that prevails towards the end of the sample.

The discount matrix for this model is

$$\boldsymbol{D} = \begin{bmatrix}1-\alpha & 1-\alpha \\ 0 & 1\end{bmatrix},$$

which has eigenvalues of 1 and $1 - \alpha$. Thus, the model is not stable as $\boldsymbol{D}^j$ does not converge to $\boldsymbol{0}$. It is, however, forecastable, provided $0 < \alpha < 2$. The model is also forecastable when $\alpha = 0$, as it then reduces to the linear regression model $y_t = \ell_0 + bt + \varepsilon_t$. Discussion of this type of discount matrix will occur in Chap. 10.

The local level model with drift is also known as "simple exponential smoothing with drift." Hyndman and Billah (2003) showed that this model is equivalent to the "Theta method" of Assimakopoulos and Nikolopoulos (2000) with b equal to half the slope of a linear regression of the observed data against their time of observation.

3.5.3 Damped Trend Model: ETS(A,A$_\text{d}$,N)

Another possibility is to take the local trend model and dampen its growth rate with a factor ϕ in the region $0 \le \phi < 1$. The resulting model is

$$\begin{aligned} y_t &= \ell_{t-1} + \phi b_{t-1} + \varepsilon_t, \\ \ell_t &= \ell_{t-1} + \phi b_{t-1} + \alpha\varepsilon_t, \\ b_t &= \phi b_{t-1} + \beta\varepsilon_t. \end{aligned}$$

The characteristics of the damped local trend model are compatible with features observed in many business and economic time series. It sometimes yields better forecasts than the local trend model. Note that the local trend model is a special case where $\phi = 1$.

The ETS(A,A_d,N) model performs remarkably well when forecasting real data (Fildes 1992).

3.5.4 Seasonal Model Based only on Seasonal Levels

If there is no trend in a time series with a seasonal pattern, the ETS(A,N,A) model can be simplified to a model that has a different level in each season. A model for a series with m periods per annum is

$$y_t = \ell_{t-m} + \varepsilon_t, \tag{3.17a}$$
$$\ell_t = \ell_{t-m} + \alpha\varepsilon_t. \tag{3.17b}$$

It conforms to a state space model where

$$\boldsymbol{w}' = \begin{bmatrix} 0\,0 \cdots 1 \end{bmatrix},$$

$$\boldsymbol{x}_t = \begin{bmatrix} \ell_t \\ \ell_{t-1} \\ \vdots \\ \ell_{t-m+1} \end{bmatrix} \quad \boldsymbol{F} = \begin{bmatrix} 0\,0 \ldots 0\,1 \\ 1\,0 \ldots 0\,0 \\ 0\,1 \ldots 0\,0 \\ \vdots\;\vdots\;\ddots\;\vdots\;\vdots \\ 0\,0 \ldots 1\,0 \end{bmatrix} \quad \text{and} \quad \boldsymbol{g} = \begin{bmatrix} \alpha \\ 0 \\ \vdots \\ 0 \end{bmatrix}.$$

The weighted average requirement is satisfied if $0 < \alpha < 1$. Because there is no link between the observations other than those m periods apart, we may consider the m sub-models separately. It follows directly that the model is stable when all the sub-models are stable, which is true provided $0 < \alpha < 2$.

3.5.5 Parsimonious Local Seasonal Model

The problem with the seasonal models (3.15) and (3.17) is that they potentially involve a large number of states, and the initial seed state $\boldsymbol{x}_0$ contains a set of parameters that need to be estimated. Modeling weekly demand data, for example, would entail 51 independent seed values for the seasonal recurrence relationships. Estimation of the seed values then makes relatively high demands on computational facilities. Furthermore, the resulting predictions may not be as robust as those from more parsimonious representations.

To emphasize the possibility of a more parsimonious approach, consider the case of a product with monthly sales that peak in December for Christmas, but which tend to be the same, on average, in the months of January to November. There are then essentially two seasonal components, one for the months of January to November, and a second for December. There is no need for 12 separate monthly components.

We require a seasonal model in a form that allows a reduced number of seasonal components. First, redefine m to denote the number of seasonal components, as distinct from the number of seasons per year. In the above example, $m = 2$ instead of 12. An m-vector z_t indicates which seasonal component applies in period t. If seasonal component j applies in period t, then the element $z_{tj} = 1$ and all other elements equal 0. It is assumed that the typical seasonal component j has its own level, which in period t is denoted by ℓ_{tj}. The levels are collected into an m-vector denoted by $\boldsymbol{\ell}_t$. Then the model is

$$y_t = z_t'\boldsymbol{\ell}_{t-1} + b_{t-1} + \varepsilon_t, \tag{3.18a}$$

$$\boldsymbol{\ell}_t = \boldsymbol{\ell}_{t-1} + \mathbf{1}b_{t-1} + (\mathbf{1}\alpha + z_t\gamma)\varepsilon_t, \tag{3.18b}$$

$$b_t = b_{t-1} + \beta\varepsilon_t, \tag{3.18c}$$

where $\mathbf{1}$ represents an m-vector of ones. The term $z_t'\boldsymbol{\ell}_{t-1}$ picks out the level of the seasonal component relevant to period t. The term $\mathbf{1}b_{t-1}$ ensures that each level is adjusted by the same growth rate. It is assumed that the random change has a common effect and an idiosyncratic effect. The term $\mathbf{1}\alpha\varepsilon_t$ represents the common effect, and the term $z_t\beta\varepsilon_t$ is the adjustment to the seasonal component associated with period t.

This model must be coupled with a method that searches systematically for months that possess common seasonal components. We discuss this problem in Chap. 14. In the special case where no common components are found (e.g., $m = 12$ for monthly data), the above model is then equivalent to the seasonal model in Sect. 3.4.3. If, in addition, there is no growth, the model is equivalent to the seasonal level model in Sect. 3.5.4.

Model (3.18) is easily adapted to handle multiple seasonal patterns. For example, daily demand may be influenced by a trading cycle that repeats itself every week, in addition to a seasonal pattern that repeats itself annually. Extensions of this kind are also considered in Chap. 14.

An important point to note is that this seasonal model does not conform to the general form (3.1), because the $\boldsymbol{g}$ and $\boldsymbol{w}$ vectors are time-dependent. A more general time-varying model must be used instead.

3.5.6 Composite Models

Two different models can be used as basic building blocks to yield even larger models. Suppose two basic innovations state space models indexed by $i = 1, 2$ are given by

$$y_t = \boldsymbol{w}_i'\boldsymbol{x}_{i,t-1} + \varepsilon_{it},$$

$$\boldsymbol{x}_{it} = \boldsymbol{F}_i\boldsymbol{x}_{i,t-1} + \boldsymbol{g}_i\varepsilon_{it},$$

where $\varepsilon_{it} \sim \text{NID}(0, v_i)$. A new model can be formed by combining them as follows:

$$y_t = \boldsymbol{w}_1' \boldsymbol{x}_{1,t-1} + \boldsymbol{w}_2' \boldsymbol{x}_{2,t-1} + \varepsilon_t,$$

$$\begin{bmatrix} \boldsymbol{x}_{1t} \\ \boldsymbol{x}_{2t} \end{bmatrix} = \begin{bmatrix} \boldsymbol{F}_1 & \boldsymbol{0} \\ \boldsymbol{0} & \boldsymbol{F}_2 \end{bmatrix} \begin{bmatrix} \boldsymbol{x}_{1,t-1} \\ \boldsymbol{x}_{2,t-1} \end{bmatrix} + \begin{bmatrix} \boldsymbol{g}_1 \\ \boldsymbol{g}_2 \end{bmatrix} \varepsilon_t.$$

For example, the local trend model (3.13) in Sect. 3.4.2 and the seasonal model (3.17) in Sect. 3.5.4 can be combined using this principle. To avoid conflict with respect to the levels, the ℓ_t in the seasonal model (3.17) is replaced by s_t. The resulting model is the local additive seasonal model (3.15) in Sect. 3.4.3.

3.6 Exercises

Exercise 3.1. Consider the local level model ETS(A,N,N). Show that the process is forecastable and stationary when $\alpha = 0$ but that neither property holds when $\alpha = 2$.

Exercise 3.2. Consider the local level model with drift, defined in Sect. 3.5.2. Define the detrended variable $z_{1t} = y_t - bt$ and the differenced variable $z_{2t} = y_t - y_{t-1}$. Show that both of these processes are stable provided $0 < \alpha < 2$ but that only z_{2t} is stationary.

Exercise 3.3. Consider the local level model ETS(A,N,N). Show that the mean and variance for $y_t|\ell_0$ are ℓ_0 and $\ell_0{}^2(1 + (t-1)\alpha^2)$ respectively.

Exercise 3.4. For the damped trend model ETS(A,A$_\text{d}$,N), find the discount matrix $\boldsymbol{D}$ and its eigenvalues.

4

Nonlinear and Heteroscedastic Innovations State Space Models

In this chapter we consider a broader class of innovations state space models, which enables us to examine multiplicative structures for any or all of the trend, the seasonal pattern and the innovations process. This general class was introduced briefly in Sect. 2.5.2. As for the linear models introduced in the previous chapter, this discussion will pave the way for a general discussion of estimation and prediction methods later in the book.

One of the intrinsic advantages of the innovations framework is that we preserve the ability to write down closed-form expressions for the recursive relationships and point forecasts. In addition, the time series may be represented as a weighted sum of the innovations, where the weights for a given innovation depend only on the initial conditions and earlier innovations, so that the weight and the innovation are conditionally independent. As before, we refer to this structure as the innovations representation of the time series. We find that these models are inherently similar to those for the linear case.

The general innovations form of the state space model is introduced in Sect. 4.1 and various special cases are considered in Sect. 4.2. We then examine seasonal models in Sect. 4.3. Finally, several variations on the core models are examined in Sect. 4.4.

4.1 Innovations Form of the General State Space Model

We employ the same basic notation as in Sect. 3.1, so that y_t denotes the element of the time series corresponding to time t. Prior to time t, y_t denotes a random variable, but it becomes a fixed value after being observed. The first n values of a time series form the n-vector $\boldsymbol{y}$.

Following the discussion in Sects. 2.5.2 and 3.1, we define the model for the variable of interest, y_t, in terms of the state variables that form the state vector, $\boldsymbol{x}_t$. We will select the elements of the state vector to describe the trend and seasonal elements of the series, using these terms as building blocks to enable us to formulate a model that captures the key components of the data generating process.

From Sect. 2.5.2, we specify the general model with state vector $\boldsymbol{x}_t = (\ell_t, b_t, s_t, s_{t-1}, \ldots, s_{t-m+1})'$ and state space equations of the form:

$$y_t = w(\boldsymbol{x}_{t-1}) + r(\boldsymbol{x}_{t-1})\varepsilon_t, \tag{4.1a}$$
$$\boldsymbol{x}_t = \boldsymbol{f}(\boldsymbol{x}_{t-1}) + \boldsymbol{g}(\boldsymbol{x}_{t-1})\varepsilon_t, \tag{4.1b}$$

where $r(\cdot)$ and $w(\cdot)$ are scalar functions, $\boldsymbol{f}(\cdot)$ and $\boldsymbol{g}(\cdot)$ are vector functions, and ε_t is a white noise process with variance σ^2. Note that we do not specify that the process is Gaussian because such an assumption may conflict with the underlying structure of the data generating process (e.g., when the series contains only non-negative values). Nevertheless, the Gaussian assumption is often a reasonable approximation when the level of the process is sufficiently far from the origin (or, more generally, the region of impossible values) and it will then be convenient to use the Gaussian assumption as a basis for inference. The functions in this model may all be time-indexed, but we shall concentrate on constant functions (the invariant form), albeit with time-varying arguments. In Chap. 3, the functions r and $\boldsymbol{g}$ were constants, whereas w and $\boldsymbol{f}$ were linear in the state vector. The simplest nonlinear scheme of interest corresponds to $\{w(\boldsymbol{x}_{t-1}) = r(\boldsymbol{x}_{t-1}) = f(\boldsymbol{x}_{t-1}) = \ell_{t-1};\ g(\boldsymbol{x}_{t-1}) = \alpha\ell_{t-1}\}$ or

$$y_t = \ell_{t-1}(1 + \varepsilon_t), \tag{4.2a}$$
$$\ell_t = \ell_{t-1}(1 + \alpha\varepsilon_t). \tag{4.2b}$$

These equations describe the ETS(M,N,N) or *local level* model given in Table 2.3 (p. 22). We may eliminate the state variable between (4.2a) and (4.2b) to arrive at a reduced form for the model:

$$y_t = y_{t-1}(1 + \varepsilon_t)(1 + \alpha\varepsilon_{t-1})/(1 + \varepsilon_{t-1}),$$
$$y_t = \ell_0(1 + \varepsilon_t)\prod_{j=1}^{t-1}(1 + \alpha\varepsilon_j).$$

We may also eliminate ε_t to arrive at the recursive relationship:

$$\ell_t = \ell_{t-1} + \alpha(y_t - \ell_{t-1}),$$
$$= \alpha y_t + (1 - \alpha)\ell_{t-1}.$$

The recursive relationship for ETS(M,N,N) is thus seen to be identical to that for ETS(A,N,N). However, the reduced form equations are clearly different, showing that the predictive distributions (and hence the prediction intervals) will differ. This difference underlies the need for a stochastic model for a time series; once a suitable model is selected, valid prediction intervals can be generated. Without an underlying model, only point forecasts are possible.

Given the insights provided by the local level model, we may approach the general model in the same way. Reduced-form expressions do not take on a useful form without additional assumptions about the various functions in the model. However, we may eliminate the error term to arrive at the recursive relationship:

$$\boldsymbol{x}_t = \boldsymbol{f}(\boldsymbol{x}_{t-1}) + \boldsymbol{g}(\boldsymbol{x}_{t-1})\frac{[y_t - w(\boldsymbol{x}_{t-1})]}{r(\boldsymbol{x}_{t-1})}. \tag{4.3}$$

Further, for convenience, we write

$$\boldsymbol{D}(\boldsymbol{x}_t) = \boldsymbol{f}(\boldsymbol{x}_t) - \frac{[\boldsymbol{g}(\boldsymbol{x}_t)w(\boldsymbol{x}_t)]}{r(\boldsymbol{x}_t)},$$

so that

$$\boldsymbol{x}_t = \boldsymbol{D}(\boldsymbol{x}_{t-1}) - \boldsymbol{g}(\boldsymbol{x}_{t-1})\frac{y_t}{r(\boldsymbol{x}_{t-1})}.$$

We may observe that $\boldsymbol{D}(\boldsymbol{x}_t)$ becomes linear in the state variables when $\boldsymbol{f}(\boldsymbol{x}_t)$ and $\boldsymbol{g}(\boldsymbol{x}_t)$ are linear in $\boldsymbol{x}_t$ and $w(\boldsymbol{x}_t) = r(\boldsymbol{x}_t)$. Further, when the vector $\boldsymbol{g}(\boldsymbol{x}_t)/r(\boldsymbol{x}_t)$ does not depend on the state variables, the transition equations given in (4.3) reduce to

$$\boldsymbol{x}_t = \boldsymbol{D}\boldsymbol{x}_{t-1} - \boldsymbol{g}y_t.$$

It then follows that the model is stable in the sense of Sect. 3.3.1. These conditions may seem restrictive, but they correspond to an important class of heteroscedastic models, as we shall see below.

The conditional expectation, which is also the one-step-ahead point forecast $\hat{y}_{t|t-1}$, is given by:

$$\mathrm{E}(y_t|y_{t-1},\ldots,y_1,\boldsymbol{x}_0) = \mathrm{E}(y_t|\boldsymbol{x}_{t-1}) = \hat{y}_{t|t-1} = w(\boldsymbol{x}_{t-1}),$$

so that the recursive relationships may be summarized as:

$$\begin{aligned}\hat{y}_{t|t-1} &= w(\boldsymbol{x}_{t-1}),\\ \varepsilon_t &= (y_t - \hat{y}_{t|t-1})/r(\boldsymbol{x}_{t-1}),\\ \boldsymbol{x}_t &= \boldsymbol{f}(\boldsymbol{x}_{t-1}) + \boldsymbol{g}(\boldsymbol{x}_{t-1})\varepsilon_t.\end{aligned}$$

Once the model has been fully specified, updating proceeds directly using these equations. This approach is in stark contrast to models based on the Kalman filter using multiple independent sources of error. In that case, no direct updating is feasible in general, and we must make use of various approximations such as the extended Kalman filter (West and Harrison 1997, pp. 496–497). The present form is too general for any meaningful discussion of particular model properties, and so we will consider these on a case-by-case basis.

As with the linear version in Sect. 3.1, the probability density function for $\boldsymbol{y}$ may be written in a relatively simple form as:

$$\begin{aligned} p(\boldsymbol{y} \mid \boldsymbol{x}_0) &= \prod_{t=1}^{n} p(y_t \mid \boldsymbol{x}_{t-1}) \\ &= \prod_{t=1}^{n} p(\varepsilon_t)/|r(\boldsymbol{x}_{t-1})|. \end{aligned}$$

If we assume that the distribution is Gaussian, this expression becomes:

$$p(\boldsymbol{y}) = (2\pi\sigma^2)^{-n/2} \left| \prod_{t=1}^{n} r(\boldsymbol{x}_{t-1}) \right|^{-1} \exp\left(-\tfrac{1}{2} \sum_{t=1}^{n} \varepsilon_t^2/\sigma^2 \right).$$

4.2 Basic Special Cases

We now explore some of the special cases that can be used to model time series, recognizing as always that such models are at best an approximation to the data generating process. We observed in Chap. 2 that models in the ETS(M,∗,∗) class give rise to the same point forecasts as those in the ETS(A,∗,∗) class, because we are deploying the same recursive relationships given in Table 2.1. The stochastic elements of the process determine whether to use the additive or multiplicative version. If the error process is homoscedastic, the constant variance assumptions in Chap. 3 are appropriate, and the predictions intervals for h-steps ahead have constant widths regardless of the current level of the process. On the other hand, if the process is heteroscedastic, and, in particular, the error variance is proportional to the current level of the process, the nonlinear schemes introduced in Sect. 4.1 are appropriate. Clearly other assumptions concerning the nature of the variance are possible, but we restrict our attention to the two options, additive or multiplicative, for the present discussion. Some extensions are considered briefly in Sect. 4.4.5.

We may justify the local models as the leading terms of a Taylor series expansion, and the only difference we would see relative to Fig. 3.1 is that the superimposed prediction intervals would widen when the observed value was high and narrow when it was low. Why does this matter? Think for the moment in terms of forecasting sales. During periods of high sales volume, the inappropriate use of a constant variance model would lead to underestimation of the level of uncertainty, and hence to a safety stock level that was too small. Similarly, in periods of low demand, the analysis would lead to carrying excess inventory. In either case, a loss of net revenue results. This effect is not always easy to recognize. For example, an empirical investigation of the coverage provided by prediction intervals might simply count the number of times the actual value fell within the prediction interval. Because the interval is too wide when the level is low, and too narrow at high levels, the overall count might well come out close to the nominal level for the interval. We need to track the coverage at a given level to be sure that the prediction intervals are constructed appropriately.

4.2.1 Local Level Model: ETS(M,N,N)

This model is described by (4.2a,b).Upon inspection of these equations, we see that the conditional variance of y_t given ℓ_{t-1} is $\sigma^2\ell_{t-1}^2$, reflecting the comments made earlier. The state equation reveals that the quantity $\alpha\ell_{t-1}\varepsilon_t$ has a persistent effect, feeding into the expected level for the next time period. When $\alpha = 0$, the mean level does not change, so that the additive and multiplicative models are then identical apart from the way the parameters were specified. When $\alpha = 1$, the model reverts to a form of random walk with the reduced form $y_t = y_{t-1}(1 + \varepsilon_t)$; the complete effect of the random error is passed on to the next period. In general, the one-step-ahead predictions are, as for the additive scheme in (3.12), given by:

$$\hat{y}_{t+1|t} = (1-\alpha)^t \ell_0 + \alpha \sum_{j=0}^{t-1} (1-\alpha)^j y_{t-j}.$$

The stability condition is satisfied provided $0 < \alpha < 2$.

A natural question to ask is what difference does it make if we select the multiplicative rather than the additive form of the local level model? Indeed, plots of simulated series look very similar to those given in Fig. 3.2, so meaningful comparisons must be sought in other ways. One approach is to look at the conditional variances given the initial conditions. Using the subscripts A and M for the additive and multiplicative schemes, we arrive at:

$$V_A(y_t|x_0) = \sigma_A^2[1 + (t-1)\alpha^2],$$
$$V_M(y_t|x_0) = x_0^2[(1 + \sigma_M^2)(1 + \alpha^2\sigma_M^2)^{t-1} - 1].$$

In order to compare the two, we set the one-step-ahead variances equal by putting $\sigma_A = x_0\sigma_M$. We may then compute the ratio V_M/V_A for different values of t and α:

σ_M	0.03	0.03	0.03	0.12	0.12	0.12
α	0.1	0.5	1.5	0.1	0.5	1.5
$t = 5$	1.000	1.001	1.004	1.001	1.010	1.058
$t = 10$	1.000	1.001	1.009	1.001	1.020	1.149
$t = 20$	1.000	1.002	1.019	1.006	1.040	1.364

Perusal of the table indicates that there are only substantial differences in the variances for longer horizons with high α and relatively high values of σ_M.

When $\sigma_M = 0.30$, the multiplicative error has a mean that is about three times its standard deviation, and the differences become noticeable quite quickly, as shown in this table for $t = 10$:

σ_M	0.03	0.12	0.30
$\alpha = 0.1$	1.000	1.001	1.008
$\alpha = 0.5$	1.001	1.020	1.134
$\alpha = 1.0$	1.004	1.067	1.519
$\alpha = 1.5$	1.009	1.149	2.473

The examination of stock price volatility represents an area where the distinction could be very important. Based on the efficient market hypothesis we would expect that $\alpha = 1$. The process might be observed at hourly or even minute intervals, yet the purpose behind the modeling would be to evaluate the volatility (essentially as measured by the variance) over much longer periods. In such circumstances, the use of an additive model when a multiplicative model was appropriate could lead to considerable under-estimation of the risks involved. Interestingly, if we start out with this form of the random walk model and consider the reduced form of (4.2), we can rewrite the measurement equation as:

$$\frac{y_t - y_{t-1}}{y_{t-1}} = \varepsilon_t.$$

That is, the one-period return on an investment follows a white noise process.

As we observed in Sect. 4.1, the Gaussian distribution is not a valid assumption for strictly positive processes such as these multiplicative models. In applications, the prediction interval appears to be satisfactory provided the h-step-ahead forecast root mean squared error is less than about one-quarter of the mean. In other cases, or for simulations, a more careful specification of the error distribution may be needed (see Chap. 15).

4.2.2 Local Trend Model: ETS(M,A,N)

We may augment the local level model by adding an evolving growth rate b_t to give the new model:

$$y_t = (\ell_{t-1} + b_{t-1})(1 + \varepsilon_t), \tag{4.4a}$$
$$\ell_t = (\ell_{t-1} + b_{t-1})(1 + \alpha\varepsilon_t), \tag{4.4b}$$
$$b_t = b_{t-1} + \beta(\ell_{t-1} + b_{t-1})\varepsilon_t. \tag{4.4c}$$

This model has a state space structure with:

$$\boldsymbol{x}_t = [l_t, b_t]', \quad w(\boldsymbol{x}_{t-1}) = r(\boldsymbol{x}_{t-1}) = \ell_{t-1} + b_{t-1},$$

$$\boldsymbol{f}(\boldsymbol{x}_{t-1}) = [\ell_{t-1} + b_{t-1}, b_{t-1}]', \quad \text{and} \quad \boldsymbol{g} = [\alpha(\ell_{t-1} + b_{t-1}), \beta(\ell_{t-1} + b_{t-1})]'.$$

Given the multiplicative nature of the model, we take the process and the underlying level to be strictly positive; the slope may be positive, zero or negative. There are now two smoothing parameters α and β, and they are scaled by the current level of the process. Because the slope is typically quite small relative to the current level of the process, the value of β will often be small. The following special cases are worth noting:

- $\beta = 0$: a global trend
- $\beta = 0$, $\alpha = 1$: random walk with a constant trend element, often known as the random walk with *drift*
- $\beta = 0$, $\alpha = 0$: fixed level and trend, thereby reducing to a classical or global linear trend model

We may represent the model in innovations form as:

$$\begin{aligned}
\hat{y}_{t|t-1} &= (\ell_{t-1} + b_{t-1}), \\
\varepsilon_t &= (y_t - \hat{y}_{t|t-1})/(\ell_{t-1} + b_{t-1}), \\
\ell_t &= (\ell_{t-1} + b_{t-1})(1 + \alpha\varepsilon_t), \\
b_t &= b_{t-1} + \beta(\ell_{t-1} + b_{t-1})\varepsilon_t.
\end{aligned}$$

An equivalent form that is more convenient for updating is:

$$\begin{aligned}
\hat{y}_{t|t-1} &= (\ell_{t-1} + b_{t-1}), \\
\varepsilon_t &= (y_t - \hat{y}_{t|t-1})/(\ell_{t-1} + b_{t-1}), \\
\ell_t &= \alpha y_t + (1-\alpha)(\ell_{t-1} + b_{t-1}), \\
b_t &= \beta^*(\ell_t - \ell_{t-1}) + (1-\beta^*)b_{t-1},
\end{aligned}$$

where $\beta^* = \beta/\alpha$. The state updates are of exactly the same algebraic form as those for the additive model in (3.14). Manipulation of the state equations provides the recursive relationships:

$$\begin{aligned}
\ell_t &= \alpha y_t + (1-\alpha)(\ell_{t-1} + b_{t-1}), \\
b_t &= \beta(y_t - \ell_{t-1}) + (1-\beta)b_{t-1}.
\end{aligned}$$

Following Sect. 3.3, we may write the general form of the stability condition as constraints on

$$\begin{aligned}
\boldsymbol{D}(\boldsymbol{x}_t) &= \boldsymbol{f}(\boldsymbol{x}_t) - \boldsymbol{g}(\boldsymbol{x}_t)w(\boldsymbol{x}_t)/r(\boldsymbol{x}_t) \\
&= \boldsymbol{f}(\boldsymbol{x}_t) - \boldsymbol{g}(\boldsymbol{x}_t)
\end{aligned}$$

when $w(\boldsymbol{x}_{t-1}) = r(\boldsymbol{x}_{t-1})$. Given the particular form of (4.4), this expression simplifies to

$$\begin{aligned}
\boldsymbol{D}(\boldsymbol{x}_{t-1}) &= [(\ell_{t-1} + b_{t-1})(1-\alpha),\ b_{t-1} - \beta(\ell_{t-1} + b_{t-1})]' \\
&= \begin{bmatrix} 1-\alpha & 1-\alpha \\ -\beta & 1-\beta \end{bmatrix} \boldsymbol{x}_{t-1}.
\end{aligned}$$

This is the same general form as for the additive models discussed in Chap. 3. So, for the local trend model, the stability conditions remain: $\alpha > 0$, $\beta > 0$ and $2\alpha + \beta < 4$.

4.2.3 Local Multiplicative Trend, Additive Error Model: ETS(A,M,N)

Some of the boxes in Tables 2.2 and 2.3 correspond to models that might not occur to the model builder as prime candidates for consideration. Nevertheless, we consider a few of these forms for several reasons. First of all, if we select a model based on some kind of automated search, it is useful to have a rich set of potential models, corresponding to the many nuances the time series might display. Second, by exploring such models we gain a better understanding of the attributes of various nonlinear configurations. Finally, if our study of their properties reveals some undesirable characteristics, we are forewarned about such possibilities before jumping to inappropriate conclusions about usable models.

If we allow an additive error to be associated with a multiplicative trend (exponential growth or decay), we have the following ETS(A,M,N) model:

$$\begin{aligned} y_t &= \ell_{t-1}b_{t-1} + \varepsilon_t, \\ \ell_t &= \ell_{t-1}b_{t-1} + \alpha\varepsilon_t, \\ b_t &= b_{t-1} + \beta\varepsilon_t/\ell_{t-1}. \end{aligned}$$

If we substitute for the error term, we arrive at the recursive relationships ($\beta = \alpha\beta^*$):

$$\begin{aligned} \ell_t &= (1-\alpha)\ell_{t-1}b_{t-1} + \alpha y_t, \\ b_t &= (1-\beta)b_{t-1} + \beta y_t/\ell_{t-1} = (1-\beta^*)b_{t-1} + \beta^*\ell_t/\ell_{t-1}. \end{aligned}$$

It is no longer possible to derive simple expressions to ensure that the stability conditions are satisfied.

4.2.4 Local Multiplicative Trend, Multiplicative Error Model: ETS(M,M,N)

In a similar vein, we can make both components multiplicative:

$$\begin{aligned} y_t &= \ell_{t-1}b_{t-1}(1 + \varepsilon_t), \\ \ell_t &= \ell_{t-1}b_{t-1}(1 + \alpha\varepsilon_t), \\ b_t &= b_{t-1}(1 + \beta\varepsilon_t). \end{aligned}$$

If we substitute for the error term, we arrive at the same updating equations:

$$\begin{aligned} \ell_t &= (1-\alpha)\ell_{t-1}b_{t-1} + \alpha y_t, \\ b_t &= (1-\beta)b_{t-1} + \beta y_t/\ell_{t-1} = (1-\beta^*)b_{t-1} + \beta^*\ell_t/\ell_{t-1}. \end{aligned}$$

This model is of interest in that we can guarantee strictly positive values[1] for the series using a set of conditions such as:

$$0 < \alpha < 1, \qquad 0 < \beta < 1 \qquad \text{and} \qquad 1 + \varepsilon_t > 0.$$

When we set $\beta = 0$, we have a constant trend term corresponding to a fixed growth rate, b. If we had $b < 1$ this would correspond to a form of damping, whereas $b > 1$ allows perpetual growth and could not satisfy the stability condition. Such a model might fit the past history of a time series, but we should be cautious about building such a relationship into the predictive recursion.

4.3 Nonlinear Seasonal Models

Much of the discussion for additive seasonal models applies equally to multiplicative models. In addition to any extant trends, we must allow for seasonal variations in a series. As in Sect. 3.4.3, we represent the seasonal factors by s_t. A common feature of such seasonal patterns is that the variability is proportional to the general level of the series. For example, discussions about the volume of air passenger traffic or about retail sales usually speak of percentage changes rather than absolute shifts in the series. Such changes also apply when the seasonal pattern corresponds to days of the week (e.g., commuter traffic) or hours of the day (e.g., electricity usage). Consideration of these examples also indicates that the seasonal pattern may change over time. Perhaps the most dramatic examples are the differences in commuter traffic on public holidays or electricity usage between weekdays and weekends. In such cases, multiple seasonal cycles may be needed to provide an effective picture (see Chap. 14). However, even in less volatile situations the need for evolving seasonal patterns is clearly evident.

4.3.1 A Multiplicative Seasonal and Error Model: ETS(M,A,M)

The seasonal variations are usually more dramatic within a short time frame than the longer-term changes in trend, so that the focus is primarily on the correct specification of the seasonal structure. We consider a model with multiplicative effects for both the seasonal and error components: ETS(M,A,M). This model may be written as:

$$y_t = (\ell_{t-1} + b_{t-1})s_{t-m}(1 + \varepsilon_t), \tag{4.5a}$$
$$\ell_t = (\ell_{t-1} + b_{t-1})(1 + \alpha\varepsilon_t), \tag{4.5b}$$
$$b_t = b_{t-1} + \beta(\ell_{t-1} + b_{t-1})\varepsilon_t, \tag{4.5c}$$
$$s_t = s_{t-m}(1 + \gamma\varepsilon_t). \tag{4.5d}$$

[1] See Chap. 15 for a detailed discussion of models for positive data.

The model has a state space structure with:

$$\boldsymbol{x}_t = \begin{bmatrix} \ell_t \\ b_t \\ s_t \\ s_{t-1} \\ \vdots \\ s_{t-m+1} \end{bmatrix},$$

$\boldsymbol{f}(\boldsymbol{x}_{t-1}) = \boldsymbol{F}\boldsymbol{x}_{t-1}$, $w(\boldsymbol{x}_{t-1}) = r(\boldsymbol{x}_{t-1}) = (\ell_{t-1} + b_{t-1})s_{t-m}$, and

$$\boldsymbol{g}(\boldsymbol{x}_{t-1}) = \begin{bmatrix} \alpha(\ell_{t-1} + b_{t-1}) \\ \beta(\ell_{t-1} + b_{t-1}) \\ \gamma s_{t-m} \\ 0 \\ \vdots \\ 0 \end{bmatrix}, \quad \text{where} \quad \boldsymbol{F} = \begin{bmatrix} 1\,1\,0\,0 \ldots 0\,0 \\ 0\,1\,0\,0 \ldots 0\,0 \\ 0\,0\,0\,0 \ldots 0\,1 \\ 0\,0\,1\,0 \ldots 0\,0 \\ 0\,0\,0\,1 \ldots 0\,0 \\ \vdots\;\vdots\;\vdots\;\vdots\;\ddots\;\vdots\;\vdots \\ 0\,0\,0\,0 \ldots 1\,0 \end{bmatrix}.$$

Just as the level and seasonal terms in the additive model are only determined up to an arbitrary additive constant, so the level, trend and seasonal terms in the present scheme are only determined up to an arbitrary multiplicative constant. To resolve the indeterminacy, we usually set $\sum_{i=0}^{m-1} s_{0-i} = m$. As before, we will find it preferable to apply this normalization throughout the series; the details are discussed in Chap. 8.

We may use (4.5) to develop the recursive relationships for this scheme and we arrive at:

$$\begin{aligned} \boldsymbol{x}_t &= \left[\boldsymbol{f}(\boldsymbol{x}_{t-1}) - \boldsymbol{g}(\boldsymbol{x}_{t-1}) \frac{w(\boldsymbol{x}_{t-1})}{r(\boldsymbol{x}_{t-1})} \right] + \frac{\boldsymbol{g}(\boldsymbol{x}_{t-1})}{r(\boldsymbol{x}_{t-1})} y_t \\ &= [\boldsymbol{F}\boldsymbol{x}_{t-1} - \boldsymbol{g}(\boldsymbol{x}_{t-1})] + \frac{\boldsymbol{g}(\boldsymbol{x}_{t-1})}{r(\boldsymbol{x}_{t-1})} y_t. \end{aligned}$$

If we substitute the specific functions derived above, we arrive at:

$$\boldsymbol{x}_t = \begin{bmatrix} \ell_t \\ b_t \\ s_t \\ s_{t-1} \\ \vdots \\ s_{t-m+1} \end{bmatrix} = \begin{bmatrix} (1-\alpha)(\ell_{t-1} + b_{t-1}) \\ b_{t-1} - \beta(\ell_{t-1} + b_{t-1}) \\ (1-\gamma)s_{t-m} \\ s_{t-1} \\ \vdots \\ s_{t-m+1} \end{bmatrix} + \begin{bmatrix} \alpha/s_{t-m} \\ \beta/s_{t-m} \\ \gamma/(\ell_{t-1} + b_{t-1}) \\ 0 \\ \vdots \\ 0 \end{bmatrix} y_t. \quad (4.6)$$

Further, we may use the first equation in (4.6) to substitute for y_t in the expression for b_t. Thus, the final recursive relationships, in agreement with Table 2.1, are:

$$\begin{aligned}
\ell_t &= (1-\alpha)(\ell_{t-1}+b_{t-1}) + \alpha y_t/s_{t-m}, \\
b_t &= \beta^*(\ell_t - \ell_{t-1}) + (1-\beta^*)b_{t-1}, \\
s_t &= (1-\gamma)s_{t-m} + \gamma y_t/(\ell_{t-1}+b_{t-1}).
\end{aligned}$$

The h-step-ahead forecasting function is:

$$\hat{y}_{t+h-1|t-1} = (\ell_{t-1} + hb_{t-1})s_{t-m-1+h^*},$$

where $h^* = h \bmod m$. This set of forecasting relationships is known as the multiplicative Holt-Winters' system for seasonal exponential smoothing; see the discussion in Sect. 1.3. Users of this method usually recommend the parameter ranges $0 < \alpha, \beta, \gamma < 1$, but the exact specification of an acceptable parameter space is extremely difficult. We defer further discussion until Chap. 8.

4.3.2 A Multiplicative Seasonal Model with Additive Errors: ETS(A,A,M)

The key feature of the seasonal model that we have just developed is the multiplicative nature of the interaction between the trend and seasonal components. Although we used a multiplicative error structure and would suggest that this assumption is more likely to be satisfied in practice, we could also develop the multiplicative seasonal structure with additive errors. The underlying model becomes:

$$\begin{aligned}
y_t &= (\ell_{t-1}+b_{t-1})s_{t-m} + \varepsilon_t, \\
\ell_t &= \ell_{t-1} + b_{t-1} + (\alpha/s_{t-m})\varepsilon_t, \\
b_t &= b_{t-1} + (\beta/s_{t-m})\varepsilon_t, \\
s_t &= s_{t-m} + (\gamma/(\ell_{t-1}+b_{t-1}))\varepsilon_t.
\end{aligned}$$

Following the same reasoning as before, we arrive back at the recursive relationships given in (4.6). There is an element of reverse engineering in this model; that is, we worked back from the recursive relationships to determine the form of the model. Making adjustments to the smoothing parameters in the manner indicated does not seem very plausible, yet logic dictates this is an implicit assumption if the ETS(A,A,M) scheme is to be used. Perhaps a better way to look at this conclusion is to recognize that one of the benefits of formulating an underlying model is that the process forces us to make our assumptions explicit. In this case, intuition guides us towards the ETS(M,A,M) model. As a practical matter, we recommend consideration of both schemes, so that resulting prediction intervals are consistent with the historical patterns in the series.

Similar adjustments enable us to consider the ETS(A,M,M) and ETS(M,M,M) schemes, and to arrive at the recursive relationships given in Table 2.1. These developments are left as end-of-chapter exercises.

4.4 Variations on the Common Models

As might be expected, the number of usable models can be expanded considerably. In particular, we may incorporate damping factors or set specific parameters to special values. We first consider models without a seasonal component and then examine special cases of the model associated with the Holt-Winters method.

4.4.1 Local Level Model with Drift

If the growth rate is steady over time, we may simplify the local trend model by setting $\beta = 0$. This modification is often effective when the forecasting horizon is fairly short and growth is positive:

$$\begin{aligned} y_t &= (\ell_{t-1} + b)(1 + \varepsilon_t), \\ \ell_t &= (\ell_{t-1} + b)(1 + \alpha\varepsilon_t). \end{aligned}$$

The principal difference between this model and the additive error version is that the prediction intervals of this model gradually widen as the mean level increases.

4.4.2 Damped Trend Model: ETS(M,A$_d$,N)

The damped trend model has the reduced growth rate ϕb_{t-1} at time t, where $0 \le \phi < 1$. The level tends to flatten out as time increases, and this feature can be useful for series whose trends are unlikely to be sustained over time. In particular, when the variable of interest is non-negative, a negative trend clearly has to flatten out sooner or later.

$$\begin{aligned} y_t &= (\ell_{t-1} + \phi b_{t-1})(1 + \varepsilon_t), \\ \ell_t &= (\ell_{t-1} + \phi b_{t-1})(1 + \alpha\varepsilon_t), \\ b_t &= \phi b_{t-1} + \beta(\ell_{t-1} + \phi b_{t-1})\varepsilon_t. \end{aligned}$$

The damped model may be combined with the previous scheme to produce a smoother trajectory towards a limiting expected value.

4.4.3 Local Multiplicative Trend with Damping: ETS(A,M$_d$,N)

If we try to introduce similar damping coefficients into multiplicative models, the models are not well-behaved. In particular, such damping forces the expected value towards a limiting value of zero. To avoid this difficulty, we follow Taylor (2003a) and raise the growth rate to a fractional power, $0 \le \phi < 1$. The relevant model, as given in Table 2.2, is:

$$y_t = \ell_{t-1} b_{t-1}^{\phi} + \varepsilon_t,$$

$$\ell_t = \ell_{t-1} b_{t-1}^{\phi} + \alpha \varepsilon_t,$$
$$b_t = b_{t-1}^{\phi} + \beta \varepsilon_t / \ell_{t-1}.$$

The state variable b_t is now a growth index with a base of 1.0. On making the appropriate substitutions, the recursive relationships in Table 2.1 follow, and are left as an exercise.

4.4.4 Various Seasonal Models

A variety of special seasonal models may be obtained as special cases of Holt-Winters' multiplicative scheme, and we present a few of these without going into great detail.

Purely Seasonal Levels

In effect, the model reduces to m distinct models with a common parameter:

$$y_t = \ell_{t-m}(1 + \varepsilon_t),$$
$$\ell_t = \ell_{t-m}(1 + \alpha \varepsilon_t).$$

The multiplicative form recognizes that periods with higher levels are likely to display greater variability.

Fixed Seasonality

Some series, particularly in the area of macroeconomics, possess quite stable seasonal patterns. In such cases, it may be desirable to set $\gamma = 0$ and treat the seasonal factors as constant. The resulting model is:

$$y_t = (\ell_{t-1} + b_{t-1}) s_j (1 + \varepsilon_t),$$
$$\ell_t = (\ell_{t-1} + b_{t-1})(1 + \alpha \varepsilon_t),$$
$$b_t = b_{t-1} + \beta(\ell_{t-1} + b_{t-1}) \varepsilon_t,$$

where $j = t \mod m$.

Parsimonious Seasonal Model

As noted in Sect. 3.5.5, series that are observed at many times within the main period, such as weeks in a year, require a very large number of starting values. Further, only a fairly short series may be available for estimation. However, many such series have a simple state space structure, such as a normal level of sales outside certain special periods. The details for the additive error versions of such models are given in Sect. 3.5.5. The multiplicative error versions follow, using the same modification as elsewhere in this chapter.

4.4.5 Other Heteroscedastic Models

The variance structure may be modified in several ways. Perhaps the simplest way is to incorporate an additional exponent in a manner reminiscent of the method employed by Box and Cox (1964) but without a transformation. We use the local trend model, ETS(M,A,N) in (4.4) by way of illustration. We separate out the error terms and modify them by using a power of the trend term, $0 \leq \theta \leq 1$:

$$\begin{aligned} y_t &= (\ell_{t-1} + b_{t-1}) + (\ell_{t-1} + b_{t-1})^{\theta}\varepsilon_t, \\ \ell_t &= (\ell_{t-1} + b_{t-1}) + \alpha(\ell_{t-1} + b_{t-1})^{\theta}\varepsilon_t, \\ b_t &= b_{t-1} + \beta(\ell_{t-1} + b_{t-1})^{\theta}\varepsilon_t. \end{aligned}$$

For example, $\theta = 1/3$ would produce a variance proportional to the $2/3$rds power of the mean, much as the cube-root transformation does. The present formulation enables us to retain the linear structure for the expectation, which in many ways is more plausible than the transformation.

4.5 Exercises

Exercise 4.1. Verify the variance expressions for ETS(M,N,N) given in Sect. 4.2.1.

Exercise 4.2. Use the approach of Sect. 4.3.1 to derive the recursive relations for the state vector of the ETS(A,M,M) model, given in Table 2.2. Extend the argument to include the ETS(A,M_d,M) model.

Exercise 4.3. Use the approach of Sect. 4.3.1 to derive the recursive relations for the state vector of the ETS(M,M,M) model, given in Table 2.2. Extend the argument to include the ETS(M,M_d,M) model.

Exercise 4.4. Evaluate the h-step-ahead forecast mean squared error for the local trend model and for the local level model with drift, given in Sect. 4.4.1. Compare the two for various combinations of h, α and β.

Exercise 4.5. Evaluate the h-step-ahead forecast mean squared error for the damped trend model ETS(M,A_d,N) and compare it with that for the local trend model for various combinations of h, ϕ, α and β.

Exercise 4.6. Show that the stability conditions for the heteroscedastic model are exactly those for the ETS(A,A,N) model.

Exercise 4.7. The data set `djiclose` contains the closing prices for the Dow Jones Index for the first day of each month from October 1928 to December 2007, along with the monthly returns for that series. Fit a heteroscedastic ETS(M,A,N) model to these data for a selected part of the series. Compare your results with the random walk with drift model for the returns series.

5

Estimation of Innovations State Space Models

For any innovations state space model, the initial (seed) states and the parameters are usually unknown, and therefore must be estimated. This can be done using maximum likelihood estimation, based on the innovations representation of the probability density function.

In Chap. 3 we outlined transformations (referred to as "general exponential smoothing") that convert a linear time series of mutually dependent random variables into an innovations series of independent and identically distributed random variables. In the heteroscedastic and nonlinear cases, such a representation remains a viable approximation in most circumstances, an issue to which we return in Chap. 15. These innovations can be used to compute the likelihood, which is then optimized with respect to the seed states and the parameters. We introduce the basic methodology in Sect. 5.1. The estimation procedures discussed in this chapter assume a finite start-up; consideration of the infinite start-up case is deferred until Chap. 12.

Any numerical optimization procedure used for this task typically requires starting values for the quantities that are to be estimated. An appropriate choice of starting values is important. The likelihood function may not be unimodal, so a poor choice of starting values can result in sub-optimal estimates. Good starting values (i.e., values that are as close as possible to the optimal estimates) not only increase the chances of finding the true optimum, but typically reduce the computational loads required during the search for the optimum solution. In Sect. 5.2 we will discuss plausible heuristics for determining the starting values.

5.1 Maximum Likelihood Estimation

The unknown model parameters and states must be estimated. Maximum likelihood (ML) estimators are sought because they are consistent and asymptotically efficient under reasonable conditions; for a general discussion see Gallant (1987, pp. 357–391). Hamilton (1994, pp. 133–149) derives the convergence properties of various numerical algorithms for computing ML estimates.

The likelihood function is based on the density of the series vector $\boldsymbol{y}$. It is a function of a p-vector $\boldsymbol{\theta}$ of parameters such as the smoothing parameters and damping factors. The likelihood also depends on the innovations variance σ^2, but for reasons that will become clearer soon, it is convenient to separate it from the main parameter vector $\boldsymbol{\theta}$. Finally, the likelihood depends on the k-vector $\boldsymbol{x}_0$ of seed states.

Under the more traditional assumptions employed in time series analysis, the generating process is presumed to have operated for an extensive period of time prior to the period of the first observation, in which case the seed state vector must be random. A likelihood function must only be based on *observable* random quantities; unobserved random variables must be averaged away. We sidestep the need for averaging, and hence simplify the task of forming the likelihood function, by assuming that the process has had no life prior to period 1, in which case the seed state vector $\boldsymbol{x}_0$ is fixed and may be treated as a vector of parameters. The case of random seed states will be considered in Chap. 12.

It was shown in Chap. 3 that any time series $\{y_t\}$ governed by a linear state space model with Gaussian innovations has a multivariate Gaussian distribution (3.2). In Sect. 4.1, the same basic result was derived as an approximation for the nonlinear version of the model, a remarkable conclusion that depends critically on the assumption of a fixed seed state. In essence, the joint density of the series was shown to be the weighted product of the densities of the individual innovations:

$$p(\boldsymbol{y} \mid \boldsymbol{\theta}, \boldsymbol{x}_0, \sigma^2) = \prod_{t=1}^{n} p(\varepsilon_t)/|r(\boldsymbol{x}_{t-1})|.$$

So the Gaussian likelihood can be written as

$$\mathcal{L}(\boldsymbol{\theta}, \boldsymbol{x}_0, \sigma^2 \mid \boldsymbol{y}) = (2\pi\sigma^2)^{-n/2} \left| \prod_{t=1}^{n} r(\boldsymbol{x}_{t-1}) \right|^{-1} \exp\left(-\tfrac{1}{2} \sum_{t=1}^{n} \varepsilon_t^2/\sigma^2 \right), \tag{5.1}$$

and the log likelihood is

$$\log \mathcal{L} = -\frac{n}{2} \log(2\pi\sigma^2) - \sum_{t=1}^{n} \log |r(\boldsymbol{x}_{t-1})| - \tfrac{1}{2} \sum_{t=1}^{n} \varepsilon_t^2/\sigma^2. \tag{5.2}$$

Then taking the partial derivative with respect to σ^2 and setting it to zero gives the maximum likelihood estimate of the innovations variance σ^2 as

$$\hat{\sigma}^2 = n^{-1} \sum_{t=1}^{n} \varepsilon_t^2.$$

This formula can be used to eliminate σ^2 from the likelihood (5.1) to give the concentrated likelihood

$$\mathcal{L}(\boldsymbol{\theta}, \boldsymbol{x}_0 \mid \boldsymbol{y}) = (2\pi e\, \hat{\sigma}^2)^{-n/2} \left| \prod_{t=1}^{n} r(\boldsymbol{x}_{t-1}) \right|^{-1}.$$

Thus, twice the negative log-likelihood is given by

$$\begin{aligned} -2 \log \mathcal{L}(\boldsymbol{\theta}, \boldsymbol{x}_0 \mid \boldsymbol{y}) &= n \log(2\pi e\, \hat{\sigma}^2) + 2 \sum_{t=1}^{n} \log |r(\boldsymbol{x}_{t-1})| \\ &= c_n + n \log \left(\sum_{t=1}^{n} \varepsilon_t^2 \right) + 2 \sum_{t=1}^{n} \log |r(\boldsymbol{x}_{t-1})|, \end{aligned}$$

where c_n is a constant depending on n but not on $\boldsymbol{\theta}$ or $\boldsymbol{x}_0$. Hence, maximum likelihood estimates of the parameters can be obtained by minimizing

$$\mathcal{L}^*(\boldsymbol{\theta}, \boldsymbol{x}_0) = n \log \left(\sum_{t=1}^{n} \varepsilon_t^2 \right) + 2 \sum_{t=1}^{n} \log |r(\boldsymbol{x}_{t-1})|. \tag{5.3}$$

Equivalently, they can be obtained by minimizing the *augmented sum of squared errors criterion*:

$$S(\boldsymbol{\theta}, \boldsymbol{x}_0) = \left[\exp(\mathcal{L}^*(\boldsymbol{\theta}, \boldsymbol{x}_0)) \right]^{1/n} = \left| \prod_{t=1}^{n} r(\boldsymbol{x}_{t-1}) \right|^{2/n} \sum_{t=1}^{n} \varepsilon_t^2. \tag{5.4}$$

In homoscedastic cases, $r(\boldsymbol{x}_{t-1}) = 1$ and (5.4) reduces to the traditional sum of squared errors.

Use of (5.4) criterion in place of the likelihood function means that the optimizer does not directly select the best value of σ^2. The number of variables being directly optimized is reduced by one, with consequent savings in computational loads. More importantly, however, it avoids a problem that sometimes arises with the likelihood function, when the optimizer chooses a trial value of the variance that is quite out-of-kilter with the errors, and a consequent numerical stability issue occurs.

For particular values of the parameters and seed states, the value of the innovation is found with $\varepsilon_t = [y_t - w(\boldsymbol{x}_{t-1})]/r(\boldsymbol{x}_{t-1})$. The state is revised with the transition

$$\boldsymbol{x}_t = \boldsymbol{f}(\boldsymbol{x}_{t-1}) + \boldsymbol{g}(\boldsymbol{x}_{t-1})\varepsilon_t.$$

In the case of homoscedastic errors and linear functional relationships, as assumed in Chap. 3, this simplifies to

$$\varepsilon_t = y_t - \boldsymbol{w}'\boldsymbol{x}_{t-1}, \tag{5.5}$$

$$\boldsymbol{x}_t = \boldsymbol{F}\boldsymbol{x}_{t-1} + \boldsymbol{g}\varepsilon_t, \tag{5.6}$$

which is the general linear form of exponential smoothing (Box et al. 1994).

Although expression (5.4) was derived from the likelihood (5.1), we could start the whole process by directly specifying that the objective is to minimize $S(\boldsymbol{\theta}, \boldsymbol{x}_0)$. This approach, known as the Augmented Least Squares (ALS)

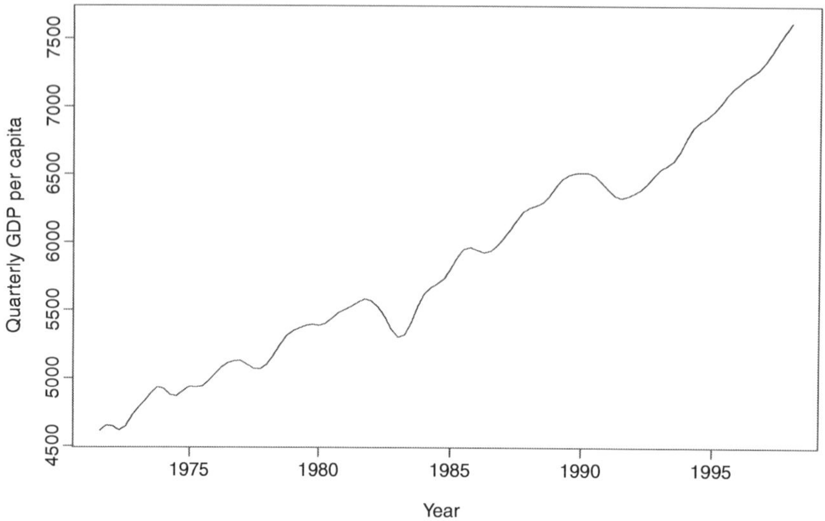

Fig. 5.1. Plot of Australian quarterly gross domestic product per capita from the September quarter of 1971 to the March quarter of 1998.

method, does not require us to make any assumptions about the distributional form of the errors. More generally, when the ML estimates are computed from (5.4) without any assumption of Gaussianity, we refer to the results as quasi-maximum likelihood estimators (Hamilton 1994, p. 126). Such estimators are often consistent, but the expressions for the standard errors of the estimators may be biased, even asymptotically.

5.1.1 Application: Australian GDP

To illustrate the method, we use the Australian quarterly real gross domestic product per capita[1] from the September quarter of 1971 to the March quarter of 1998. The deseasonalized series, which consists of 107 observations, is depicted in Fig. 5.1. We fitted a local linear trend model (3.13):

$$y_t = \ell_{t-1} + b_{t-1} + \varepsilon_t \tag{5.7a}$$
$$\ell_t = \ell_{t-1} + b_{t-1} + \alpha\varepsilon_t \tag{5.7b}$$
$$b_t = b_{t-1} + \beta\varepsilon_t \tag{5.7c}$$

by minimizing (5.4).

The results obtained depend on the constraints imposed on the parameter values during estimation. As seen in Sect. 3.4.2, the constraints $0 < \alpha < 1$ and $0 < \beta < \alpha$ are typically imposed. They ensure that the states can be

[1] Source: Australian Bureau of Statistics.

Table 5.1. Maximum likelihood results: the local trend model applied to the Australian quarterly gross domestic product.

	Constraints	
	Conventional	Stable
α	1.00	0.61
β	1.00	2.55
ℓ_0	4571.3	4568.7
b_0	36.5	35.1
MSE	639	291
MAPE	0.36%	0.24%

interpreted as averages. However, another set of constraints arises from the stability criterion (Sect. 3.3.1, p. 36). They ensure that the observations have a diminishing effect as they get older. This second set of constraints (derived in Chap. 10) is $\alpha \geq 0$, $\beta \geq 0$ and $2\alpha + \beta \leq 4$. It contains the first constraint set and is much larger.

The results are summarized in Table 5.1. The parameter estimates with the conventional constraints lie on the boundary of the parameter space. In the second case, the estimates lie in the interior of the parameter space.

The MSE is more than halved by relaxing the conventional constraints to the stability conditions. The MAPEs indicate that both approaches provide local linear trends with a remarkably good fit. The lower MAPE of 0.24% for the stability approach is consistent with the MSE results.

The optimal value of 2.55 for β in the second estimation may seem quite high. It ensures, however, that the growth rate is very responsive to unanticipated changes in the series. A plot of the estimated growth rates is shown in Fig. 5.2. The effect is to ensure that the local trend adapts quickly to changes in the direction of the series values.

5.2 A Heuristic Approach to Estimation

The method of estimation described in the previous section seeks values of the seed vector $\boldsymbol{x}_0$ and the parameters $\boldsymbol{\theta}$ that *jointly* minimize the augmented sum of squared errors. The inclusion of the seed state vector can be a source of relatively high computational loads. For example, if it is applied to a weekly time series, a linear trend with seasonal effects has 54 seed states, but only three smoothing parameters. In such cases, it is common practice to approximate the seed values using a heuristic method, and simply minimize with respect to the parameters $\boldsymbol{\theta}$ alone.

Heuristic methods are typically devised on a case by case basis. For example, the seed level in a local level model is often approximated by the first value of a series, or sometimes by a simple average of the first few values of a series (see Makridakis et al. 1998).

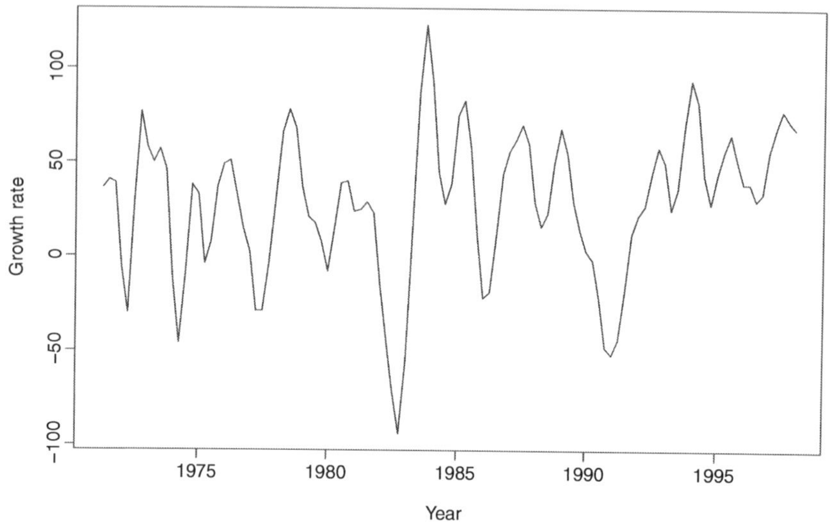

Fig. 5.2. Estimation of the local trend model for the Australian quarterly gross domestic product per capita from the September quarter of 1971 to the March quarter of 1998: growth rate estimates.

In Sect. 2.6.1 we described a heuristic method that works well for almost all series. For non-seasonal data, we fit a straight line $a + bt$ to the first ten observations and set $\ell_0 = a$. We use $b_0 = b$ when the model assumes an additive trend, and for a model with a multiplicative trend we set $b_0 = 1 + b/a$. For seasonal data, the seed seasonal components are obtained using a classical decomposition method (Makridakis et al. 1998) applied to the first few years of data. Then, as for non-seasonal data, we fit a straight line to the first ten deseasonalized observations to obtain ℓ_0 and b_0.

These are simply *starting values* for the full optimization. We do not generally use them in forecasting, unless the seasonal period m is very large (as with weekly data).

Such heuristics can be very successful, but they are not failsafe. The exponential decay of the effect of the seed vector in (3.6) is supposed to ensure that the effect on the sum of squared errors of any additional error introduced by the use of a heuristic is negligible. This is only true, however, when the smoothing parameters are relatively large, and so the states change rapidly over time. Then the structure that prevailed at the start of the series has little impact on the series close to the end. However, when the smoothing parameters are small, the end states are unlikely to be very different from those at the start of the observational period. In this case, early seed values are not discounted heavily, so the effects of any extra error arising from a heuristic are unlikely to disappear. Heuristics may potentially lead to poor results in some circumstances. Hyndman et al. (2002) report that optimization of the seed states improves forecasts of the M3 series by about 2.8%. It appears that full optimization is to be recommended where it is practicable.

When heuristics are used, they should be designed to apply to the model when the smoothing parameters are small. Fortunately, this is not difficult to achieve. For example, as α and β approach zero, model (5.4) reduces to the global linear trend model, so that our heuristic based on fitting the linear trend to the first ten observations can be expected to perform well in such circumstances. When α and β are large, the heuristic will perform less well, but the initial conditions are then discounted more rapidly, so the effect is reduced.

Heuristics originated in an era when computers were very much slower than those available today. They were devised to avoid what were then quite daunting computational loads. Nowadays, full optimization of the likelihood can be undertaken in a fraction of the time. For example, Hyndman et al. (2002) found that it took only 16 min to fit a collection of models, similar to those from Tables 2.2 and 2.3, to the 3,003 time series from the M3 forecasting competition. Much faster times should now be possible with modern computers.

It may be argued that the above example with weekly seasonal effects suggests that there remain cases where full optimization may still not be practicable. However, in this type of situation, it is advisable to reduce the number of optimizable variables. For example, weeks with similar seasonal characteristics could be grouped to reduce the number of seasonal indexes. Alternatively, Fourier representations based on sine and cosine functions, might be used. Models with a reduced number of states are likely to yield more robust forecasts.

Heuristics, however, still have a useful place in forecasting. For example, they can be used to provide starting values for the optimizer. This way the seed values are still optimized, but the optimizer begins its search from a point that is likely to be closer to the optimal solution. An advantage of this approach is that it reduces the chances of sub-optimal solutions with multimodal augmented sum of squared errors functions. Moreover, with series containing slow changes to states, the optimizer automatically prevents any deleterious effects of poor starting values, should they occur with the use of a heuristic. Finally, the time required for optimization is typically shortened by the use of heuristics.

5.3 Exercises

The following exercises should be completed using the quarterly US gross domestic product series available in the data set usgdp.

Exercise 5.1. Fit the local level model ETS(A,N,N) to the data. This will require calculating the sum of squared errors criterion and minimizing it with respect to the value of ℓ_0 and α. Do the estimation using your own **R** code. Then compare the results with those obtained using the `ets()` function in the `forecast` package for **R**.

Exercise 5.2. Fit the local level model with drift (Sect. 3.5.2) to the log-transformed data.

Exercise 5.3. The multiplicative version of a local level model with drift model is

$$\begin{aligned} y_t &= \ell_{t-1}b(1+\varepsilon_t) \\ \ell_t &= \ell_{t-1}b(1+\alpha\varepsilon_t) \end{aligned}$$

where b is a multiplicative drift term. Fit this model to the raw data using the augmented sum of squared errors criterion. Contrast the results with those from Exercise 5.2.

6

Prediction Distributions and Intervals

Point forecasts for each of the state space models were given in Table 2.1 (p. 18). It is also useful to compute the associated prediction distributions and prediction intervals for each model. In this chapter, we discuss how to compute these distributions and intervals.

There are several sources of uncertainty when forecasting a future value of a time series (Chatfield 1993):

1. The uncertainty in model choice—maybe another model is correct, or maybe none of the candidate models is correct.
2. The uncertainty in the future innovations $\varepsilon_{n+1}, \dots, \varepsilon_{n+h}$.
3. The uncertainty in the estimates of the parameters: α, β, γ, ϕ and $\boldsymbol{x}_0$.

Ideally, the prediction distribution and intervals should take all of these into account. However, this is a difficult problem, and in most time series analysis only the uncertainty in the future innovations is taken into account.

If we assume that the model and its parameters (including $\boldsymbol{x}_0$) are known, then we also know $\boldsymbol{x}_n$, the state vector at the last period of observation, because the error in the transition equation can be calculated from the observations up to time n. Consequently, we define the prediction distribution as the distribution of a future value of the series given the model, its estimated parameters, and $\boldsymbol{x}_n$. A short-hand way of writing this is $y_{n+h|n} \equiv y_{n+h} \mid \boldsymbol{x}_n$.

We briefly discuss how to allow for parameter estimation uncertainty in Sect. 6.1. We do not address how to allow for model uncertainty, although this is an important issue. Hyndman (2001) showed that model uncertainty is likely to be a much bigger source of error than parameter uncertainty.

The mean of the prediction distribution is called the *forecast mean* and is denoted by $\mu_{n+h|n} = \mathrm{E}(y_{n+h} \mid \boldsymbol{x}_n)$. The corresponding *forecast variance* is given by $v_{n+h|n} = \mathrm{V}(y_{n+h} \mid \boldsymbol{x}_n)$. We will find expressions for these quantities for many of the models discussed in this book.

We are also interested in "lead-time demand" forecasting, where we predict the *aggregate* of the next h observations rather than each of the next h observations individually. We discuss this briefly here and in more detail in Chap. 18.

The most direct method of obtaining prediction distributions is to simulate many possible future sample paths from the fitted model, and to estimate the distributions from the simulated data. This approach will work for any time series model, including all of the models discussed in this book. We describe the simulation method in more detail in Sect. 6.1.

While the simulation approach is simple and can be applied to any well-specified time series model, the computations can be time-consuming. Furthermore, the resulting prediction intervals are only available numerically rather than algebraically. Therefore, the approach does not allow for algebraic analysis of the prediction distributions.

An alternative approach is to derive the distributions analytically. Analytical results on prediction distributions can provide additional insight and can be much quicker to compute. These results are relatively easy to derive for some models (particularly the linear models), but very difficult for others. In fact, there are analytical results on prediction distributions for only 15 of the 30 models in our exponential smoothing framework.

When discussing the analytical prediction distributions, it is helpful to divide the thirty state space models given in Tables 2.2 and 2.3 (pp. 21–22) into five classes; Classes 1–4 are shown in Table 6.1.

For each of Classes 1–3, we give expressions for the forecast means and variances. Class 1 consists of the linear models with homoscedastic errors; these are discussed in Sect. 6.2. In Sect. 6.3 we discuss Class 2, which contains the linear models with heteroscedastic errors. Class 3 models are discussed

Table 6.1. The models separated in the exponential smoothing framework split into Classes 1–5.

Class 1 →	A,N,N	A,N,A				
	A,A,N	A,A,A				
	A,A_d,N	A,A_d,A				
Class 2 →	M,N,N	M,N,A	M,N,M			← Class 3
	M,A,N	M,A,A	M,A,M			
	M,A_d,N	M,A_d,A	M,A_d,M			
Class 4 →	M,M,N		M,M,M			
	M,M_d,N		M,M_d,M			
Class 5 →		M,M,A	A,N,M	A,M,N	A,M_d,N	
		M,M_d,A	A,A,M	A,M,A	A,M_d,A	
			A,A_d,M	A,M,M	A,M_d,M	

in Sect. 6.4; these are the models with multiplicative errors and multiplicative seasonality but additive trend.

Class 4 consists of the models with multiplicative errors, multiplicative trend, and either no seasonality or multiplicative seasonality. For Class 4, there are no available analytical expressions for forecast means or variances, and so we recommend using simulation to find prediction intervals.

The remaining 11 models are in Class 5. For these models, we also recommend using simulation to obtain prediction intervals. However, Class 5 models are those that can occasionally lead to numerical difficulties with very long forecast horizons. Specifically, the forecast variances are infinite, although this does not usually matter in practice for short- or medium-term forecasts. This issue is explored in Chap. 15.

Section 6.5 discusses the use of the forecast mean and variance formulae to construct prediction intervals even in cases where the prediction distributions are not Gaussian. In Sect. 6.6, we discuss lead-time demand forecasting for Class 1 models.

Most of the results in this chapter are based on Hyndman et al. (2005) and Snyder et al. (2004), although we use a slightly different parameterization in this book, and we extend the results in some new directions.

To simplify some of the expressions, we introduce the following notation:

$$h = mh_m + h_m^+,$$

where[1] h is the forecast horizon, m is the number of periods in each season, $h_m = \lfloor (h-1)/m \rfloor$ and $h_m^+ = \big[(h-1) \bmod m\big] + 1$. In other words, h_m is the number of complete years in the forecast period *prior* to time h, and h_m^+ is the number of remaining times in the forecast period up to and including time h. Thus, h_m^+ can take values $1, 2, \ldots, m$.

6.1 Simulated Prediction Distributions and Intervals

Recall from Chap. 4 that the general model with state vector

$$\boldsymbol{x}_t = (\ell_t, b_t, s_t, s_{t-1}, \ldots, s_{t-m+1})'$$

has the form

$$\begin{aligned} y_t &= w(\boldsymbol{x}_{t-1}) + r(\boldsymbol{x}_{t-1})\varepsilon_t, \\ \boldsymbol{x}_t &= \boldsymbol{f}(\boldsymbol{x}_{t-1}) + \boldsymbol{g}(\boldsymbol{x}_{t-1})\varepsilon_t, \end{aligned}$$

where $w(\cdot)$ and $r(\cdot)$ are scalar functions, $\boldsymbol{f}(\cdot)$ and $\boldsymbol{g}(\cdot)$ are vector functions, and $\{\varepsilon_t\}$ is a white noise process with variance σ^2.

One simple approach to obtaining the prediction distribution is to simulate sample paths from the models, conditional on the final state $\boldsymbol{x}_n$. This

[1] The notation $\lfloor u \rfloor$ means the integer part of u.

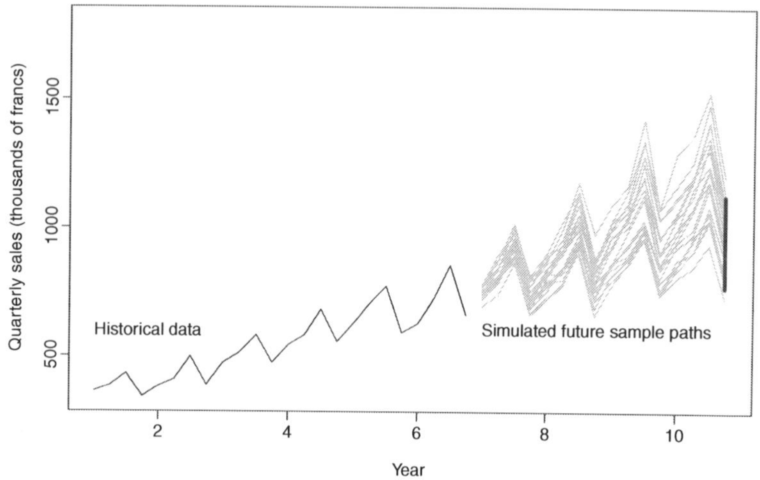

Fig. 6.1. Quarterly French exports data with 20 simulated future sample paths generated using the ETS(M,A,M) model assuming Gaussian innovations. The *solid vertical line* on the right shows a 90% prediction interval for the 16-step forecast horizon, calculated from the 0.05 and 0.95 quantiles of the 5,000 simulated values.

was the approach taken by Ord et al. (1997) and Hyndman et al. (2002). That is, we generate observations $\{y_t^{(i)}\}$, for $t = n+1,\ldots,n+h$, starting with $\boldsymbol{x}_n$ from the fitted model. Each ε_t value is obtained from a random number generator assuming a Gaussian or other appropriate distribution. This procedure is repeated for $i = 1,\ldots,M$, where M is a large integer. (In practice, we often use $M = 5{,}000$.)

Figure 6.1 shows a series of quarterly exports of a French company (in thousands of francs) taken from Makridakis et al. (1998, p. 162). We fit an ETS(M,A,M) model to the data. Then the model is used to simulate 5,000 future sample paths of the data. Twenty of these sample paths are shown in Fig. 6.1.

Characteristics of the prediction distribution of $y_{n+h|n}$ can then be estimated from the simulated values at a specific forecast horizon: $\boldsymbol{y}_{n+h|n} = \{y_{n+h}^{(1)},\ldots,y_{n+h}^{(M)}\}$. For example, prediction intervals can be obtained using quantiles of the simulated sample paths. An approximate $100(1-\alpha)\%$ prediction interval for forecast horizon h is given by the $\alpha/2$ and $1-\alpha/2$ quantiles of $\boldsymbol{y}_{n+h|n}$. The solid vertical line on the right of Fig. 6.1 is a 90% prediction interval computed in this way from the 0.05 and 0.95 quantiles of the simulated values at the 16-step horizon.

The full prediction density can be estimated using a kernel density estimator (Silverman 1986) applied to $\boldsymbol{y}_{n+h|n}$. Figure 6.2 shows the prediction

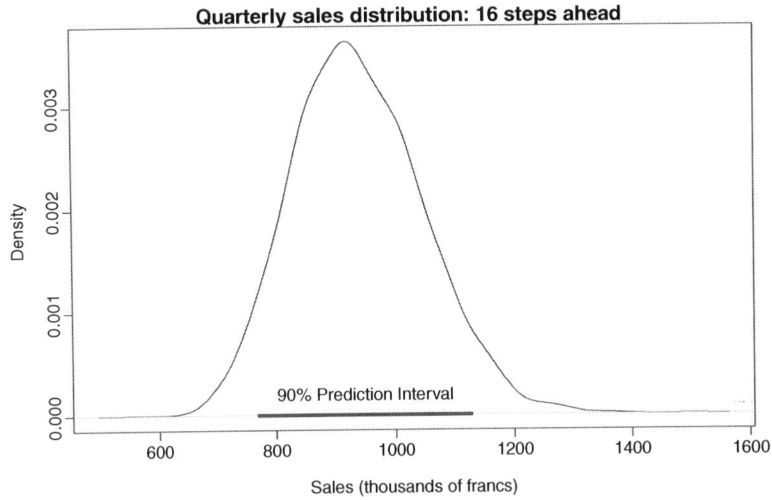

Fig. 6.2. The 16-step forecast density estimated from 5,000 simulated future sample paths. The 90% prediction interval is calculated from the 0.05 and 0.95 quantiles.

density for the data in Fig. 6.1 obtained in this way, along with the 90% prediction interval.

There are several advantages in computing prediction distributions and intervals in this way:

- If the distribution of ε_t is not Gaussian, another distribution can be used to generate the ε_t values when simulating the future sample paths.
- The historical ε_t values can be resampled to give bootstrap prediction distributions without making any distributional assumptions.
- The method can be used for nonlinear models where ε_t may be Gaussian but y_t is not Gaussian.
- The method avoids the complex formulae that are necessary to compute analytical prediction intervals for some nonlinear models.
- For some models (those in Classes 4 and 5), simulation is the only method available for computing prediction distributions and intervals.
- It is possible to take into account the error in estimating the model parameters. In this case, the simulated sample paths are generated using the same model but with randomly varying parameters, reflecting the parameter uncertainty in the fitted model. This was done in Ord et al. (1997) for models with multiplicative error, and in Snyder et al. (2001) for models with additive error.
- The increasing speed of computers makes the simulation approach more viable every year.

6.1.1 Lead-Time Forecasting

In inventory control, forecasts of the sum of the next h observations are often required. These are used for determination of ordering requirements such as reorder levels, order-up-to levels and reorder quantities.

Suppose that a replenishment decision is to be made at the beginning of period $n+1$. Any order placed at this time is assumed to arrive a lead-time later, at the start of period $n+h+1$. Thus, we need to forecast the aggregate of unknown future values y_{n+j}, defined by

$$Y_n(h) = \sum_{j=1}^{h} y_{n+j}.$$

The problem is to make inferences about the distribution of $Y_n(h)$ which (in the inventory context) is known as the "lead-time demand." The results from the simulation of single periods give the prediction distributions and intervals for individual forecast horizons, but for re-ordering purposes it is more useful to have the lead-time prediction distribution and interval. Because $Y_n(h)$ involves a summation, the central limit theorem states that its distribution will tend towards Gaussianity as h increases. However, for small to moderate h, we need to estimate the distribution.

The simulation approach can easily be used here by computing values of $Y_n(h)$ from the simulated future sample paths. For example, to get the distribution of $Y_n(3)$ for the quarterly French exports data, we sum the first three values of the simulated future sample paths shown in Fig. 6.1. This gives us 5,000 values from the distribution of $Y_n(3)$ (assuming the model is correct). Figure 6.3 shows the density computed from these 5,000 values along with a 90% prediction interval.

Here we have assumed that the lead-time h is fixed. Fixed lead-times are relevant when suppliers make regular deliveries, an increasingly common situation in supply chain management. For stochastic lead-times, we could randomly generate h from a Poisson distribution (or some other count distribution) when simulating values of $Y_n(h)$. This would be used when suppliers make irregular deliveries.

6.2 Class 1: Linear Homoscedastic State Space Models

We now derive some analytical results for the prediction distributions of the linear homoscedastic (Class 1) models. These provide additional insight and can be much quicker to compute than the simulation approach. Derivations of the results in this section are given in Appendix "Derivation of Results for Class 1."

The linear ETS models are (A,N,N), (A,A,N), (A,A$_\text{d}$,N), (A,N,A), (A,A,A) and (A,A$_\text{d}$,A). The forecast means are given in Table 6.2. Because of the linear

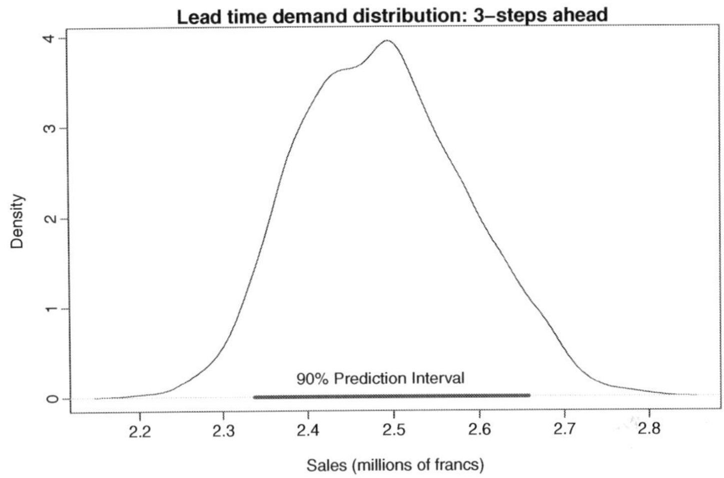

Fig. 6.3. The 3-step lead-time demand density estimated from 5,000 simulated future sample paths assuming Gaussian innovations. The 90% prediction interval is calculated from the 0.05 and 0.95 quantiles.

Table 6.2. Forecast means and c_j values for the linear homoscedastic (Class 1) and linear heteroscedastic (Class 2) state space models.

Model	Forecast mean: $\mu_{n+h\|n}$	c_j
(A,N,N)/(M,N,N)	ℓ_n	α
(A,A,N)/(M,A,N)	$\ell_n + hb_n$	$\alpha + \beta j$
(A,A$_d$,N)/(M,A$_d$,N)	$\ell_n + \phi_h b_n$	$\alpha + \beta\phi_j$
(A,N,A)/(M,N,A)	$\ell_n + s_{n-m+h_m^+}$	$\alpha + \gamma d_{j,m}$
(A,A,A)/(M,A,A)	$\ell_n + hb_n + s_{n-m+h_m^+}$	$\alpha + \beta j + \gamma d_{j,m}$
(A,A$_d$,A)/(M,A$_d$,A)	$\ell_n + \phi_h b_n + s_{n-m+h_m^+}$	$\alpha + \beta\phi_j + \gamma d_{j,m}$

The values of c_j are used in the forecast variance expressions (6.1) and (6.2). Here, $d_{j,m} = 1$ if $j = 0 \pmod m$ and 0 otherwise, and $\phi_j = \phi + \phi^2 + \cdots + \phi^j$.

structure of the models, the forecast means are identical to the point forecasts given in Table 2.1 (p. 18).

The forecast variances are given by

$$v_{n+h|n} = \mathrm{V}(y_{n+h} \mid \boldsymbol{x}_n) = \begin{cases} \sigma^2 & \text{if } h = 1; \\ \sigma^2 \left[1 + \sum_{j=1}^{h-1} c_j^2\right] & \text{if } h \geq 2; \end{cases} \tag{6.1}$$

where c_j is given in Table 6.2. Note that $v_{n+h|n}$ does not depend on $\boldsymbol{x}_n$ or n, but only on h and the smoothing parameters.

Table 6.3. Forecast variance expressions for each linear homoscedastic state space model, where $v_{n+h|n} = \mathrm{V}(y_{n+h} \mid \boldsymbol{x}_n)$.

Model	Forecast variance: $v_{n+h\|n}$
(A,N,N)	$v_{n+h\|n} = \sigma^2\big[1+\alpha^2(h-1)\big]$
(A,A,N)	$v_{n+h\|n} = \sigma^2\Big[1+(h-1)\big\{\alpha^2+\alpha\beta h+\tfrac{1}{6}\beta^2 h(2h-1)\big\}\Big]$
(A,A$_\text{d}$,N)	$v_{n+h\|n} = \sigma^2\Big[1+\alpha^2(h-1)+\frac{\beta\phi h}{(1-\phi)^2}\{2\alpha(1-\phi)+\beta\phi\} - \frac{\beta\phi(1-\phi^h)}{(1-\phi)^2(1-\phi^2)}\big\{2\alpha(1-\phi^2)+\beta\phi(1+2\phi-\phi^h)\big\}\Big]$
(A,N,A)	$v_{n+h\|n} = \sigma^2\Big[1+\alpha^2(h-1)+\gamma h_m(2\alpha+\gamma)\Big]$
(A,A,A)	$v_{n+h\|n} = \sigma^2\Big[1+(h-1)\big\{\alpha^2+\alpha\beta h+\tfrac{1}{6}\beta^2 h(2h-1)\big\} + \gamma h_m\big\{2\alpha+\gamma+\beta m(h_m+1)\big\}\Big]$
(A,A$_\text{d}$,A)	$v_{n+h\|n} = \sigma^2\Big[1+\alpha^2(h-1)+\frac{\beta\phi h}{(1-\phi)^2}\{2\alpha(1-\phi)+\beta\phi\} - \frac{\beta\phi(1-\phi^h)}{(1-\phi)^2(1-\phi^2)}\big\{2\alpha(1-\phi^2)+\beta\phi(1+2\phi-\phi^h)\big\} + \gamma h_m(2\alpha+\gamma) + \frac{2\beta\gamma\phi}{(1-\phi)(1-\phi^m)}\big\{h_m(1-\phi^m)-\phi^m(1-\phi^{mh_m})\big\}\Big]$

Because the models are linear and ε_t is assumed to be Gaussian, $y_{n+h} \mid \boldsymbol{x}_n$ is also Gaussian. Therefore, prediction intervals are easily obtained from the forecast means and variances.

In practice, we would normally substitute the numerical values of c_j from Table 6.2 into (6.1) to obtain numerical values for the variance. However, it is sometimes useful to expand (6.1) algebraically by substituting in the *expressions* for c_j from Table 6.2. The resulting variance expressions are given in Table 6.3.

We note in passing that $v_{n+h|n}$ is linear in h when $\beta = 0$, but cubic in h when $\beta > 0$. Thus, models with non-zero β tend to have prediction intervals that widen rapidly as h increases.

Traditionally, prediction intervals for the linear exponential smoothing methods have been found through heuristic approaches or by employing equivalent or approximate ARIMA models. Where an equivalent ARIMA model exists (see Chap. 11), the results in Table 6.3 provide identical forecast variances to those from the ARIMA model.

State space models with multiple sources of error have also been used to find forecast variances for SES and Holt's method (Harrison 1967; Johnston and Harrison 1986). With these models, the variances are limiting values, although the convergence is rapid. The variance formulae arising from these two cases are the same as in our results.

Prediction intervals for the additive Holt-Winters method have previously been considered by Yar and Chatfield (1990). They assumed that the one-period ahead forecast errors are independent, but they did not assume any particular underlying model for the smoothing methods. The formulae presented here for the ETS(A,A,A) model are equivalent to those given by Yar and Chatfield (1990).

6.3 Class 2: Linear Heteroscedastic State Space Models

Derivations of the results in this section are given in Appendix "Derivation of Results for Class 2."

The ETS models in Class 2 are (M,N,N), (M,A,N), (M,A_d,N), (M,N,A), (M,A,A) and (M,A_d,A). These are similar to those in Class 1 except that multiplicative rather than additive errors are used. Consequently, the forecast means of Class 2 models are identical to the forecast means of the analogous Class 1 model (assuming the same parameters), but the prediction intervals and distributions will be different. The forecast means for Class 2 also coincide with the usual point forecasts. Specific values of the forecast means are given in Table 6.2.

The forecast variance is given by

$$v_{n+h|n} = (1+\sigma^2)\theta_h - \mu_{n+h|n}^2, \tag{6.2}$$

where

$$\theta_1 = \mu_{n+1|n}^2 \qquad \text{and} \qquad \theta_h = \mu_{n+h|n}^2 + \sigma^2 \sum_{j=1}^{h-1} c_j^2 \theta_{h-j}, \tag{6.3}$$

where each c_j is identical to that for the corresponding additive error model from Class 1 in Table 6.2.

For most models, there is no non-recursive expression for the variance, and we simply substitute the relevant c_j values into (6.2) and (6.3) to obtain numerical expressions for the variance. However, for the ETS(M,N,N) model, we can go a little further (Exercise 6.1).

6.4 Class 3: Some Nonlinear Seasonal State Space Models

Derivations of the results in this section are given in Appendix "Derivation of results for Class 3."

The Class 3 models are (M,N,M), (M,A,M) and (M,A_d,M). These are similar to the seasonal models in Class 2 except that the seasonal component is multiplicative rather than additive.

Table 6.4. Values of $\mu_{n+h|n}$, $\tilde{\mu}_{n+h|n}$ and c_j for the Class 3 models.

	Approx $\mu_{n+h\|n}$	$\tilde{\mu}_{n+h\|n}$	c_j
ETS(M,N,M)	$\ell_n s_{n-m+h_m^+}$	ℓ_n	α
ETS(M,A,M)	$(\ell_n + hb_n)s_{n-m+h_m^+}$	$\ell_n + hb_n$	$\alpha + \beta j$
ETS(M,A$_d$,M)	$(\ell_n + \phi_h b_n)s_{n-m+h_m^+}$	$\ell_n + \phi_h b_n$	$\alpha + \beta\phi_j$

Here, $\phi_j = \phi + \phi^2 + \cdots + \phi^j$. Values of c_j are used in the forecast variance expressions (6.5).

6.4.1 Approximate Forecast Means and Variances

For these models, the exact forecast means and variances are complicated to compute when $h \geq m$. However, by noting that σ^2 is usually small (much less than 1), we can obtain approximate expressions for the mean and variance which are often useful. Let $\hat{y}_{n+h|n}$ be the usual point forecast as given in Table 2.1. Then,

$$\mu_{n+h|n} \approx \hat{y}_{n+h|n} \tag{6.4}$$

and

$$v_{n+h|n} \approx s^2_{n-m+h_m^+}\left[\theta_h(1+\sigma^2)(1+\gamma^2\sigma^2)^{h_m} - \tilde{\mu}^2_{n+h|n}\right], \tag{6.5}$$

where

$$\tilde{\mu}_{n+h|n} = \hat{y}_{n+h|n}/s_{n-m+h_m^+}$$

is the seasonally adjusted point forecast, $\theta_1 = \tilde{\mu}^2_{n+1|n}$, and

$$\theta_h = \tilde{\mu}^2_{n+h|n} + \sigma^2 \sum_{j=1}^{h-1} c_j^2 \theta_{h-j}, \qquad h \geq 2. \tag{6.6}$$

These expressions are exact for $h \leq m$, but are only approximate for $h > m$. The variance formula (6.5) agrees with those in Koehler et al. (2001) and Chatfield and Yar (1991) (who only considered the first year of forecasts).

Specific values for $\mu_{n+h|n}$, $\tilde{\mu}_{n+h|n}$ and c_j for the particular models in Class 3 are given in Table 6.4.

Example 6.1: ETS(M,N,M) model

For the ETS(M,N,M) model, $\theta_1 = \ell_n^2$, and for $h \geq 2$,

$$\begin{aligned}\theta_h &= \ell_n^2 + \alpha^2\sigma^2 \sum_{j=1}^{h-1} \theta_{h-j} \\ &= \ell_n^2 + \alpha^2\sigma^2(\theta_1 + \theta_2 + \cdots + \theta_{h-1}).\end{aligned}$$

Then, by induction, we can show that $\theta_h = \ell_n^2(1+\alpha^2\sigma^2)^{h-1}$. Plugging this into (6.5) gives the following simpler expression for $v_{n+h|n}$:

$$v_{n+h|n} \approx s^2_{n-m+h_m^+}\ell_n^2\Big[(1+\sigma^2)(1+\alpha^2\sigma^2)^{h-1}(1+\gamma^2\sigma^2)^{h_m}-1\Big].$$

The expression is exact for $h \leq m$.

6.4.2 Exact Forecast Means and Variances

To obtain the exact formulae for $h > m$, we first write the models in Class 3 using the following nonlinear state space model:

$$\begin{aligned} y_t &= \boldsymbol{w}_1'\boldsymbol{x}_{t-1}\boldsymbol{w}_2'\boldsymbol{z}_{t-1}(1+\varepsilon_t), \\ \boldsymbol{x}_t &= (\boldsymbol{F}_1 + \boldsymbol{G}_1\varepsilon_t)\boldsymbol{x}_{t-1}, \\ \boldsymbol{z}_t &= (\boldsymbol{F}_2 + \boldsymbol{G}_2\varepsilon_t)\boldsymbol{z}_{t-1}, \end{aligned}$$

where $\boldsymbol{F}_1$, $\boldsymbol{F}_2$, $\boldsymbol{G}_1$, $\boldsymbol{G}_2$, $\boldsymbol{w}_1'$ and $\boldsymbol{w}_2'$ are all matrix or vector coefficients, and $\boldsymbol{x}_t$ and $\boldsymbol{z}_t$ are unobserved state vectors at time t. As for Class 2, $\{\varepsilon_t\}$ is $\text{NID}(0,\sigma^2)$, where the lower tail of the distribution is truncated so that $1+\varepsilon_t$ is positive.

Let k be the length of vector $\boldsymbol{x}_t$ and q be the length of vector $\boldsymbol{z}_t$. Then the orders of the above matrices are as follows:

$$\begin{array}{lll} \boldsymbol{F}_1\ (k\times k) & \boldsymbol{G}_1\ (k\times k) & \boldsymbol{w}_1'\ (1\times k) \\ \boldsymbol{F}_2\ (q\times q) & \boldsymbol{G}_2\ (q\times q) & \boldsymbol{w}_2'\ (1\times q) \end{array}$$

- For the ETS(M,N,M) model, $\boldsymbol{x}_t = \ell_t$, $\boldsymbol{z}_t = (s_t,\ldots,s_{t-m+1})'$, and the matrix coefficients are $\boldsymbol{w}_1 = 1$, $\boldsymbol{w}_2' = [0,\ldots,0,1]$,

$$\boldsymbol{F}_1 = 1, \quad \boldsymbol{F}_2 = \begin{bmatrix} \boldsymbol{0}'_{m-1} & 1 \\ \boldsymbol{I}_{m-1} & \boldsymbol{0}_{m-1} \end{bmatrix}, \quad \boldsymbol{G}_1 = \alpha, \quad \text{and} \quad \boldsymbol{G}_2 = \begin{bmatrix} \boldsymbol{0}'_{m-1} & \gamma \\ \boldsymbol{O}_{m-1} & \boldsymbol{0}_{m-1} \end{bmatrix}.$$

- For the ETS(M,A$_\text{d}$,M) model, $\boldsymbol{x}_t = (\ell_t, b_t)'$, $\boldsymbol{w}_1' = [1,1]$,

$$\boldsymbol{F}_1 = \begin{bmatrix} 1 & \phi \\ 0 & \phi \end{bmatrix}, \quad \boldsymbol{G}_1 = \begin{bmatrix} \alpha & \alpha \\ \beta & \beta \end{bmatrix},$$

and $\boldsymbol{z}_2$, $\boldsymbol{w}_2$, $\boldsymbol{F}_2$ and $\boldsymbol{G}_2$ are the same as for the ETS(M,N,M) model.
- The ETS(M,A,M) model is equivalent to the ETS(M,A$_\text{d}$,M) model with $\phi = 1$.

For models in this class,

$$\mu_{n+h|n} = \boldsymbol{w}_1' \boldsymbol{M}_{h-1} \boldsymbol{w}_2 \tag{6.7}$$

and

$$v_{n+h|n} = (1+\sigma^2)(\boldsymbol{w}_2' \otimes \boldsymbol{w}_1') \boldsymbol{V}_{n+h-1|n} (\boldsymbol{w}_2' \otimes \boldsymbol{w}_1')' + \sigma^2 \mu_{n+h|n}^2, \tag{6.8}$$

where $\otimes$ denotes a Kronecker product (Schott 2005, Sect. 8.2), $\boldsymbol{M}_0 = \boldsymbol{x}_n \boldsymbol{z}_n'$, $\boldsymbol{V}_0 = \boldsymbol{O}_{2m}$, and for $h \geq 1$,

$$\boldsymbol{M}_h = \boldsymbol{F}_1 \boldsymbol{M}_{h-1} \boldsymbol{F}_2' + \boldsymbol{G}_1 \boldsymbol{M}_{h-1} \boldsymbol{G}_2' \sigma^2 \tag{6.9}$$

and

$$\begin{aligned}
\boldsymbol{V}_{n+h|n} &= (\boldsymbol{F}_2 \otimes \boldsymbol{F}_1) \boldsymbol{V}_{n+h-1|n} (\boldsymbol{F}_2 \otimes \boldsymbol{F}_1)' \\
&+ \sigma^2 \Big[(\boldsymbol{F}_2 \otimes \boldsymbol{F}_1) \boldsymbol{V}_{n+h-1|n} (\boldsymbol{G}_2 \otimes \boldsymbol{G}_1)' + (\boldsymbol{G}_2 \otimes \boldsymbol{G}_1) \boldsymbol{V}_{n+h-1|n} (\boldsymbol{F}_2 \otimes \boldsymbol{F}_1)' \Big] \\
&+ \sigma^2 (\boldsymbol{G}_2 \otimes \boldsymbol{F}_1 + \boldsymbol{F}_2 \otimes \boldsymbol{G}_1) \Big[\boldsymbol{V}_{n+h-1|n} + \overrightarrow{\boldsymbol{M}}_{h-1} \overrightarrow{\boldsymbol{M}}_{h-1}' \Big] (\boldsymbol{G}_2 \otimes \boldsymbol{F}_1 + \boldsymbol{F}_2 \otimes \boldsymbol{G}_1)' \\
&+ \sigma^4 (\boldsymbol{G}_2 \otimes \boldsymbol{G}_1) \Big[3 \boldsymbol{V}_{n+h-1|n} + 2 \overrightarrow{\boldsymbol{M}}_{h-1} \overrightarrow{\boldsymbol{M}}_{h-1}' \Big] (\boldsymbol{G}_2 \otimes \boldsymbol{G}_1)',
\end{aligned} \tag{6.10}$$

where $\overrightarrow{\boldsymbol{M}}_{h-1} = \text{vec}(\boldsymbol{M}_{h-1})$. (That is, the columns of $\boldsymbol{M}_{h-1}$ are stacked to form a vector.) Note, in particular, that $\mu_{n+1|n} = (\boldsymbol{w}_1' \boldsymbol{x}_n)(\boldsymbol{w}_2' \boldsymbol{z}_n)$ and $v_{n+1|n} = \sigma^2 \mu_{n+1|n}^2$. While these expressions look complicated and provide little insight, it is relatively easy to compute them using computer matrix languages such as **R** and Matlab.

In Appendix "Derivation of results for Class 3," we show that the approximations (6.4) and (6.5) follow from the exact expressions (6.7) and (6.8). Note that the usual point forecasts for these models are given by (6.4) rather than (6.7).

6.4.3 The Accuracy of the Approximations

In order to investigate the accuracy of the approximations (6.4) and (6.5) for the exact mean and variance given by (6.7) and (6.8), we provide some comparisons for the ETS(M,A,M) model in Class 3.

These comparisons are done for quarterly data, where the values for the components are assumed to be the following: $\ell_n = 100$, $b_n = 2$, $s_n = 0.80$, $s_{n-1} = 1.20$, $s_{n-2} = 0.90$ and $s_{n-3} = 1.10$. We use the following base level values for the parameters: $\alpha = 0.2$, $\beta = 0.06$, $\gamma = 0.1$, and $\sigma = 0.05$. We vary these parameters one at a time as shown in Table 6.5.

The results in Table 6.5 show that the mean and approximate mean are always very close, and that the percentage difference in the standard

Table 6.5. Comparison of exact and approximate means and standard deviations for the ETS(M,A,M) model in Class 3.

Period ahead h	Exact mean (6.7) $\mu_{n+h\|n}$	Approximate mean (6.4)	Exact SD (6.8) $\sqrt{v_{n+h\|n}}$	Approximate SD (6.5)	SD percent difference
$\sigma = 0.05, \alpha = 0.2, \beta = 0.06, \gamma = 0.1$					
5	121.01	121.00	7.53	7.33	2.69
6	100.81	100.80	6.68	6.52	2.37
7	136.81	136.80	9.70	9.50	2.07
8	92.81	92.80	7.06	6.93	1.80
9	129.83	129.80	10.85	10.45	3.68
10	108.03	108.00	9.65	9.34	3.21
11	146.44	146.40	13.99	13.60	2.81
12	99.22	99.20	10.13	9.88	2.47
$\sigma = \mathbf{0.1}, \alpha = 0.2, \beta = 0.06, \gamma = 0.1$					
5	121.05	121.00	15.09	14.68	2.73
6	100.84	100.80	13.39	13.07	2.40
7	136.86	136.80	19.45	19.04	2.11
8	92.84	92.80	14.15	13.89	1.84
9	129.93	129.80	21.77	20.96	3.75
10	108.11	108.00	19.39	18.75	3.29
11	146.55	146.40	28.11	27.30	2.89
12	99.30	99.20	20.35	19.83	2.55
$\sigma = 0.05, \alpha = \mathbf{0.6}, \beta = 0.06, \gamma = 0.1$					
5	121.02	121.00	10.87	10.60	2.47
6	100.82	100.80	9.96	9.76	2.04
7	136.83	136.80	14.76	14.51	1.72
8	92.82	92.80	10.86	10.70	1.47
9	129.86	129.80	16.64	16.19	2.71
10	108.05	108.00	14.83	14.48	2.37
11	146.46	146.40	21.45	21.00	2.09
12	99.24	99.20	15.45	15.16	1.86
$\sigma = 0.05, \alpha = 0.2, \beta = \mathbf{0.18}, \gamma = 0.1$					
5	121.03	121.00	10.19	9.87	3.08
6	100.82	100.80	9.88	9.66	2.27
7	136.83	136.80	15.55	15.29	1.69
8	92.82	92.80	12.14	11.98	1.28
9	129.87	129.80	19.67	19.16	2.56
10	108.06	108.00	18.41	18.04	2.03
11	146.48	146.40	27.86	27.41	1.64
12	99.26	99.20	20.93	20.65	1.35
$\sigma = 0.05, \alpha = 0.2, \beta = 0.06, \gamma = \mathbf{0.3}$					
5	121.04	121.00	8.10	7.53	7.12
6	100.83	100.80	7.13	6.68	6.36
7	136.84	136.80	10.28	9.70	5.64
8	92.83	92.80	7.42	7.05	4.97
9	129.90	129.80	11.89	10.77	9.46
10	108.08	108.00	10.47	9.59	8.42
11	146.51	146.40	15.04	13.91	7.49
12	99.27	99.20	10.79	10.07	6.67

deviations only becomes substantial when we increase γ. This result for the standard deviation is not surprising because the approximation is exact if $\gamma = 0$. In fact, we recommend that the approximation not be used if the smoothing parameter for γ exceeds 0.10.

6.5 Prediction Intervals

The prediction distributions for Class 1 are clearly Gaussian, as the models are linear and the errors are Gaussian. Consequently, $100(1-\alpha)\%$ prediction intervals can be calculated from the forecast means and variances in the usual way, namely $\mu_{n+h|n} \pm z_{\alpha/2}\sqrt{v_{n+h|n}}$, where z_q denotes the qth quantile of a standard Gaussian distribution.

In applying these formulae, the maximum likelihood estimator for σ^2 (see p. 68) is simply

$$\hat{\sigma}^2 = n^{-1}\sum_{t=1}^{n} \hat{\varepsilon}_t^2,$$

where $\hat{\varepsilon}_t = y_t - \mu_{t|t-1}$.

The prediction distributions for Classes 2 and 3 are non-Gaussian because of the nonlinearity of the state space equations. However, prediction intervals based on the above (Gaussian) formula will usually give reasonably accurate results, as the following example shows. In cases where the Gaussian approximation may be unreasonable, it is necessary to use the simulation approach of Sect. 6.1.

6.5.1 Application: Quarterly French Exports

As a numerical example, we consider the quarterly French exports data given in Fig. 6.1, and use the ETS(M,A,M) model. We estimate the parameters to be $\alpha = 0.8185$, $\beta = 0.01$, $\gamma = 0.01$ and $\sigma = 0.0352$, with the final states $\ell_n = 757.3$, $b_n = 15.7$, and $z_n = (0.873, 1.141, 1.022, 0.964)'$.

Figure 6.4 shows the forecast standard deviations calculated exactly using (6.8) and approximately using (6.5). The approximate values are so close to the exact values in this case (because σ^2 and γ are both very small) that it is almost impossible to distinguish the two lines.

The data with three years of forecasts are shown in Fig. 6.5. In this case, the conditional mean forecasts obtained from model ETS(M,A,M) are virtually indistinguishable from the usual forecasts because σ is so small (they are identical up to $h = m$). The solid lines show prediction intervals calculated as $\mu_{n+h|n} \pm 1.96\sqrt{v_{n+h|n}}$, and the dotted lines show prediction intervals computed by generating 20,000 future sample paths from the fitted model and finding the 2.5 and 97.5% quantiles at each forecast horizon.

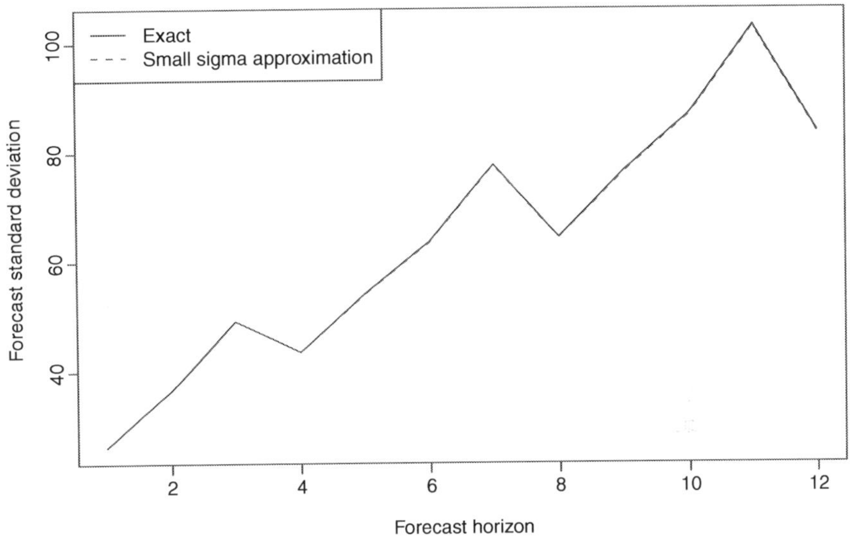

Fig. 6.4. Forecast standard deviations calculated (**a**) exactly using (6.8); and (**b**) approximately using (6.5).

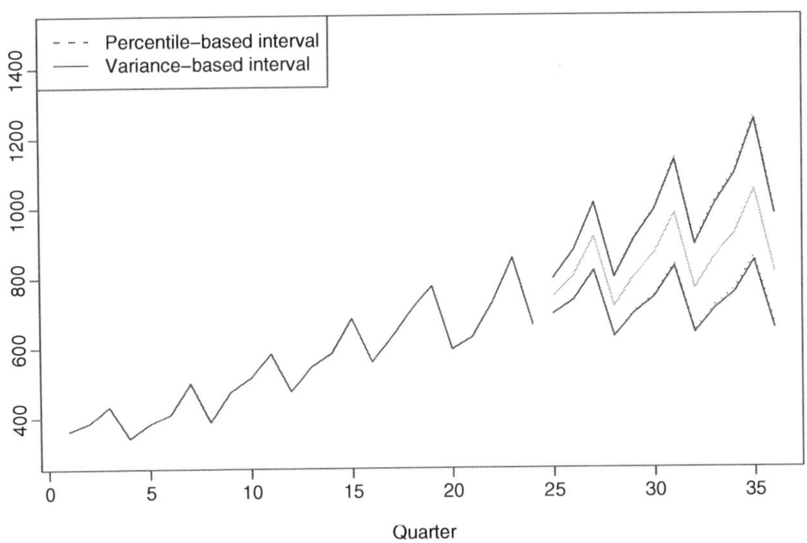

Fig. 6.5. Quarterly French exports data with 3 years of forecasts. The *solid lines* show prediction intervals calculated as $\mu_{n+h|n} \pm 1.96\sqrt{v_{n+h|n}}$, and the *dotted lines* show prediction intervals computed by generating 20,000 future sample paths from the fitted model and finding the 2.5 and 97.5% quantiles at each forecast horizon.

Clearly, the variance-based intervals are a good approximation despite the non-Gaussianity of the prediction distributions.

6.6 Lead-Time Demand Forecasts for Linear Homoscedastic Models

For Class 1 models, it is also possible to obtain some analytical results on the distribution of lead-time demand, defined by

$$Y_n(h) = \sum_{j=1}^{h} y_{n+j}. \tag{6.11}$$

In particular, the variance of lead-time demand can be used when implementing an inventory strategy, although the basic exponential smoothing procedures originally provided only point forecasts, and rather ad hoc formulae were the vogue in inventory control software.

Harrison (1967) and Johnston and Harrison (1986) derived a variance formula for lead-time demand based on simple exponential smoothing using a state space model with independent error terms. They utilized the fact that simple exponential smoothing emerges as the steady state form of the model predictions in large samples. Adopting a different model, Snyder et al. (1999) were able to obtain the same formula without recourse to a restrictive large sample assumption. Around the same time, Graves (1999) obtained the formula using an ARIMA(0,1,1) model.

Harrison (1967) and Johnston and Harrison (1986) also obtained a variance formula for lead-time demand when trend-corrected exponential smoothing is employed. Yar and Chatfield (1990), however, suggested a slightly different formula. They also provide a formula that incorporates seasonal effects for use with the additive Holt-Winters method.

The approach we adopt here is based on Snyder et al. (2004), although the parameterization in this book is slightly different from that used in Snyder et al. (2004). The results obtained subsume those in Harrison (1967), Johnston and Harrison (1986), Yar and Chatfield (1990), Graves (1999) and Snyder et al. (1999). In addition, for ETS(A,A,A), the recursive variance formula in Yar and Chatfield (1990) has been replaced with a closed-form counterpart.

6.6.1 Means and Variances of Lead-Time Demand

In Appendix "Derivation of C_j values" we show that

$$y_{n+j} = \mu_{n+j|n} + \sum_{i=1}^{j-1} c_{j-i}\varepsilon_{n+i} + \varepsilon q_{n+j},$$

where $\mu_{n+j|n}$ and c_k are given in Table 6.2. Substitute this into (6.11) to give

$$Y_n(h) = \sum_{j=1}^{h} \left(\mu_{n+j|n} + \sum_{i=1}^{j-1} c_{j-i}\varepsilon_{n+i} + \varepsilon_{n+j}\right) = \sum_{j=1}^{h} \mu_{n+j|n} + \sum_{j=1}^{h} C_{j-1}\varepsilon_{n+h-j+1}, \tag{6.12}$$

where

$$C_0 = 1 \quad \text{and} \quad C_j = 1 + \sum_{i=1}^{j} c_i \quad \text{for} \quad j = 1, \dots, h-1. \tag{6.13}$$

Thus, lead-time demand can be resolved into a linear function of the uncorrelated level and error components.

From (6.12), it is easy to see that the point forecast (conditional mean) is simply

$$\hat{Y}_n(h) = \mathrm{E}(Y_n(h) \mid \boldsymbol{x}_n) = \sum_{j=1}^{h} \mu_{n+j|n} \tag{6.14}$$

and the conditional variance is given by

$$\mathrm{V}\left(Y_n(h) \mid \boldsymbol{x}_n\right) = \sigma^2 \sum_{j=0}^{h-1} C_j^2. \tag{6.15}$$

The value of C_j for each of the models is given in Table 6.6. These expressions are derived in Appendix "Derivation of C_j values."

As with the equations for forecast variance at a specific forecast horizon, we can substitute these expressions into (6.15) to derive a specific formula for each model. This leads to a lot of tedious algebra that is of limited value. Therefore we only give the result for model ETS(A,N,N):

Table 6.6. Values of C_j to be used in computing the lead-time variance in (6.15).

Model	C_j
(A,N,N)	$1 + j\alpha$
(A,A,N)	$1 + j\left[\alpha + \frac{1}{2}\beta(j+1)\right]$
(A,A$_d$,N)	$1 + j\alpha + \frac{\beta\phi}{(1-\phi)^2}\left[(j+1)(1-\phi) - (1-\phi^{j+1})\right]$
(A,N,A)	$1 + j\alpha + \gamma j_m$
(A,A,A)	$1 + j\left[\alpha + \frac{1}{2}\beta(j+1)\right] + \gamma j_m$
(A,A$_d$,A)	$1 + j\alpha + \frac{\beta\phi}{(1-\phi)^2}\left[(j+1)(1-\phi) - (1-\phi^{j+1})\right] + \gamma j_m$

Here m is the number of periods in each season and $j_m = \lfloor j/m \rfloor$ is the number of complete seasonal cycles that occur within j time periods.

$$\begin{aligned}\mathrm{V}(Y_n(h) \mid \boldsymbol{x}_n) &= \sum_{j=0}^{h-1}(1+j\alpha)^2 \\ &= \sigma^2 h\Big[1+\alpha(h-1)+\tfrac{1}{6}\alpha^2(h-1)(2h-1)\Big]. \end{aligned} \tag{6.16}$$

6.6.2 Matrix Calculation of Means and Variances

The mean and variance of the lead-time demand, and the forecast mean and variance for a single period, can also be computed recursively using matrix equations. From Chap. 3, we know that the form of the Class 1 models is

$$\begin{aligned} y_t &= \boldsymbol{w}'\boldsymbol{x}_{t-1} + \varepsilon_t, \\ \boldsymbol{x}_t &= \boldsymbol{F}\boldsymbol{x}_{t-1} + \boldsymbol{g}\varepsilon_t, \end{aligned}$$

where $\boldsymbol{w}'$ is a row vector, $\boldsymbol{g}$ is a column vector, $\boldsymbol{F}$ is a matrix, $\boldsymbol{x}_t$ is the unobserved state vector at time t, and $\{\varepsilon_t\}$ is $\mathrm{NID}(0,\sigma^2)$.

Observe that the lead-time demand can be determined recursively by

$$Y_n(j) = Y_n(j-1) + y_{n+j}, \tag{6.17}$$

where $Y_n(0) = 0$ and $Y_n(j) = \sum_{i=1}^{j} y_{n+i}$. Consequently, (6.17) can be written as

$$Y_n(j) = Y_n(j-1) + \boldsymbol{w}'\boldsymbol{x}_{n+j-1} + \varepsilon_{n+j}. \tag{6.18}$$

So, if the state vector $\boldsymbol{x}_{n+j}$ is augmented with $Y_n(j)$, the first-order recurrence relationship

$$\begin{bmatrix}\boldsymbol{x}_{n+j} \\ Y_n(j)\end{bmatrix} = \begin{bmatrix}\boldsymbol{F} & \boldsymbol{0} \\ \boldsymbol{w}' & 1\end{bmatrix}\begin{bmatrix}\boldsymbol{x}_{n+j-1} \\ Y_n(j-1)\end{bmatrix} + \begin{bmatrix}\boldsymbol{g} \\ 1\end{bmatrix}\varepsilon_{n+j}$$

is obtained. This has the general form $\boldsymbol{z}_{n+j} = \boldsymbol{A}\boldsymbol{z}_{n+j-1} + \boldsymbol{b}\varepsilon_{n+j}$. If the mean and variance of the $\boldsymbol{z}_{n+j}$ are denoted by $\boldsymbol{m}^z_{n+j|n} = \mathrm{E}(\boldsymbol{z}_{n+j} \mid \boldsymbol{x}_n)$ and $\boldsymbol{V}^z_{n+j|n} = \mathrm{V}(\boldsymbol{z}_{n+j} \mid \boldsymbol{x}_n)$, then they can be computed recursively using the equations

$$\begin{aligned} \boldsymbol{m}^z_{n+j|n} &= \boldsymbol{A}\boldsymbol{m}^z_{n+j-1|n}, \\ \boldsymbol{V}^z_{n+j|n} &= \boldsymbol{A}\boldsymbol{V}^z_{n+j-1|n}\boldsymbol{A}' + \sigma^2\boldsymbol{b}\boldsymbol{b}'. \end{aligned}$$

The mean of the lead-time demand $Y_n(h)$ is the last element in $\boldsymbol{m}^z_{n+h|n}$, and the variance of $Y_n(h)$ is the bottom right element of $\boldsymbol{V}^z_{n+h|n}$.

This same procedure of using an augmented matrix can also be applied to find the forecast mean and variance of y_{n+h} for any single future time period $t = n+h$. In this case, the state vector $\boldsymbol{x}_{n+j}$ is augmented with y_{n+j} in place of $Y_n(j)$, and

$$\boldsymbol{A} = \begin{bmatrix}\boldsymbol{F} & \boldsymbol{0} \\ \boldsymbol{w}' & 0\end{bmatrix}.$$

Then, the mean and variance of y_{n+h} are the last elements in $\boldsymbol{m}^z_{n+h|n}$ and $\boldsymbol{V}^z_{n+h|n}$ respectively. Of course, one can use $\boldsymbol{A} = [\boldsymbol{F}, \boldsymbol{w}']'$ and the general form $\boldsymbol{z}_{n+j} = \boldsymbol{A}\boldsymbol{x}_{n+j-1} + \boldsymbol{b}\varepsilon_{n+j}$ to remove the unnecessary multiplications by 0 in an actual implementation.

6.6.3 Stochastic Lead-Times

In practice, lead-times are often stochastic, depending on various factors including demand in the previous time periods. We explore the effect of stochastic lead-times on forecast variances in the case of the ETS(A,N,N) model for simple exponential smoothing.

Let the lead-time, T, be stochastic with mean $\mathrm{E}(T) = h$. The mean lead-time demand, given the level at time n, is

$$\mathrm{E}(Y_n(T) \mid \ell_n) = \mathrm{E}_T[\mathrm{E}(Y_n(T) \mid T, \ell_n)] = h\ell_n,$$

as in the case of a fixed lead-time. The variance of the lead-time demand reduces to

$$\begin{aligned}
\mathrm{V}(Y_n(T) \mid \ell_n) &= \mathrm{V}_T[\mathrm{E}(Y_n(T) \mid T, \ell_n)] + \mathrm{E}_T[\mathrm{V}(Y_n(T) \mid T, \ell_n)] \\
&= \mathrm{V}_T(\ell_n T) + \mathrm{E}_T\left[\sigma^2 \sum_{j=1}^{T} c_{j,T}^2\right] \\
&= \ell_n^2 \mathrm{V}(T) + \sigma^2 \mathrm{E}_T\left[\sum_{j=1}^{T} \left\{1 + 2\alpha(T-j) + \alpha^2(T-j)^2\right\}\right] \\
&= \ell_n^2 \mathrm{V}(T) + \sigma^2 h + \sigma^2\alpha\left[(1 + \tfrac{1}{2}\alpha)h_{[2]} + \tfrac{1}{3}\alpha h_{[3]}\right],
\end{aligned}$$

where $h_{[j]} = \mathrm{E}[T(T-1)\dots(T-j+1)]$, $j = 1, 2, \dots$, is known as the jth factorial moment of the distribution of T.

For example, when the lead-time is fixed, $h_{[j]} = h(h-1)\dots(h-j+1)$. When the lead-time is Poisson with mean h, then $h_{[j]} = h^j$. Therefore, the lead-time demand variance becomes

$$\mathrm{V}(Y_n(T) \mid \ell_n) = (\ell_n^2 + \sigma^2)h + \sigma^2\alpha\left[(1 + \tfrac{1}{2}\alpha)h^2 + \tfrac{1}{3}\alpha h^3\right].$$

Compare this with the variance for a fixed lead-time as given in (6.16). The two variances are plotted in Fig. 6.6 for $\alpha = 0.1$, $\sigma = 1$ and $\ell_n = 2$, showing that a stochastic lead-time can substantially increase the lead-time demand variance.

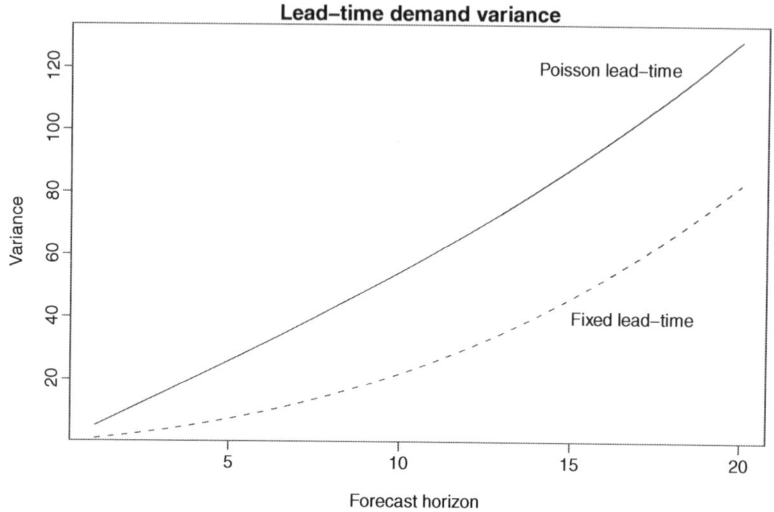

Fig. 6.6. Lead-time demand variance for an ETS(A,N,N) model with fixed and stochastic lead-times. Here, $\alpha = 0.1$, $\sigma = 1$ and $\ell_n = 2$.

6.7 Exercises

Exercise 6.1. For the ETS(M,N,N) model, show that

$$\theta_j = \ell_n^2(1+\alpha^2\sigma^2)^{j-1}$$

and

$$v_{n+h|n} = \ell_n^2\left[(1+\alpha^2\sigma^2)^{h-1}(1+\sigma^2)-1\right].$$

Exercise 6.2. For the ETS(A,A,A) model, use (6.23) replacing ϕ_j by j to show that

$$v_{n+h|n} = \sigma^2\Big[1+(h-1)\big\{\alpha^2+\alpha\beta h+\tfrac{1}{6}\beta^2 h(2h-1)\big\}$$
$$+\gamma h_m\left\{2\alpha+\gamma+\beta m(h_m+1)\right\}\Big].$$

Exercise 6.3. Monthly US 10-year bonds data were forecast with an ETS(A,A$_\text{d}$,N) model in Sect. 2.8.1 (p. 28). Find the 95% prediction intervals for this model algebraically and compare the results obtained by simulating 5,000 future sample paths using **R**.

Exercise 6.4. Quarterly UK passenger vehicle production data were forecast with an ETS(A,N,A) model in Sect. 2.8.1 (p. 28). Find the 95% prediction intervals for this model algebraically and compare the results obtained by simulating 5,000 future sample paths using **R**.

Appendix: Derivations

Derivation of Results for Class 1

The results for Class 1 models are obtained by first noting that all of the linear, homoscedastic ETS models can be written using the following linear state space model, introduced in Chap. 3:

$$y_t = \boldsymbol{w}'\boldsymbol{x}_{t-1} + \varepsilon_t \tag{6.19}$$
$$\boldsymbol{x}_t = \boldsymbol{F}\boldsymbol{x}_{t-1} + \boldsymbol{g}\varepsilon_t, \tag{6.20}$$

where $\boldsymbol{w}'$ is a row vector, $\boldsymbol{g}$ is a column vector, $\boldsymbol{F}$ is a matrix, and $\boldsymbol{x}_t$ is the unobserved state vector at time t. In each case, $\{\varepsilon_t\}$ is NID$(0, \sigma^2)$.

Let $\boldsymbol{I}_k$ denote the $k \times k$ identity matrix, and $\boldsymbol{0}_k$ denote a zero vector of length k. Then

- The ETS(A,N,N) model has $\boldsymbol{x}_t = \ell_t$, $\boldsymbol{w} = \boldsymbol{F} = 1$ and $\boldsymbol{g} = \alpha$;
- The ETS(A,A$_d$,N) model has $\boldsymbol{x}_t = (\ell_t, b_t)'$, $\boldsymbol{w}' = [1\ \phi]$,

$$\boldsymbol{F} = \begin{bmatrix} 1 & \phi \\ 0 & \phi \end{bmatrix} \quad \text{and} \quad \boldsymbol{g} = \begin{bmatrix} \alpha \\ \beta \end{bmatrix};$$

- The ETS(A,N,A) model has $\boldsymbol{x}_t = (\ell_t, s_t, s_{t-1}, \ldots, s_{t-(m-1)})'$, $\boldsymbol{w}' = [1\ \boldsymbol{0}'_{m-1}\ 1]$,

$$\boldsymbol{F} = \begin{bmatrix} 1 & \boldsymbol{0}'_{m-1} & 0 \\ 0 & \boldsymbol{0}'_{m-1} & 1 \\ \boldsymbol{0}_{m-1} & \boldsymbol{I}_{m-1} & \boldsymbol{0}_{m-1} \end{bmatrix} \quad \text{and} \quad \boldsymbol{g} = \begin{bmatrix} \alpha \\ \gamma \\ \boldsymbol{0}_{m-1} \end{bmatrix};$$

- The ETS(A,A$_d$,A) model has $\boldsymbol{x}_t = (\ell_t, b_t, s_t, s_{t-1}, \ldots, s_{t-(m-1)})'$, $\boldsymbol{w}' = [1\ \phi\ \boldsymbol{0}'_{m-1}\ 1]$,

$$\boldsymbol{F} = \begin{bmatrix} 1 & \phi & \boldsymbol{0}'_{m-1} & 0 \\ 0 & \phi & \boldsymbol{0}'_{m-1} & 0 \\ 0 & 0 & \boldsymbol{0}'_{m-1} & 1 \\ \boldsymbol{0}_{m-1} & \boldsymbol{0}_{m-1} & \boldsymbol{I}_{m-1} & \boldsymbol{0}_{m-1} \end{bmatrix} \quad \text{and} \quad \boldsymbol{g} = \begin{bmatrix} \alpha \\ \beta \\ \gamma \\ \boldsymbol{0}_{m-1} \end{bmatrix}.$$

The matrices for (A,A,N) and (A,A,A) are the same as for (A,A$_d$,N) and (A,A$_d$,A) respectively, but with $\phi = 1$.

Forecast Mean

Let $\boldsymbol{m}_{n+h|n} = \mathrm{E}(\boldsymbol{x}_{n+h} \mid \boldsymbol{x}_n)$. Then $\boldsymbol{m}_{n|n} = \boldsymbol{x}_n$ and

$$\boldsymbol{m}_{n+h|n} = \boldsymbol{F}\boldsymbol{m}_{n+h-1|n} = \boldsymbol{F}^2\boldsymbol{m}_{n+h-2|n} = \cdots = \boldsymbol{F}^h\boldsymbol{m}_{n|n} = \boldsymbol{F}^h\boldsymbol{x}_n.$$

Therefore

$$\mu_{n+h|n} = \mathrm{E}(y_{n+h}|\boldsymbol{x}_n) = \boldsymbol{w}'\boldsymbol{m}_{n+h-1|n} = \boldsymbol{w}'\boldsymbol{F}^{h-1}\boldsymbol{x}_n.$$

Example 6.2: Forecast mean of the ETS(A,A$_d$,A) model

For the ETS(A,A$_\text{d}$,A) model, $\boldsymbol{w}' = [1\ \phi\ \boldsymbol{0}'_{m-1}\ 1]$ and

$$\boldsymbol{F}^j = \begin{bmatrix} 1 & \phi_j & 0 & 0 & \dots & 0 \\ 0 & \phi^j & 0 & 0 & \dots & 0 \\ 0 & 0 & d_{j+m,m} & d_{j+m+1,m} & \dots & d_{j+2m-1,m} \\ 0 & 0 & d_{j+m-1,m} & d_{j+m,m} & \dots & d_{j+2m-2,m} \\ \vdots & \vdots & \vdots & \vdots & \ddots & \vdots \\ 0 & 0 & d_{j+1,m} & d_{j+2,m} & \dots & d_{j+m,m} \end{bmatrix},$$

where $\phi_j = \phi + \phi^2 + \dots + \phi^j$, and $d_{k,m} = 1$ if $k = 0 \pmod{m}$ and $d_{k,m} = 0$ otherwise. Therefore,

$$\boldsymbol{w}'\boldsymbol{F}^j = [1, \phi_{j+1}, d_{j+1,m}, d_{j+2,m}, \dots, d_{j+m,m}] \tag{6.21}$$

and

$$\mu_{n+h|n} = \ell_n + \phi_h b_n + s_{n-m+h_m^+}.$$

The forecast means for the other models can be derived similarly, and are listed in Table 6.2

Forecast Variance

Define the state forecast variance as $\boldsymbol{V}_{n+h|n} = \mathrm{V}(\boldsymbol{x}_{n+h} \mid \boldsymbol{x}_n)$. Note that $\boldsymbol{V}_{n|n} = \boldsymbol{O}$, where $\boldsymbol{O}$ denotes a matrix of zeros. Then, from (6.20),

$$\boldsymbol{V}_{n+h|n} = \boldsymbol{F}\boldsymbol{V}_{n+h-1|n}\boldsymbol{F}' + \boldsymbol{g}\boldsymbol{g}'\sigma^2,$$

and therefore

$$\boldsymbol{V}_{n+h|n} = \sigma^2 \sum_{j=0}^{h-1} \boldsymbol{F}^j\boldsymbol{g}\boldsymbol{g}'(\boldsymbol{F}^j)'.$$

Hence, using (6.19), the forecast variance for h periods ahead is

$$\begin{aligned} v_{n+h|n} &= \mathrm{V}(y_{n+h} \mid \boldsymbol{x}_n) \\ &= \boldsymbol{w}'\boldsymbol{V}_{n+h-1|n}\boldsymbol{w} + \sigma^2 = \begin{cases} \sigma^2 & \text{if } h = 1; \\ \sigma^2\left[1 + \sum_{j=1}^{h-1} c_j^2\right] & \text{if } h \geq 2; \end{cases} \end{aligned} \tag{6.22}$$

where $c_j = \boldsymbol{w}'\boldsymbol{F}^{j-1}\boldsymbol{g}$.

Example 6.3: Forecast variance for the ETS(A,A$_d$,A) model

Using (6.21), we find that $c_j = \boldsymbol{w}'\boldsymbol{F}^{j-1}\boldsymbol{g} = \alpha + \beta\phi_j + \gamma d_{j,m}$. Consequently, from (6.22) we obtain

$$\begin{aligned} v_{n+h|n} &= \sigma^2 \left[1 + \sum_{j=1}^{h-1} (\alpha + \beta\phi_j + \gamma d_{j,m})^2\right] \\ &= \sigma^2 \left[1 + \sum_{j=1}^{h-1} \left(\alpha^2 + 2\alpha\beta\phi_j + \beta^2\phi_j^2 + \{\gamma^2 + 2\alpha\gamma + 2\beta\gamma\phi_j\} d_{j,m}\right)\right]. \end{aligned} \tag{6.23}$$

In order to expand this expression, first recall the following well known results for arithmetic and geometric series (Morgan 2005):

$$\sum_{j=1}^{p} j = \tfrac{1}{2}p(p+1), \quad \sum_{j=1}^{p} j^2 = \tfrac{1}{6}p(p+1)(2p+1) \quad \text{and} \quad \sum_{j=1}^{p} a^j = \frac{a(1-a^p)}{1-a},$$

where $a \neq 1$, from which it is easy to show that

$$\sum_{j=1}^{p} ja^j = \frac{a[1-(p+1)a^p + pa^{p+1}]}{(1-a)^2}, \qquad \sum_{j=1}^{p} j(p-j+1) = \tfrac{1}{6}p(p+1)(p+2)$$

and $\phi_j = \phi(1-\phi^j)/(1-\phi)$ when $\phi < 1$. Then the following expressions also follow for $\phi < 1$:

$$\sum_{j=1}^{h-1} \phi_j = \frac{\phi}{(1-\phi)^2}\left[h(1-\phi) - (1-\phi^h)\right]$$

and
$$\begin{aligned} \sum_{j=1}^{h-1} \phi_j^2 &= \frac{\phi^2}{(1-\phi)^2} \sum_{j=1}^{h-1} (1 - 2\phi^j + \phi^{2j}) \\ &= \frac{\phi^2}{(1-\phi)^2(1-\phi^2)}\left[h(1-\phi^2) - (1+2\phi-\phi^h)(1-\phi^h)\right]. \end{aligned}$$

Furthermore, $\sum_{j=1}^{h-1} d_{j,m} = h_m$. If $h-1 < m$ (i.e., $h_m = 0$), then $\sum_{j=1}^{h-1} \phi_j d_{j,m} = 0$, and if $h-1 \geq m$ (i.e., $h_m \geq 1$), then

$$\begin{aligned} \sum_{j=1}^{h-1} \phi_j d_{j,m} &= \sum_{\ell=1}^{h_m} \phi_{\ell m} = \frac{\phi}{1-\phi} \sum_{\ell=1}^{h_m} (1-\phi^{\ell m}) \\ &= \frac{\phi}{(1-\phi)(1-\phi^m)}\left[h_m(1-\phi^m) - \phi^m(1-\phi^{m h_m})\right]. \end{aligned}$$

(continued)

Using the above results, we can rewrite (6.23) as

$$\begin{aligned} v_{n+h|n} = \sigma^2\Big[1 + \alpha^2(h-1) + \frac{\beta\phi h}{(1-\phi)^2}\{2\alpha(1-\phi)+\beta\phi\} \\ - \frac{\beta\phi(1-\phi^h)}{(1-\phi)^2(1-\phi^2)}\left\{2\alpha(1-\phi^2)+\beta\phi(1+2\phi-\phi^h)\right\} \\ + \gamma h_m(2\alpha+\gamma) + \frac{2\beta\gamma\phi}{(1-\phi)(1-\phi^m)}\left\{h_m(1-\phi^m)-\phi^m(1-\phi^{mh_m})\right\}\Big]. \end{aligned} \tag{6.24}$$

This is the forecast variance for the ETS(A,A_d,A) model when $h \geq 2$.

Example 6.4: Forecast variance for the ETS(A,A,A) model

To obtain the forecast variance for the ETS(A,A,A) model, we could take the limit as $\phi \to 1$ in (6.24) and apply L'Hospital's rule. However, in many ways it is simpler to go back to (6.23) and replace ϕ_j with j. This yields (Exercise 6.2)

$$\begin{aligned} v_{n+h|n} = \sigma^2\Big[1 + (h-1)\{\alpha^2 + \alpha\beta h + \tfrac{1}{6}\beta^2 h(2h-1)\} \\ + \gamma h_m\{2\alpha+\gamma+\beta m(h_m+1)\}\Big]. \end{aligned} \tag{6.25}$$

The forecast variance expressions for all other models can be obtained as special cases of either (6.24) or (6.25):

- For (A,A_d,N), we use the results of (A,A_d,A) with $\gamma = 0$ and $s_t = 0$ for all t.
- For (A,A,N), we use the results of (A,A,A) with $\gamma = 0$ and $s_t = 0$ for all t.
- The results for (A,N,N) are obtained from (A,A,N) by further setting $\beta = 0$ and $b_t = 0$ for all t.
- The results for (A,N,A) are obtained as a special case of (A,A,A) with $\beta = 0$ and $b_t = 0$ for all t.

Derivation of Results for Class 2

The models in Class 2 can all be written using the following state space model:

$$y_t = \boldsymbol{w}'\boldsymbol{x}_{t-1}(1+\varepsilon_t), \tag{6.26}$$

$$\boldsymbol{x}_t = (\boldsymbol{F} + \boldsymbol{g}\boldsymbol{w}'\varepsilon_t)\boldsymbol{x}_{t-1}, \tag{6.27}$$

where $\boldsymbol{w}$, $\boldsymbol{g}$, $\boldsymbol{F}$, $\boldsymbol{x}_t$ and ε_t are the same as for the corresponding Class 1 model. The lower tail of the error distribution is truncated so that $1+\varepsilon_t$ is positive.

The truncation is usually negligible as σ is usually relatively small for these models.

Let $\boldsymbol{m}_{n+h|n} = \mathrm{E}(\boldsymbol{x}_{n+h} \mid \boldsymbol{x}_n)$ and $\boldsymbol{V}_{n+h|n} = \mathrm{V}(\boldsymbol{x}_{n+h} \mid \boldsymbol{x}_n)$ as in Sect. 6.2. The forecast means for Class 2 have the same form as for Class 1, namely

$$\mu_{n+h|n} = \boldsymbol{w}'\boldsymbol{m}_{n+h-1|n} = \boldsymbol{w}'\boldsymbol{F}^{h-1}\boldsymbol{x}_n.$$

From (6.26), it can be seen that the forecast variance is given by

$$\begin{aligned} v_{n+h|n} &= \boldsymbol{w}'\boldsymbol{V}_{n+h-1|n}\boldsymbol{w}(1+\sigma^2) + \sigma^2\boldsymbol{w}'\boldsymbol{m}_{n+h-1|n}\boldsymbol{m}'_{n+h-1|n}\boldsymbol{w} \\ &= \boldsymbol{w}'\boldsymbol{V}_{n+h-1|n}\boldsymbol{w}(1+\sigma^2) + \sigma^2\mu^2_{n+h|n}. \end{aligned}$$

To obtain $\boldsymbol{V}_{n+h-1|n}$, first note that $\boldsymbol{x}_t = \boldsymbol{F}\boldsymbol{x}_{t-1} + \boldsymbol{g}e_t$, where $e_t = y_t - \boldsymbol{w}'\boldsymbol{x}_{t-1} = \boldsymbol{w}'\boldsymbol{x}_{t-1}\varepsilon_t$. Then it is readily seen that $\boldsymbol{V}_{n+h|n} = \boldsymbol{F}\boldsymbol{V}_{n+h-1|n}\boldsymbol{F}' + \boldsymbol{g}\boldsymbol{g}'\mathrm{V}(e_{n+h} \mid \boldsymbol{x}_n)$. Now let θ_h be defined such that $\mathrm{V}(e_{n+h} \mid \boldsymbol{x}_n) = \theta_h\sigma^2$. Then, by repeated substitution,

$$\boldsymbol{V}_{n+h|n} = \sigma^2 \sum_{j=0}^{h-1} \boldsymbol{F}^j\boldsymbol{g}\boldsymbol{g}'(\boldsymbol{F}^j)'\theta_{h-j}.$$

Therefore,

$$\boldsymbol{w}'\boldsymbol{V}_{n+h-1|n}\boldsymbol{w} = \sigma^2 \sum_{j=1}^{h-1} c_j^2\theta_{h-j}, \tag{6.28}$$

where $c_j = \boldsymbol{w}'\boldsymbol{F}^{j-1}\boldsymbol{g}$. Now

$$e_{n+h} = \left[\boldsymbol{w}'(\boldsymbol{x}_{n+h-1} - \boldsymbol{m}_{n+h-1|n}) + \boldsymbol{w}'\boldsymbol{m}_{n+h-1|n}\right]\varepsilon_{n+h},$$

which we square and take expectations to give $\theta_h = \boldsymbol{w}'\boldsymbol{V}_{n+h-1|n}\boldsymbol{w} + \mu^2_{n+h|n}$. Substituting (6.28) into this expression for θ_h gives

$$\theta_h = \sigma^2 \sum_{j=1}^{h-1} c_j^2\theta_{h-j} + \mu^2_{n+h|n}, \tag{6.29}$$

where $\theta_1 = \mu^2_{n+1|n}$. The forecast variance is then given by

$$v_{n+h|n} = (1+\sigma^2)\theta_h - \mu^2_{n+h|n}. \tag{6.30}$$

Derivation of Results for Class 3

Note that we can write (see p. 85)

$$y_t = \boldsymbol{w}_1'\boldsymbol{x}_{t-1}\boldsymbol{z}_{t-1}'\boldsymbol{w}_2(1+\varepsilon_t).$$

So let $Q_h = x_{n+h}z'_{n+h}$, $M_h = \mathrm{E}(Q_h \mid x_n, z_n)$ and $V_{n+h|n} = \mathrm{V}(\vec{Q}_h \mid x_n, z_n)$ where $\vec{Q}_h = \mathrm{vec}(Q_h)$. Note that

$$\begin{aligned}Q_h &= (F_1 x_{n+h-1} + G_1 x_{n+h-1}\varepsilon_{n+h})(z'_{n+h-1}F'_2 + z'_{n+h-1}G'_2\varepsilon_{n+h}) \\ &= F_1 Q_{h-1}F'_2 + (F_1 Q_{h-1}G'_2 + G_1 Q_{h-1}F'_2)\varepsilon_{n+h} + G_1 Q_{h-1}G'_2\varepsilon^2_{n+h}.\end{aligned}$$

It follows that $M_0 = x_n z'_n$ and

$$M_h = F_1 M_{h-1} F'_2 + G_1 M_{h-1} G'_2 \sigma^2. \tag{6.31}$$

For the variance of Q_h , we find $V_0 = 0$, and

$$\begin{aligned}V_{n+h|n} &= \mathrm{V}\big[\mathrm{vec}(F_1 Q_{h-1}F'_2) + \mathrm{vec}(F_1 Q_{h-1}G'_2 + G_1 Q_{h-1}F'_2)\varepsilon_{n+h} \\ &\qquad + \mathrm{vec}(G_1 Q_{h-1}G'_2)\varepsilon^2_{n+h}\big] \\ &= (F_2 \otimes F_1)V_{n+h-1|n}(F_2 \otimes F_1)' \\ &\quad + (G_2 \otimes F_1 + F_2 \otimes G_1)\mathrm{V}(\vec{Q}_{h-1}\varepsilon_{n+h})(G_2 \otimes F_1 + F_2 \otimes G_1)' \\ &\quad + (G_2 \otimes G_1)\mathrm{V}(\vec{Q}_{h-1}\varepsilon^2_{n+h})(G_2 \otimes G_1)' \\ &\quad + (F_2 \otimes F_1)\mathrm{Cov}(\vec{Q}_{h-1}, \vec{Q}_{h-1}\varepsilon^2_{n+h})(G_2 \otimes G_1)' \\ &\quad + (G_2 \otimes G_1)\mathrm{Cov}(\vec{Q}_{h-1}\varepsilon^2_{n+h}, \vec{Q}_{h-1})(F_2 \otimes F_1)'.\end{aligned}$$

Next we find that

$$\begin{aligned}\mathrm{V}(\vec{Q}_{h-1}\varepsilon_{n+h}) &= \mathrm{E}[\vec{Q}_{h-1}(\vec{Q}_{h-1})'\varepsilon^2_{n+h}] \\ &= \sigma^2\big[V_{n+h-1|n} + \vec{M}_{h-1}(\vec{M}_{h-1})'\big], \\ \mathrm{V}(\vec{Q}_{h-1}\varepsilon^2_{n+h}) &= \mathrm{E}\big[\vec{Q}_{h-1}(\vec{Q}_{h-1})'\varepsilon^4_{n+h}\big] - \mathrm{E}(\vec{Q}_{h-1})\mathrm{E}(\vec{Q}_{h-1})'\sigma^4 \\ &= 3\sigma^4\big[V_{n+h-1|n} + \vec{M}_{h-1}(\vec{M}_{h-1})'\big] - \vec{M}_{h-1}(\vec{M}_{h-1})'\sigma^4 \\ &= \sigma^4\big[3V_{n+h-1|n} + 2\vec{M}_{h-1}(\vec{M}_{h-1})'\big],\end{aligned}$$

and

$$\begin{aligned}\mathrm{Cov}(\vec{Q}_{h-1}, \vec{Q}_{h-1}\varepsilon^2_{n+h}) &= \mathrm{E}\big[\vec{Q}_{h-1}(\vec{Q}_{h-1})'\varepsilon^2_{n+h}\big] - \mathrm{E}(\vec{Q}_{h-1})\mathrm{E}(\vec{Q}_{h-1})'\sigma^2 \\ &= \sigma^2(V_{n+h-1|n} + \vec{M}_{h-1}(\vec{M}_{h-1})') - \sigma^2\vec{M}_{h-1}(\vec{M}_{h-1})' \\ &= \sigma^2 V_{n+h-1|n}.\end{aligned}$$

It follows that

$$\begin{aligned}V_{n+h|n} &= (F_2 \otimes F_1)V_{n+h-1|n}(F_2 \otimes F_1)' \\ &\quad + \sigma^2\Big[(F_2 \otimes F_1)V_{n+h-1|n}(G_2 \otimes G_1)' + (G_2 \otimes G_1)V_{n+h-1|n}(F_2 \otimes F_1)'\Big]\end{aligned}$$

$$+\sigma^2(G_2 \otimes F_1 + F_2 \otimes G_1)\Big[V_{n+h-1|n} + \vec{M}_{h-1}(\vec{M}_{h-1})'\Big] \\ \times (G_2 \otimes F_1 + F_2 \otimes G_1)' \\ +\sigma^4(G_2 \otimes G_1)\Big[3V_{n+h-1|n} + 2\vec{M}_{h-1}(\vec{M}_{h-1})'\Big](G_2 \otimes G_1)'.$$

The forecast mean and variance are given by

$$\mu_{n+h|n} = \mathrm{E}(y_{n+h} \mid x_n, z_n) = w_1' M_{h-1} w_2$$

and

$$\begin{aligned} v_{n+h|n} &= \mathrm{V}(y_{n+h} \mid x_n, z_n) = \mathrm{V}[\mathrm{vec}(w_1' Q_{h-1} w_2 + w_1' Q_{h-1} w_2' \varepsilon_{n+h})] \\ &= \mathrm{V}[(w_2' \otimes w_1')\vec{Q}_{h-1} + (w_2' \otimes w_1')\vec{Q}_{h-1}\varepsilon_{n+h}] \\ &= (w_2' \otimes w_1')[V_{n+h-1|n}(1+\sigma^2) + \sigma^2 \vec{M}_{h-1}(\vec{M}_{h-1})'](w_2 \otimes w_1) \\ &= (1+\sigma^2)(w_2' \otimes w_1')V_{n+h-1|n}(w_2' \otimes w_1')' + \sigma^2 \mu_{n+h|n}^2. \end{aligned}$$

When σ is sufficiently small (much less than 1), it is possible to obtain some simpler but approximate expressions. The second term in (6.31) can be dropped to give $M_h = F_1^{h-1} M_0 (F_2^{h-1})'$, and so

$$\mu_{n+h|n} \approx w_1' F_1^{h-1} x_n (w_2' F_2^{h-1} z_n)'.$$

The order of this approximation can be obtained by noting that the observation equation may be written as $y_t = u_{1,t} u_{2,t} u_{3,t}$, where $u_{1,t} = w_1' x_{t-1}$, $u_{2,t} = w_2' z_{t-1}$ and $u_{3,t} = 1 + \varepsilon_t$. Then

$$\mathrm{E}(y_t) = \mathrm{E}(u_{1,t} u_{2,t} u_{3,t}) = \mathrm{E}(u_{1,t} u_{2,t})\mathrm{E}(u_{3,t}),$$

because $u_{3,t}$ is independent of $u_{1,t}$ and $u_{2,t}$. Therefore, because $\mathrm{E}(u_{1,t}u_{2,t}) = \mathrm{E}(u_{1,t})\mathrm{E}(u_{2,t}) + \mathrm{Cov}(u_{1,t}, u_{2,t})$, we have the approximation:

$$\mu_{n+h|n} = \mathrm{E}(y_{n+h} \mid x_n, z_n) = \mathrm{E}(u_{1,n+h} \mid x_n)\mathrm{E}(u_{2,n+h} \mid z_n)\mathrm{E}(u_{3,n+h}) + O(\sigma^2).$$

When $u_{2,n+h}$ is constant the result is exact. Now let

$$\begin{aligned} \mu_{1,h} &= \mathrm{E}(u_{1,n+h+1} \mid x_n) = \mathrm{E}(w_1' x_{n+h} \mid x_n) = w_1' F_1^h x_n, \\ \mu_{2,h} &= \mathrm{E}(u_{2,n+h+1} \mid z_n) = \mathrm{E}(w_2' z_{n+h} \mid z_n) = w_2' F_2^h z_n, \\ v_{1,h} &= \mathrm{V}(u_{1,n+h+1} \mid x_n) = \mathrm{V}(w_1' x_{n+h} \mid x_n), \\ v_{2,h} &= \mathrm{V}(u_{2,n+h+1} \mid z_n) = \mathrm{V}(w_2' z_{n+h} \mid z_n), \\ \text{and} \quad v_{12,h} &= \mathrm{Cov}(u_{1,n+h+1}^2, u_{2,n+h+1}^2 \mid x_n, z_n) \\ &= \mathrm{Cov}([w_1' x_{n+h}]^2, [w_2' z_{n+h}]^2 \mid x_n, z_n). \end{aligned}$$

Then

$$\mu_{n+h|n} = \mu_{1,h-1}\mu_{2,h-1} + O(\sigma^2) = \boldsymbol{w}_1'\boldsymbol{F}_1^{h-1}\boldsymbol{x}_n\boldsymbol{w}_2'\boldsymbol{F}_2^{h-1}\boldsymbol{z}_n + O(\sigma^2).$$

By the same arguments, we have

$$\mathrm{E}(y_t^2) = \mathrm{E}(u_{1,t}^2 u_{2,t}^2 u_{3,t}^2) = \mathrm{E}(u_{1,t}^2 u_{2,t}^2)\mathrm{E}(u_{3,t}^2),$$

and

$$\begin{aligned}\mathrm{E}(y_{n+h}^2 \mid \boldsymbol{z}_n, \boldsymbol{x}_n) &= \mathrm{E}(u_{1,n+h}^2 u_{2,n+h}^2 \mid \boldsymbol{x}_n, \boldsymbol{z}_n)\mathrm{E}(u_{3,n+h}^2) \\ &= \left[\mathrm{Cov}(u_{1,n+h}^2, u_{2,n+h}^2 \mid \boldsymbol{x}_n, \boldsymbol{z}_n) + \mathrm{E}(u_{1,n+h}^2 \mid \boldsymbol{x}_n)\mathrm{E}(u_{2,n+h}^2 \mid \boldsymbol{z}_n)\right] \mathrm{E}(u_{3,n+h}^2) \\ &= (1+\sigma^2)[v_{12,h-1} + (v_{1,h-1} + \mu_{1,h-1}^2)(v_{2,h-1} + \mu_{2,h-1}^2)].\end{aligned}$$

Assuming that the covariance $v_{12,h-1}$ is small compared to the other terms, we obtain

$$v_{n+h|n} \approx (1+\sigma^2)(v_{1,h-1} + \mu_{1,h-1}^2)(v_{2,h-1} + \mu_{2,h-1}^2) - \mu_{1,h-1}^2\mu_{2,h-1}^2.$$

We now simplify these results for the ETS(M,A$_d$,M) case where $\boldsymbol{x}_t = (\ell_t, b_t)'$ and $\boldsymbol{z}_t = (s_t, \ldots, s_{t-m+1})'$, and the matrix coefficients are $\boldsymbol{w}_1' = [1, \phi]$, $\boldsymbol{w}_2' = [0, \ldots, 0, 1]$,

$$\boldsymbol{F}_1 = \begin{bmatrix} 1 & \phi \\ 0 & \phi \end{bmatrix}, \quad \boldsymbol{F}_2 = \begin{bmatrix} \boldsymbol{0}_{m-1}' & 1 \\ \boldsymbol{I}_{m-1} & \boldsymbol{0}_{m-1} \end{bmatrix},$$

$$\boldsymbol{G}_1 = \begin{bmatrix} \alpha & \alpha \\ \beta & \beta \end{bmatrix}, \quad \text{and} \quad \boldsymbol{G}_2 = \begin{bmatrix} \boldsymbol{0}_{m-1}' & \gamma \\ \boldsymbol{O}_{m-1} & \boldsymbol{0}_{m-1} \end{bmatrix}.$$

Many terms will be zero in the formulae for the expected value and the variance because of the following relationships: $\boldsymbol{G}_2^2 = \boldsymbol{O}_m$, $\boldsymbol{w}_2'\boldsymbol{G}_2 = \boldsymbol{0}_m'$, and $(\boldsymbol{w}_2' \otimes \boldsymbol{w}_1')(\boldsymbol{G}_2 \otimes \boldsymbol{X}) = \boldsymbol{0}_{2m}'$ where $\boldsymbol{X}$ is any 2×2 matrix. For the terms that remain, $\boldsymbol{w}_2' \otimes \boldsymbol{w}_1'$ and its transpose will only use the terms from the last two rows of the last two columns of the large matrices because $\boldsymbol{w}_2' \otimes \boldsymbol{w}_1' = [\boldsymbol{0}_{2m-2}', 1, 1]$.

Using the small σ approximations and exploiting the structure of the ETS(M,A$_d$,M) model, we can obtain simpler expressions that approximate $\mu_{n+h|n}$ and $v_{n+h|n}$.

Note that $\boldsymbol{w}_2'\boldsymbol{F}_2^j\boldsymbol{G}_2 = \gamma d_{j+1,m}\boldsymbol{w}_2'$. So, for $h < m$, we have

$$\boldsymbol{w}_2'\boldsymbol{z}_{n+h} \mid \boldsymbol{z}_n = \boldsymbol{w}_2'\prod_{j=1}^{h}(\boldsymbol{F}_2 + \boldsymbol{G}_2\varepsilon_{n+h-j+1})\boldsymbol{z}_n = \boldsymbol{w}_2'\boldsymbol{F}_2^h\boldsymbol{z}_n = s_{n-m+h+1}$$

Furthermore,

$$\mu_{2,h} = s_{n-m+h_m^+}$$

$$\text{and} \qquad v_{2,h} = \left[(1+\gamma^2\sigma^2)^{h_m} - 1\right]s_{n-m+h_m^+}^2.$$

Also note that $\boldsymbol{x}_n$ has the same properties as for ETS(M,A$_\text{d}$,N) in Class 2. Thus

$$\begin{aligned} \mu_{1,h} &= \ell_n + \phi_h b_n \\ \text{and} \qquad v_{1,h} &= (1+\sigma^2)\theta_h - \mu_{1,h}^2. \end{aligned}$$

Combining all of the terms, we arrive at the approximations

$$\begin{aligned} \mu_{n+h|n} &= \tilde{\mu}_{n+h|n} s_{n-m+h_m^+} + O(\sigma^2) \\ \text{and} \qquad v_{n+h|n} &\approx s_{n-m+h_m^+}^2 \Big[\theta_h (1+\sigma^2)(1+\gamma^2\sigma^2)^{h_m} - \tilde{\mu}_{n+h|n}^2\Big], \end{aligned}$$

where $\tilde{\mu}_{n+h|n} = \ell_n + \phi_h b_n$, $\theta_1 = \tilde{\mu}_{n+1|n}^2$, and

$$\theta_h = \tilde{\mu}_{n+h|n}^2 + \sigma^2 \sum_{j=1}^{h-1} (\alpha + \beta\phi_j)^2 \theta_{h-j}, \qquad h \geq 2.$$

These expressions are exact for $h \leq m$. The other cases of Class 3 can be derived as special cases of ETS(M,A$_\text{d}$,M).

Derivation of C_j Values

We first demonstrate that for Class 1 models, lead-time demand can be resolved into a linear function of the uncorrelated level and error components. Back-solve the transition equation (6.20) from period $n+j$ to period n, to give

$$\boldsymbol{x}_{n+j} = \boldsymbol{F}^j \boldsymbol{x}_n + \sum_{i=1}^{j} \boldsymbol{F}^{j-i} \boldsymbol{g} \varepsilon_{n+i}.$$

Now from (6.19) and (6.20) we have

$$\begin{aligned} y_{n+j} &= \boldsymbol{w}'\boldsymbol{x}_{n+j-1} + \varepsilon_{n+j} \\ &= \boldsymbol{w}'\boldsymbol{F}\boldsymbol{x}_{n+j-2} + \boldsymbol{w}'\boldsymbol{g}\varepsilon_{n+j-1} + \varepsilon_{n+j} \\ &\;\;\vdots \\ &= \boldsymbol{w}'\boldsymbol{F}^{j-1}\boldsymbol{x}_n + \sum_{i=1}^{j-1} \boldsymbol{w}'\boldsymbol{F}^{j-i-1}\boldsymbol{g}\varepsilon_{n+i} + \varepsilon_{n+j} \\ &= \mu_{n+j|n} + \sum_{i=1}^{j-1} c_{j-i}\varepsilon_{n+i} + \varepsilon_{n+j}, \end{aligned}$$

where $c_k = \boldsymbol{w}'\boldsymbol{F}^{k-1}\boldsymbol{g}$. Substituting this into (6.11) gives (6.15).

To derive the value of C_j for the ETS(A,A$_d$,A) model, we plug the value of c_i from Table 6.2 into (6.13) to obtain

$$\begin{aligned}
C_j &= 1 + \sum_{i=1}^{j}(\alpha + \beta\phi_i + \gamma d_{i,m}) \\
&= 1 + \alpha j + \beta\sum_{i=1}^{j}\phi_i + \gamma\sum_{i=1}^{j} d_{i,m} \\
&= 1 + \alpha j + \frac{\beta\phi}{(1-\phi)^2}\Big[(j+1)(1-\phi) - (1-\phi^{j+1})\Big] + \gamma j_m,
\end{aligned}$$

where $j_m = \lfloor j/m \rfloor$ is the number of complete seasonal cycles that occur within j time periods.

A similar derivation for the ETS(A,A,A) model leads to

$$C_j = 1 + \sum_{i=1}^{j}(\alpha + i\beta + \gamma d_{i,m}) = 1 + j\Big[\alpha + \tfrac{1}{2}\beta(j+1)\Big] + \gamma j_m.$$

The expressions for C_j for the other linear models are obtained as special cases of either ETS(A,A$_d$,A) or ETS(A,A,A) and are given in Table 6.6.

7

Selection of Models

One important step in the forecasting process is the selection of a model that could have generated the time series and would, therefore, be a reasonable choice for producing forecasts and prediction intervals. As we have seen in Chaps. 2–4, there are many specific models within the general innovations state space model (2.12). There are also many approaches that one might implement in a model selection process. In Sect. 7.1, we will describe the use of information criteria for selecting among the innovations state space models. These information criteria have been developed specifically for time series data and are based on maximized likelihoods. We will consider four commonly recommended information criteria and one relatively new information criterion. Then, in Sect. 7.2, we will use the MASE from Chap. 2 to develop measures for comparing model selection procedures. These measures will be used in Sects. 7.2.2 and 7.2.3 to compare the five information criteria with each other, and the commonly applied prediction validation method for model selection using the M3 competition data (Makridakis and Hibon 2000) and a hospital data set. We also compare the results with the application of damped trend models for all time series. Finally, some implications of these comparisons will be given in Sect. 7.3.

7.1 Information Criteria for Model Selection

The goal in model selection is to pick the model with the best predictive ability on average. Finding the model with the smallest within-sample one-step-ahead forecast errors, or even the one with the maximum likelihood, does not assure us that the model will be the best one for forecasting.

One approach is to use an information criterion which penalizes the likelihood to compensate for the potential overfitting of data. The general form of the information criteria for an innovations state space model is

$$\text{IC} = -2\log \mathcal{L}(\hat{\boldsymbol{\theta}}, \hat{\boldsymbol{x}}_0 \mid \boldsymbol{y}) + q\zeta(n), \tag{7.1}$$

Table 7.1. Penalties in the information criteria.

Criterion	$\zeta(n)$	Penalty	Source
AIC	2	$2q$	Akaike (1974)
BIC	$\log(n)$	$q\log(n)$	Schwarz (1978)
HQIC	$2\log(\log(n))$	$2q\log(\log(n))$	Hannan and Quinn (1979)
AICc	$2n/(n-q-1)$	$2qn/(n-q-1)$	Sugiura (1978)
LEIC	Empirical c	qc	Billah et al. (2003)

where $\mathcal{L}(\hat{\boldsymbol{\theta}}, \hat{\boldsymbol{x}}_0 \mid \boldsymbol{y})$ is the maximized likelihood function, q is the number of parameters in $\hat{\boldsymbol{\theta}}$ plus the number of free states in $\hat{\boldsymbol{x}}_0$, and $\zeta(n)$ is a function of the sample size. Thus, $q\zeta(n)$ is the penalty assigned to a model for the number of parameters and states in the model. (We also require that the state space model has no redundant states—see Sect. 10.1, p. 149.) The information criteria that will be introduced in this chapter are summarized in Table 7.1.

For the Gaussian likelihood, we can drop the additive constants in $-2\log(\mathcal{L}(\hat{\boldsymbol{\theta}}, \hat{\boldsymbol{x}}_0 \mid \boldsymbol{y}))$ and replace the expression by $\mathcal{L}^*(\boldsymbol{\theta}, \boldsymbol{x}_0)$ from (5.3) to obtain

$$\mathrm{IC} = n\log\left(\sum_{t=1}^{n}\varepsilon_t^2\right) + 2\sum_{t=1}^{n}\log|r(x_{t-1})| + q\zeta(n). \tag{7.2}$$

Recall from Chap. 5 that $\varepsilon_t = [y_t - w(\boldsymbol{x}_{t-1})]/r(\boldsymbol{x}_{t-1})$. Also, the likelihood function is based on a fixed seed state $\boldsymbol{x}_0$. Not only is the fixed seed state critical for this form of the Gaussian likelihood in the nonlinear version, it is essential in both the linear and nonlinear case for comparing models that differ by a nonstationary state (see Chap. 12 for a discussion of this problem).

The procedure for using an information criterion in model selection is to compute the IC for each model and choose the model with the minimum IC. Of course, we do not believe that there is an absolutely correct model for a time series, but this process should find a reasonable model for forecasting.

In the *Akaike Information Criterion* (AIC) (Akaike 1974), $\zeta(n) = 2$, and hence the penalty is $2q$. The AIC is derived by considering the principles of maximum likelihood and negative entropy. Suppose future values of a time series $\boldsymbol{y}^* = [y_{n+1}, \ldots, y_{n+h}]$ are to be predicted from present and past values $\boldsymbol{y} = [y_1, \ldots, y_n]$. Model selection can be viewed as the problem of approximating $f(\boldsymbol{y}^*|\boldsymbol{y})$, the true conditional density of $\boldsymbol{y}^*$, given that $\boldsymbol{y}$ is observed. If $g(\boldsymbol{y}^*|\boldsymbol{y})$ is an estimate of f, its goodness in approximating f can be measured by its negative entropy

$$I_{\boldsymbol{y}^*|\boldsymbol{y}}(f,g) = \int f(\boldsymbol{y}^*|\boldsymbol{y})\log\left(\frac{f(\boldsymbol{y}^*|\boldsymbol{y})}{g(\boldsymbol{y}^*|\boldsymbol{y})}\right)d\boldsymbol{y}^*.$$

This measure is also known as the Kullback-Leibler conditional discriminant information, and its size reflects the model approximation error. The negative entropy principle is to select the approximating density g which minimizes the expected negative entropy $\mathrm{E}_{\boldsymbol{y}}[I_{\boldsymbol{y}^*|\boldsymbol{y}}(f,g)]$ (Akaike 1977). Because the true density f is not known, the negative entropies of various competing models must be estimated. Akaike's criterion estimates twice the negative entropy and is designed to produce an approximate asymptotically unbiased estimator as n increases. Thus, the model having the minimum AIC should have the minimum prediction error for y^*, at least asymptotically.

In the *Schwarz Bayesian information criterion* (BIC) (Schwarz 1978), $\zeta(n) = \log(n)$, and the penalty is $q\log(n)$. Schwarz derived his criterion as the Bayes solution to the problem of model identification. Asymptotically, the BIC is minimized at the model order having the highest posterior probability. The BIC is "order consistent" under suitable conditions. A criterion is order consistent if, as the sample size increases, the criterion is minimized at the true order with a probability that approaches unity. For our models the order is the number of parameters and free states. In contrast, the AIC has been criticized because it is inconsistent and tends to overfit models. Geweke and Meese (1981) showed this for regression models, Shibata (1976) for autoregressive models, and Hannan (1980) for ARMA models.

In the *Hannan–Quinn information criterion* (HQIC) (Hannan and Quinn 1979), $\zeta(n) = 2\log(\log(n))$ and the penalty is $2q\log(\log(n))$. For the purpose of understanding the objective of Hannan and Quinn in the HQIC, we divide the Gaussian IC in (7.2) by n to put the information criterion into the following form:

$$\mathrm{IC} = \log\left(\sum_{t=1}^{n}\varepsilon_t^2\right) + (2/n)\sum_{t=1}^{n}\log|r(x_{t-1})| + qC_n,$$

where $C_n = n^{-1}\zeta(n)$. Hannan and Quinn's goal was to find an information criterion based on the minimization of the IC that is order consistent and for which C_n decreases as fast as possible. Thus, HQIC has the property that, like the BIC, it is order consistent and yet comes closer to achieving the optimal forecasting performance of the AIC.

In the *bias corrected AIC* that is denoted by AICc (Sugiura 1978; Hurvich and Tsai 1989), $\zeta(n) = n/(n-q-1)$, and the penalty is $2qn/(n-q-1)$. While the BIC and HQIC are order consistent, they are not asymptotically efficient like the AIC. In addition, the AIC is only approximately an unbiased estimator. In fact, it has a negative bias that becomes more pronounced as n/q decreases. The AICc is an asymptotically efficient information criterion that does an approximate correction for this negative bias, and has been shown to provide better model order choices for small samples.

In the *linear empirical information criterion* (LEIC) (Billah et al. 2003, 2005), $\zeta(n) = c$, where c is estimated empirically for an ensemble of N similar time series with M competing models, and the penalty is qc. The procedure for estimating c requires that a specified number of time periods H be withheld from each time series. The forecasting errors for these withheld time periods are used to compare the competing models and determine a value for c.

The details of estimating c are provided in Appendix "The linear empirical information criterion."

7.2 Choosing a Model Selection Procedure

The potential models to be used in our consideration of the model selection process are listed in Tables 2.2 and 2.3. The first question to be considered is whether each model in those tables is the best model for forecasting some time series. We will use the data from the M3 competition (Makridakis and Hibon 2000) as an example to see that each model is "best" for a reasonable number of time series in that data set. Another interesting question is whether using one model to forecast all time series might be better than employing a model selection method. Some evidence that one could do well by always using damped trend models is provided by the M3 data. However, examination of a set of hospital data containing monthly time series shows that this is not always the case. The M3 data and the hospital data will also be used to compare model selection methods that include the information criteria from Sect. 7.1 and a prediction validation method that is explained in Appendix "Prediction validation method of model selection." It will be seen in both cases that it is reasonable to choose the AIC as the model selection method.

7.2.1 Measures for Comparing Model Section Procedures

In our comparisons, we will include the following procedures for choosing forecasting models for N time series:

- A single model for all time series
- Minimum IC (AIC, BIC, AICc, HQIC, LEIC)
- Prediction validation (VAL) (see Appendix "Prediction validation method of model selection")

Thus, we consider seven model selection procedures, which may be labeled procedure 1 to procedure 7.

The mean absolute scaled error (MASE) proposed by Hyndman and Koehler (2006) is used to determine the success of a model selection procedure. Consider the situation in which we have a collection of N time series for which there are M potential models for forecasting future values. The set of observations for the time series $\{y_t^{(j)}\}$ $(j = 1, \ldots, N)$ is split into two parts: a *fitting set* of the first n_j values and a *forecasting set* of the last H values. The forecasting accuracy of model i $(i = 1, \ldots, M)$, for time series $\{y_t^{(j)}\}$ will be measured by the mean absolute scaled forecast error, defined by

$$\text{MASE}(H,i,j) = \frac{1}{H}\sum_{h=1}^{H}\frac{|y_{n_j+h}^{(j)} - \hat{y}_{n_j}^{(i,j)}(h)|}{\text{MAE}_j}, \tag{7.3}$$

where $\text{MAE}_j = (n_j - 1)^{-1} \sum_{t=2}^{n_j} |y_t^{(j)} - y_{t-1}^{(j)}|$, and $\hat{y}_{i,n_j}^{(i,j)}(h)$ is the h-step-ahead forecast when model i is used for the jth time series.

We define three measures for comparing model selection procedures for forecasting. All of the measures are based on the mean absolute scaled forecast error MASE(H,i,j), as defined in (7.3). The models are numbered from 1 to M, and for model selection procedure k, we denote the number of the model selected for time series $\{y_t^{(j)}\}$ by k_j. The rank $r(H,k_j,j)$ for procedure k and time series j is the rank of MASE(H,k_j,j) among the values of MASE(H,i,j), $i = 1,\dots,M$, when they are ranked in ascending order. Note that this ranking is out of the M models for the model selected by procedure k, and is not a ranking of the procedures.

For a specified model selection procedure k and number of forecasting horizons H, the following measures will be computed:

$$\text{Mean rank MASE}(H,k) = \frac{1}{N}\sum_{j=1}^{N} r(H,k_j,j),$$

$$\text{Mean MASE}(H,k) = \frac{1}{N}\sum_{j=1}^{N} \text{MASE}(H,k_j,j),$$

$$\text{Median MASE}(H,k) = \text{median}\{\text{MASE}(H,k_j,j);\ j = 1,\dots,N\}.$$

A model is fitted to a time series using maximum likelihood estimation (see 5.2 for the logarithm of the likelihood function). A check should always be carried out to ensure that the maximum likelihood for a model does not exceed that of an encompassing model. A violation of this condition indicates that the solution for the encompassing model is not a global optimum. In this case, the likelihood for the encompassing model should be seeded with the optimal values for the smaller model.

7.2.2 Comparing Selection Procedures on the M3 Data

In this section we will use the M3 competition data (Makridakis and Hibon 2000) to compare the model selection procedures listed in the beginning of Sect. 7.2.1. First, we examine the annual time series from the M3 competition to determine how frequently a model is best for forecasting a time series in that data set. The ten non-seasonal models from Tables 2.2 and 2.3 are fitted to the 645 annual time series in the M3 data set using the maximum likelihood method of Chap. 5.

Table 7.2 contains the number and percentage of series (out of the 645 annual time series) for which each model is defined to be the best model for forecasting. In this table, model i^* is defined to be the best model if $r(H,i^*,j) = 1$; that is, if MASE$(H,i^*,j) = \min\{\text{MASE}(H,i,j);\ i = 1,\dots,M\}$. Here, the number of forecasting horizons is $H = 6$ and the number of models

Table 7.2. Number and percentage of 645 annual M3 time series with minimum MASE(H,i,j).

Model	Count	Percent
ETS(A,N,N)	141	21.86
ETS(A,M,N)	84	13.02
ETS(M,M,N)	74	11.47
ETS(A,M_d,N)	72	11.16
ETS(M,A,N)	56	8.68
ETS(A,A,N)	54	8.37
ETS(M,M_d,N)	52	8.06
ETS(A,A_d,N)	40	6.20
ETS(M,N,N)	37	5.74
ETS(M,A_d,N)	35	5.43

Table 7.3. Number and percentage of 1428 monthly M3 time series with minimum MASE(H,i,j).

Model	Count	Percent	Model	Count	Percent
ETS(M,M,N)	92	6.44	ETS(A,M,A)	40	2.80
ETS(M,A,N)	81	5.67	ETS(A,M_d,N)	39	2.73
ETS(M,A,M)	78	5.46	ETS(A,M_d,M)	39	2.73
ETS(A,M,N)	76	5.32	ETS(A,N,A)	38	2.66
ETS(A,N,N)	69	4.83	ETS(M,A_d,M)	37	2.59
ETS(A,A,M)	63	4.41	ETS(M,A_d,N)	36	2.52
ETS(M,M,M)	60	4.20	ETS(M,M_d,M)	35	2.45
ETS(A,N,M)	58	4.06	ETS(A,A_d,N)	34	2.38
ETS(A,A,N)	57	3.99	ETS(M,M_d,A)	33	2.31
ETS(A,M,M)	54	3.78	ETS(M,M_d,N)	33	2.31
ETS(M,N,M)	49	3.43	ETS(A,A_d,M)	32	2.24
ETS(M,N,A)	48	3.36	ETS(M,M,A)	30	2.10
ETS(M,N,N)	47	3.29	ETS(A,A,A)	30	2.10
ETS(M,A,A)	44	3.08	ETS(A,A_d,A)	30	2.10
ETS(M,A_d,A)	43	3.01	ETS(A,M_d,A)	23	1.61

is $M = 10$. The model for simple exponential smoothing has the largest percentage (21.9%) of time series for which a single model is the best model for forecasting. The smallest percentage of time series for any model is 5.4%. We will see later that the high percentage for the ETS(A,N,N) model does not indicate that it is the best model for forecasting all of the annual time series. However, this table does indicate that every model is best for some of the time series, and it might be beneficial to have a procedure for choosing from among all these non-seasonal models.

Table 7.3 contains the analogous results for the 1,428 monthly time series from the M3 data set. All 30 models from Tables 2.2 and 2.3 were applied to

Table 7.4. The ten non-seasonal models for annual M3 time series.

Model	Mean rank	Mean MASE	Median MASE
ETS(A,A_d,N)	4.97	2.92	1.82
ETS(M,A_d,N)	5.23	2.97	1.95
ETS(A,A,N)	5.25	2.99	1.97
ETS(A,M_d,N)	5.29	3.57	1.75
ETS(M,M_d,N)	5.31	3.24	1.89
ETS(M,A,N)	5.37	2.96	2.01
ETS(A,M,N)	5.77	4.18	1.96
ETS(M,M,N)	5.87	3.63	2.05
ETS(A,N,N)	5.93	3.17	2.26
ETS(M,N,N)	6.02	3.19	2.26

the monthly time series. The counts and percentages in this table also support the notion of trying to find a model selection procedure.

Consider the procedure where a single model is selected to forecast all time series in a collection of N time series. Each row of Table 7.4 displays the three measures (mean rank, mean MASE, and median MASE) when a specified model i^* is applied to all of the $N = 645$ annual time series in the M3 data set. The three measures are based on MASE$(6, i^*, j)$ because $H = 6$ and $k_j = i^*$ for all time series, $j = 1, \ldots, N$. Each of the specified models is one of the $M = 10$ non-seasonal models in Tables 2.2 and 2.3. A chi-square statistic (KFHS) for the mean ranks, as proposed in Koning et al. (2005), shows that we can reject the hypothesis that the mean ranks are equal at less than a 0.001 level of significance (KFHS $= 82.9$ with 9 degrees of freedom). The model with the smallest mean rank of 4.97 out of 10 is the additive damped trend model with additive error, ETS(A,A_d,N). While Table 7.2 shows that the ETS(A,A_d,N) model is ranked first for only 6.2% of the 645 time series, it has the smallest mean rank. It also has the smallest mean MASE and the second smallest median MASE. The ETS(A,N,N) model, which was the best model (i.e., $r(H, i, j) = 1$) for the most time series in Table 7.4, is poor with respect to all three measures. Thus, it is the best model for the largest number of series, but does poorly on other series. On the other hand ETS(A,A_d,N) is not the best as often, but it is more robust in that it does not do so poorly when applied to all time series.

A similar comparison of mean ranks for the $M = 30$ models in Tables 2.2 and 2.3 on the $N = 756$ quarterly time series in the M3 data, showed that the additive damped trend model with multiplicative seasonality and error, ETS(M,A_d,M), has the smallest mean rank of 12.84. For the $N = 1{,}428$ monthly time series in the M3 data, the model that has the smallest mean rank out of $M = 30$ is ETS(A,A_d,M), with a rank of 14.09.

Now we turn to the question of whether we can improve the forecasts by using a procedure that allows the choice of model to be different for different time series rather than using one model for all time series. We will compare the following seven model selection procedures: a single model for forecasting all time series, the five IC methods, and the prediction validation method. Based on the results obtained using the mean rank of the MASE(H, i, j) in the preceding paragraphs of this section, we will consider damped trend models for the choice of the single model. In particular, the ETS(A,A_d,N) model will be used when choosing among the three linear models and when choosing among all ten non-seasonal models for the annual M3 data. The ETS(A,A_d,A) model will be used when choosing among six linear models for quarterly and monthly data, and the ETS(M,A_d,M) and ETS(A,A_d,M) models when choosing among all 30 models for quarterly and monthly data, respectively. The potential models for these categories are listed in Tables 2.2 and 2.3. In a linear model, the error term and any trend or seasonal components are additive.

We continue to use the M3 data set in our comparisons of the model selection procedures. Each time series $\{y_t^{(j)}\}$ is divided into two parts: the fitting set of n_j values and the forecasting set of H values. For the LEIC and VAL selection methods, the fitting set is divided further into two segments of n_j^* and H values. The values of H are 6, 8, and 18 for annual, quarterly, and monthly data, respectively.

The results of comparing the seven procedures are summarized in Table 7.5. In the table, we refer to the procedures that employ one of the five ICs or prediction validation as *model selection methods* and the procedure that picks a single model as *damped trend*. By looking at this table, we can compare a specified damped trend model, the AIC, and the best model selection method with respect to each of the three measures: mean rank, mean MASE, and median MASE. "Best Method(s)" indicates the model selection methods that have the minimum value for the specified measure and type of data on that row.

Examination of Table 7.5a provides some interesting insights. For this table, the potential models for selection included only the linear models from Tables 2.2 and 2.3. There are three potential non-seasonal linear models for the annual time series and six potential linear models for the quarterly and monthly data. The last two columns in the table indicate that, among the model selection methods, the AIC always has the minimum, or nearly the minimum, value for each measure. On the other hand, applying the indicated damped trend model to the entire data type is almost always equally satisfactory with respect to the three measures. The two damped trend models are encompassing in that all the other possible model choices for the type of data are special cases. Thus, it is not surprising that the ETS(A,A_d,N) model performs well for annual data, and the ETS(A,A_d,A) model does well for

Table 7.5. Comparisons using MASE and MAPE for models in Tables 2.2 and 2.3.

Measure	Data type	Damped trend	AIC	Best method(s)
(a) Comparison using MASE for linear models				
Mean Rank	Annual	1.86/ETS(A,A_d,N)	1.84	1.84/AIC
	Quarterly	2.96/ETS(A,A_d,A)	3.08	3.08/AIC
	Monthly	3.29/ETS(A,A_d,A)	3.07	3.03/AICc
Mean MASE	Annual	2.92/ETS(A,A_d,N)	2.94	2.94/AIC
	Quarterly	2.14/ETS(A,A_d,A)	2.15	2.15/AIC, LEIC
	Monthly	2.09/ETS(A,A_d,A)	2.06	2.05/AICc
Median MASE	Annual	1.82/ETS(A,A_d,N)	1.82	1.82/AIC
	Quarterly	1.46/ETS(A,A_d,A)	1.47	1.47/AIC
	Monthly	1.12/ETS(A,A_d,A)	1.08	1.07/AICc
(b) Comparison using MASE for all models				
Mean Rank	Annual	4.97/ETS(A,A_d,N)	5.42	5.29/BIC
	Quarterly	12.84/ETS(M,A_d,M)	13.97	13.97/AIC
	Monthly	14.09/ETS(A,A_d,M)	13.50	13.29/AICc
Mean MASE	Annual	2.92/ETS(A,A_d,N)	3.30	2.91/LEIC
	Quarterly	2.13/ETS(M,A_d,M)	2.27	2.27/AIC
	Monthly	2.10/ETS(A,A_d,M)	2.07	2.08/AIC, AICc, HQIC
Median MASE	Annual	1.82/ETS(A,A_d,N)	1.98	1.92/LEIC
	Quarterly	1.50/ETS(M,A_d,M)	1.54	1.54/AIC
	Monthly	1.10/ETS(A,A_d,M)	1.10	1.07/HQIC
(c) Comparison using MAPE for linear models				
Mean Rank	Annual	1.86/ETS(A,A_d,N)	1.83	1.83/AIC
	Quarterly	2.98/ETS(A,A_d,A)	3.07	3.07/AIC
	Monthly	3.22/ETS(A,A_d,A)	3.08	3.06/AICc
Mean MAPE	Annual	22.66/ETS(A,A_d,N)	22.00	21.33/AICc
	Quarterly	12.06/ETS(A,A_d,A)	11.95	11.94/LEIC
	Monthly	22.01/ETS(A,A_d,A)	21.75	21.23/AICc
Median MAPE	Annual	10.92/ETS(A,A_d,N)	11.18	11.16/AICc, LEIC
	Quarterly	5.32/ETS(A,A_d,A)	5.46	5.46/AIC
	Monthly	9.30/ETS(A,A_d,A)	9.29	9.29/AIC, AICc
(d) Comparison using MAPE for all models				
Mean Rank	Annual	4.98/ETS(A,A_d,N)	5.45	5.26/LEIC
	Quarterly	12.86/ETS(M,A_d,M)	13.87	13.87/AIC
	Monthly	13.76/ETS(A,A_d,M)	13.62	13.54/AICc
Mean MAPE	Annual	22.66/ETS(A,A_d,N)	25.42	20.71/LEIC
	Quarterly	11.96/ETS(M,A_d,M)	12.23	12.15/HQIC
	Monthly	20.02/ETS(A,A_d,M)	21.63	21.62/HQIC
Median MAPE	Annual	10.92/ETS(A,A_d,N)	11.54	11.16/LEIC
	Quarterly	5.22/ETS(M,A_d,M)	5.62	5.54/VAL
	Monthly	9.15/ETS(A,A_d,M)	9.03	8.96/VAL

quarterly and monthly data. One could decide to choose these models rather than use the AIC model selection method. However, one should be reassured that the AIC will do as well as or better than the encompassing model, and it would lead to the selection of simpler models when possible.

The information in Table 7.5b is more complicated. The potential models for the selection methods in this table include the ten non-seasonal models for the annual time series and all 30 models for quarterly and monthly time series from Tables 2.2 and 2.3. The results for quarterly and monthly data are similar to those for the linear models. The AIC does nearly as well as or better than the other model selection methods. As in the case of the linear models, a single damped trend model performs well. Unlike the case of the linear models, neither the ETS(M,A_d,M) model nor the ETS(A,A_d,M) model is an encompassing model for the 30 models, and therefore, it is not as clear which damped trend model to pick. Another observation is that the mean MASE and median MASE do not decrease when the number of models in the selection process is increased from six to 30, as we would hope. For both monthly and quarterly time series, one should consider using the AIC with an expanded set of linear models, but far fewer than all 30 models.

The annual data in Table 7.5b tend to be shorter than the quarterly and monthly data. The longest annual series had 41 observations, and there are many very short time series. Hence, it is not unexpected to find that the model selection methods did not do as well as a single model. The same comparisons as in Table 7.5b were done for annual time series that were greater than or equal to 20 in length, with essentially no change to the results. In fact, model selection comparisons were also done for quarterly and monthly data of length greater than or equal to 28 and 72, respectively, with no change to the general implications in Table 7.5b. For annual time series, the comparisons on the annual data indicate that one should either use the damped trend model ETS(A,A_d,N) or limit the AIC to the three linear models. These findings match and help to explain the poor performance of the AIC for choosing among innovations state space models in Hyndman et al. (2002) for annual data from the M3 competition.

All of the comparisons in Table 7.5a, b were repeated using the mean absolute percentage error (MAPE) from Sect. 2.7.2, and again the implications were the same. See Table 7.5c, d.

For a comparison of the individual methods (five ICs and VAL) using the MASE, see Table 7.6. This table allows the reader to see more detail for the model selection methods that are summarized in the last two columns of Table 7.5a, b. In the next section, we will examine model selection for a set of hospital data and will present the results in the same form as Table 7.6.

Table 7.6. Comparisons of methods on the M3 data using MASE for the models in Tables 2.2 and 2.3.

Measure	Data type	AIC	BIC	HQIC	AICc	LEIC	VAL
(a) Comparison of methods using MASE for linear models							
Mean Rank	Annual	1.84	1.86	1.86	1.88	1.86	1.97
	Quarterly	3.08	3.24	3.14	3.16	3.12	3.26
	Monthly	3.07	3.15	3.05	3.03	3.23	3.20
Mean MASE	Annual	2.94	2.96	2.95	2.96	2.95	3.04
	Quarterly	2.15	2.21	2.16	2.17	2.15	2.19
	Monthly	2.06	2.13	2.09	2.05	2.19	2.17
Median MASE	Annual	1.82	1.85	1.85	1.85	1.85	1.95
	Quarterly	1.47	1.58	1.50	1.49	1.49	1.53
	Monthly	1.08	1.11	1.08	1.07	1.12	1.10
(b) Comparison of methods using MASE for all models							
Mean Rank	Annual	5.42	5.29	5.39	5.33	5.31	5.55
	Quarterly	13.97	14.75	14.20	14.47	15.14	14.87
	Monthly	13.50	13.60	13.33	13.29	14.78	13.92
Mean MASE	Annual	3.30	3.28	3.29	3.26	2.91	3.37
	Quarterly	2.27	2.38	2.29	2.29	2.40	2.29
	Monthly	2.08	2.10	2.08	2.08	2.19	2.20
Median MASE	Annual	1.98	1.95	1.97	1.97	1.92	2.00
	Quarterly	1.54	1.57	1.55	1.56	1.61	1.55
	Monthly	1.10	1.11	1.07	1.09	1.14	1.09

7.2.3 Comparing Selection Procedures on a Hospital Data Set

For another comparison of the model selection procedures, we used time series from a hospital data set.[1] Each time series comprises a monthly patient count for one of 20 products that are related to medical problems. We included only time series that had a mean patient count of at least ten and no individual values of 0. There were 767 time series that met these conditions. As in the comparisons using the M3 data, we withheld $H = 18$ time periods from the fitting set for the LEIC and the VAL model selection methods; similarly, we set $H = 18$ time periods for the comparisons in the forecasting set. The length of every time series was 7 years of monthly observations. Thus, the length n_j of all fitting sets had the same value of 66 time periods (i.e., $84 - 18 = 66$).

[1] The data were provided by Paul Savage of Healthcare, LLC Intelligence and Hans Levenbach of Delphus, Inc.

Table 7.7. Comparisons of methods on the hospital data set using MASE for models in Tables 2.2 and 2.3.

Measure	Data type	AIC	BIC	HQIC	AICc	LEIC	VAL
(a) Comparison of methods using MASE for linear models							
Mean Rank	Monthly	3.10	3.07	3.01	3.10	3.07	3.36
Mean MASE	Monthly	0.94	0.91	0.92	0.94	0.91	0.96
Median MASE	Monthly	0.83	0.83	0.83	0.83	0.83	0.84
(b) Comparison of methods using MASE for all models							
Mean Rank	Monthly	13.66	13.25	13.22	13.52	13.53	14.80
Mean MASE	Monthly	0.98	0.96	0.97	0.98	0.95	1.00
Median MASE	Monthly	0.84	0.84	0.83	0.84	0.85	0.86

By examining Table 7.7, we see that the results for the hospital data set using the MASE are somewhat similar to those for monthly time series in the M3 data set. Because there are definitely time series with values near 0, we believe that in this case the MASE is a more reliable measure than the MAPE for comparing forecasts. For the selection from only linear models in Table 7.7a, the three measures (mean rank, mean MASE, and median MASE) indicate that there is not much difference between the five IC methods (AIC, BIC, HQIC, AICc, and LEIC). The VAL method seems to be clearly the worst choice. In Table 7.7b, where the selection is among all 30 models, the VAL method remains the poorest choice, and similar to the results with the M3 data, there is no improvement with the increase in potential models. A difference from the findings with the M3 data is that we found that it is not a good idea to use a single damped trend model for forecasting.

7.3 Implications for Model Selection Procedures

The comparisons of the model selection procedures in Sects. 7.2.2 and 7.2.3 provide us with some interesting information on how to select models, even though the study was limited to the M3 data and the hospital data. First, the AIC model selection method was shown to be a reasonable choice among the six model selection methods for the three types of data (annual, quarterly, and monthly) in the M3 data and for the monthly time series in the hospital data. The number of observations for annual data is always likely to be small (i.e., less than or equal to 40), and thus the IC procedures may not have sufficient data to compete with simply choosing a single model such as the ETS(A,A_d,N) model when all ten non-seasonal models are considered. However, using the AIC on the three linear non-seasonal models fared as well as the ETS(A,A_d,N) and would allow the possibility of choosing simpler models, especially when there is little trend in the data. Thus, for annual time series we recommend using the AIC and choosing among the three linear non-seasonal models.

For the monthly data, the AIC is better than the choice of selecting a single damped trend model in both the M3 data and the hospital data. Because it is definitely not clear which single model to use, we suggest using the AIC. One might also consider limiting the choice of models to a set that includes the linear models but is smaller than the complete set of 30 models. We make the same recommendations for the quarterly time series, with additional emphasis on reducing the number of models from 30.

In common with other studies of model selection, our focus has been exclusively on selection methods that relate to point forecasts. Model selection procedures designed to produce good interval forecasts are likely to have similar properties to those discussed in this chapter, but the issue is one to be addressed in future research.

7.4 Exercises

Exercise 7.1. Select a data set with monthly time series, and write some **R** code to do the following:

a. Find the maximum likelihood estimates, forecasts for $h = 1, \dots, 18$, forecast errors for $h = 1, \dots, 18$, and $\text{MASE}(18, i, j)$ for each time series j in the data set and each linear model i from Table 2.1.
b. Use the AIC to pick a model k_j for each time series j and identify the $\text{MASE}(18, k_j, j)$ for each time series from values in part a above.
c. Use the BIC to pick a model k_j for each time series j and identify the $\text{MASE}(18, k_j, j)$ for each time series from values in part a above.
d. Compare the forecast accuracy obtained when selecting a model with the AIC or BIC, and when using the ETS(A,A_d,A) model for all series. (See Sect. 7.2.1 for suggested measures).

Exercise 7.2. Repeat Exercise 7.1 with the set of potential models in part a expanded to include the ETS(M,A_d,M) model and its submodels, and with the ETS(M,A_d,M) model added to the comparison in part c.

Appendix: Model Selection Algorithms

The Linear Empirical Information Criterion

In the *linear empirical information criterion* (LEIC) (Billah et al. 2003, 2005), $\zeta(n) = c$, where is c is estimated empirically from an ensemble of N similar time series for M competing models. The number of observations in the fitting set for time series $\{y_t^{(j)}\}$, $j = 1\ldots, N$, is denoted by n_j. Each of the N sets of observations is divided into two segments: the first segment consists of $n_j^* = n_j - H$ observations and the second segment consists of the last H observations. Let $n = \max\{n_{j^*};\ j = 1,\ldots,N\}$. Then, values of c between 0.25 and $2\log(n)$ in steps of δ provide a range of values wide enough to contain all the commonly used penalty functions. The value of $\delta = 0.25$ has worked well in practice. The procedure for estimating c for the LEIC is as follows:

Model estimation

1. For each of the N series, use the first n_j^* observations to estimate the parameters and initial state vector in each of the M competing models by maximum likelihood estimation.
2. Record the maximized log-likelihoods for all estimated models.

Penalty estimation

1. For each trial value of c do the following
   (a) For each time series $\{y_t^{(j)}\}$, $j = 1\ldots, N$, select a model with the minimum LEIC using (7.2) and $\zeta(n) =$ the trial value for c.
   (b) For each forecast horizon h, $h = 1,\ldots, H$, and time series $\{y_t^{(j)}\}$, $j = 1\ldots, N$, compute the absolute scaled error

$$\text{ASE}(h, c, j) = \frac{|y_{n_j^*+h}^{(j)} - \hat{y}_{n_j^*}^{(c,j)}(h)|}{\text{MAE}_j},$$

   where $\text{MAE}_j = (n_j^* - 1)^{-1}\sum_{t=2}^{n_j^*} |y_t^{(j)} - y_{t-1}^{(j)}|$, and $\hat{y}_{n_j^*}^{(c,j)}(h)$ is the h-step-ahead forecast using the model selected for the jth series.
2. For each value of c and for each forecast horizon h, calculate the mean absolute scaled error MASE across the N time series to obtain

$$\text{MASE}(h, c) = \frac{1}{N}\sum_{j=1}^{N} \text{ASE}(h, c, j). \tag{7.4}$$

3. Select a value of $c^{(h)}$ by minimizing the MASE(h,c) over the grid of c values. Thus, a value of $c^{(h)}$ is selected for each forecast horizon h, $h = 1,\dots,H$.
4. Compute the final value of c by averaging the H values of $c^{(h)}$:

$$c = \frac{1}{H}\sum_{h=1}^{H} c^{(h)}.$$

Observe that MASE(H,i,j) in (7.3) is an average across H forecast horizons for a specified model i and time series j, while MASE(h,c) in (7.4) is an average across N time series for a fixed forecasting horizon h and model determined by trial value c. Also, it is important to re-estimate the parameters and initial state vector for the selected model by using all of the n_j values.

Prediction Validation Method of Model Selection

The *prediction validation method* (VAL) is a method that has frequently been used in practice. In this method, a model is chosen from M models for time series $\{y_t^{(j)}\}$ as follows:

1. Divide the fitting set for time series $\{y_t^{(j)}\}$ of length n_j into two segments: the first segment consists of $n_j^* = n_j - H$ observations, and the second segment consists of the last H observations.
2. Using $y_1^{(j)}$ to $y_{n_j^*}^{(j)}$, find the maximum likelihood estimates for each model i, $i = 1,\dots,M$.
3. For each model i, compute the forecasts $\hat{y}_{n_j^*}^{(i,j)}(h)$, $h = 1,\dots,H$.
4. Compute the MASE(H,i,j), as defined in (7.3), with n_j replaced by n_{j^*}.
5. Choose model k_j, where

$$\text{MASE}(H,k_j,j) = \min\{\text{MASE}(H,i,j);\ i = 1,\dots,M\}.$$

The parameters and initial state vector for the selected model must be re-estimated using all n_j values.

Part III

Further Topics

8

Normalizing Seasonal Components

In exponential smoothing methods, the m seasonal components are combined with level and trend components to indicate changes to the time series that are caused by seasonal effects. It is sometimes desirable to report the value of these m seasonal components, and then it is important for them to make intuitive sense. For example, in the additive seasonal model ETS(A,A,A), the seasonal components are added to the other components of the model. If one seasonal component is positive, there must be at least one other seasonal component that is negative, and the average of the m seasonal components should be 0. When the average value of the m additive seasonal components at time t is 0, the seasonal components are said to be *normalized*. Similarly, we say that multiplicative seasonal components are normalized if the average of the m multiplicative seasonal components at time t is 1.

Normalized seasonal components can be used to seasonally adjust the data. To calculate the seasonally adjusted data when the model contains an additive seasonal component, it is necessary to subtract the seasonal component from the data. For a multiplicative seasonal component, the data should be divided by the seasonal component.

Thus far, the specified models have not had normalized seasonal components. This is because normalization is only necessary when the seasonal component is to be analyzed separately or used for seasonal adjustment. If one is only interested in the point forecasts and forecast variances for prediction intervals, then it is not necessary to normalize the seasonal components. In most cases, the forecasts and forecast variances obtained with the non-normalized models are identical to those obtained with the normalized models. As we will see, the only exception to this equivalence is when the trend is multiplicative and the seasonality is additive.

In Sect. 8.1, we discuss normalizing models with additive seasonal components. Normalization of models with multiplicative seasonal components is covered in Sect. 8.2. An example to show the potential importance of normalization is presented in Sect. 8.3.

8.1 Normalizing Additive Seasonal Components

The additive seasonal components are said to be *normalized* when the sum of any m consecutive components sum to zero:

$$\sum_{i=0}^{m-1} s_{t-i} = 0 \qquad \text{for } t \geq 0. \tag{8.1}$$

A common practice is to impose (8.1) for $t = 0$ so that estimates of the m initial seasonal components in $\boldsymbol{x}_0$ sum to 0. (The simplest way to impose this constraint is to estimate only $m - 1$ of the initial seasonal components and set the final component to be minus the sum of the others.)

However, for the models with additive seasonality that we have seen in previous chapters, the normalization property (8.1) for the estimates of the seasonal components is lost for $t > 0$ as they are revised in the exponential smoothing process. To illustrate this point, we examine the case of additive errors. Note that the seasonal component is defined by

$$\begin{aligned} s_t &= s_{t-m} + \gamma\varepsilon_t \qquad (8.2)\\ &= s_{t-2m} + \gamma(\varepsilon_t + \varepsilon_{t-m})\\ &\ \ \vdots\\ &= s_{t_m^+} + \gamma\sum_{i=0}^{t_m} \varepsilon_{t-im}, \end{aligned}$$

where $t_m = \lfloor (t-1)/m \rfloor$ and $t_m^+ = \lceil (t-1) \bmod m \rceil + 1 - m$. Therefore,

$$\sum_{i=0}^{m-1} s_{t-i} = \sum_{k=0}^{m-1} s_{-k} + \gamma\sum_{i=1}^{t} \varepsilon_i = \gamma\sum_{i=1}^{t} \varepsilon_i.$$

So the sum of m consecutive components behaves like a random walk and, over time, will range a long way from zero, particularly if γ is large.

One solution that has been suggested for the normalizing problem in the additive error situation is to replace (8.2) with the following equation:

$$s_t = -\sum_{i=1}^{m-1} s_{t-i} + \gamma\varepsilon_t. \tag{8.3}$$

Although the expected value of the sum $\sum_{i=0}^{m-1} s_{t-i}$ is 0, this proposed model does not have the property that the sum of any m consecutive seasonal components is 0. Furthermore, estimates of the individual seasonal components can (and frequently do) become quite unrealistic with this formulation of the seasonal state equation. If $\sum_{i=0}^{m-1} s_{-i} = 0$, as is required in practice for initial estimates, then (8.3) is equivalent to $s_t = s_{t-m} + \gamma(\varepsilon_t - \varepsilon_{t-1})$. Looking at (8.3) in this latter form makes it seem even more unreasonable as a substitute for (8.2).

8.1.1 Roberts-McKenzie Normalization

Roberts (1982) and McKenzie (1986) showed how to normalize the seasonal estimates of additive seasonal components in the case when both the trend and the errors are also additive. In their method, every seasonal component has to be revised in each time period. In order to describe and extend their method, we first introduce a new notation. The seasonal component at time t for the season that corresponds to time period $t-i$ will be denoted by $s_t^{(i)}$, $i = 0,\ldots,m-1$. Note that $s_t^{(i)}$ and $s_{t-1}^{(i-1)}$ represent seasonal components for the same season at two different time periods, t and $t-1$. Then, for all models in Tables 2.2 and 2.3 with an additive seasonal component, the state equations for the seasonal components can be written as follows:

$$s_t^{(0)} = s_{t-1}^{(m-1)} + \gamma q(\boldsymbol{x}_{t-1})\varepsilon_t, \tag{8.4a}$$

$$s_t^{(i)} = s_{t-1}^{(i-1)}, \qquad i = 1,\ldots,m-1, \tag{8.4b}$$

where $q(\boldsymbol{x}_{t-1}) = 1$ for models with additive error, $q(\boldsymbol{x}_{t-1}) = \hat{y}_{t|t-1}$ for models with multiplicative error, and $\boldsymbol{x}_t = [\ell_t, b_t, s_t^{(0)}, s_t^{(1)}, \ldots, s_t^{(m-1)}]'$. In the observation equation, we replace the seasonal component s_{t-m} by $s_{t-1}^{(m-1)}$.

To obtain normalized seasonal components that correspond to the Roberts (1982) and McKenzie (1986) normalization, we simply subtract a small term a_t, called the *additive normalizing factor*, from each $s_t^{(i)}$ to ensure that the seasonal components at each time period sum to zero. Thus, (8.4) is replaced by

$$\tilde{s}_t^{(0)} = \tilde{s}_{t-1}^{(m-1)} + \gamma q(\tilde{\boldsymbol{x}}_{t-1})\varepsilon_t - a_t, \tag{8.5a}$$

$$\tilde{s}_t^{(i)} = \tilde{s}_{t-1}^{(i-1)} - a_t, \qquad i = 1,\ldots,m-1, \tag{8.5b}$$

where a tilde is used to denote the normalized components, and the additive normalizing factor is

$$a_t = (\gamma/m)q(\tilde{\boldsymbol{x}}_{t-1})\varepsilon_t.$$

Observe that for these normalized seasonal components

$$\begin{aligned}\sum_{i=0}^{m-1} \tilde{s}_t^{(i)} &= \sum_{i=0}^{m-2} \left[\tilde{s}_{t-1}^{(i)} - (\gamma/m)q(\tilde{\boldsymbol{x}}_{t-1})\varepsilon_t\right] + \left[\tilde{s}_{t-1}^{(m-1)} + (\gamma - \gamma/m)q(\tilde{\boldsymbol{x}}_{t-1})\varepsilon_t\right] \\ &= \sum_{i=0}^{m-1} \tilde{s}_{t-1}^{(i)}.\end{aligned}$$

Thus, if the initial values of the seasonal components at $t = 0$ sum to 0, this property is maintained at all time periods.

When both the trend and the seasonal components are additive, we will show that an additional adjustment in the level equation will maintain the same forecast means and variances for the model. However, for models with multiplicative trend but additive seasonality, the normalized model will give different forecasts from the non-normalized model.

8.1.2 Adjusted Components

In almost all cases, the Roberts-McKenzie scheme outlined above can be implemented simply by adjusting the usual state components to obtain normalized components. That is, we can use the original models for forecasting and recover the normalized factors later if required.

With the new notation for the seasonal components, the damped trend additive seasonal models, ETS(A,A$_d$,A) and ETS(M,A$_d$,A) from Tables 2.2 and 2.3, have the following form:

$$\begin{aligned} y_t &= \ell_{t-1} + \phi b_{t-1} + s_{t-1}^{(m-1)} + q(\boldsymbol{x}_{t-1})\varepsilon_t, &(8.6a)\\ \ell_t &= \ell_{t-1} + \phi b_{t-1} + \alpha q(\boldsymbol{x}_{t-1})\varepsilon_t, &(8.6b)\\ b_t &= \phi b_{t-1} + \beta q(\boldsymbol{x}_{t-1})\varepsilon_t, &(8.6c)\\ s_t^{(0)} &= s_{t-1}^{(m-1)} + \gamma q(\boldsymbol{x}_{t-1})\varepsilon_t, &(8.6d)\\ s_t^{(i)} &= s_{t-1}^{(i-1)}, \qquad i = 1, \ldots, m-1. &(8.6e) \end{aligned}$$

The only models in Tables 2.2 and 2.3 with additive seasonal components that are not represented as special cases of (8.6) are those with multiplicative trend.

The normalized form of the models represented in (8.6) is given below:

$$\begin{aligned} y_t &= \tilde{\ell}_{t-1} + \phi \tilde{b}_{t-1} + \tilde{s}_{t-1}^{(m-1)} + q(\tilde{\boldsymbol{x}}_{t-1})\varepsilon_t, &(8.7a)\\ \tilde{\ell}_t &= \tilde{\ell}_{t-1} + \phi \tilde{b}_{t-1} + \alpha q(\tilde{\boldsymbol{x}}_{t-1})\varepsilon_t + a_t, &(8.7b)\\ \tilde{b}_t &= \phi \tilde{b}_{t-1} + \beta q(\tilde{\boldsymbol{x}}_{t-1})\varepsilon_t, &(8.7c)\\ \tilde{s}_t^{(0)} &= \tilde{s}_{t-1}^{(m-1)} + \gamma q(\tilde{\boldsymbol{x}}_{t-1})\varepsilon_t - a_t, &(8.7d)\\ \tilde{s}_t^{(i)} &= \tilde{s}_{t-1}^{(i-1)} - a_t, \qquad i = 1, \ldots, m-1. &(8.7e) \end{aligned}$$

We now turn to examining how the states, the forecast means, the forecast variances, and the prediction distributions for the original model in (8.6) are related to those for the normalized model in (8.7). The *cumulative additive normalizing factor* is defined as

$$A_t = \frac{1}{m} \sum_{i=0}^{m-1} s_t^{(i)}, \qquad t \geq 0.$$

Assume that we have observed $y_1, y_2, \ldots, y_t$, and $\boldsymbol{x}_0 = \tilde{\boldsymbol{x}}_0$, with $\sum_{i=0}^{m-1} \tilde{s}_0^{(i)} = 0$. Then the following relationships between the normalized model (8.7) and the original model (8.6) are valid (see Appendix):

- Recursive formula for the cumulative additive normalizing factor:

$$A_t = A_{t-1} + a_t. \tag{8.8}$$

- Normalized states at time t:

$$\tilde{\ell}_t = \ell_t + A_t, \tag{8.9a}$$

$$\tilde{b}_t = b_t, \tag{8.9b}$$

$$\tilde{s}_t^{(i)} = s_t^{(i)} - A_t \qquad \text{for } i = 0, \ldots, m-1. \tag{8.9c}$$

- Point forecasts and forecast errors are equal for all $t \geq 0$ and $h \geq 1$:

$$\tilde{y}_{t+h|t} = \hat{y}_{t+h|t} = \ell_t + \phi_h b_t + s_t^{(m-h_m)}, \tag{8.10}$$

where $\tilde{y}_{t+h|t}$ is the forecast using (8.7) and $h_m = \lfloor (h-1)/m \rfloor$.
- Forecast variances are equal for all $t \geq 0$ and $h \geq 1$:
  - Class 1 models from Chap. 6 where $q(\boldsymbol{x}_t) = \mu_{t+h|t} = 1$,

$$\tilde{v}_{t+h|t} = v_{t+h|t} = \begin{cases} \sigma^2 & \text{if } h = 1 \\ \sigma^2 \left[1 + \sum_{j=1}^{h-1} c_j^2 \right] & \text{if } h \geq 2. \end{cases} \tag{8.11}$$

  - Class 2 models from Chap. 6 where $q(\boldsymbol{x}_t) = \mu_{t+h|t} = \hat{y}_{t+1|t}$

$$\tilde{v}_{t+h|t} = v_{t+h|t} = (1+\sigma^2)\theta_h - \mu_{t+h|t}^2, \tag{8.12a}$$

$$\theta_h = \begin{cases} \mu_{t+1|t}^2 & \text{if } h = 1 \\ \mu_{t+h|t}^2 + \sigma^2 \sum_{j=1}^{h-1} c_j^2 \theta_{h-j} & \text{if } h \geq 2, \end{cases} \tag{8.12b}$$

  where $c_j = \boldsymbol{w}' \boldsymbol{F}^{j-1} \boldsymbol{g}$ (see Table 6.2 on page 81 for values of c_j corresponding to specific models).
- Simulated prediction distributions are the same:

$$\tilde{y}_{t+h}^{(i)} = y_{t+h}^{(i)} \quad \text{for the } i\text{th simulated value at time } t, \tag{8.13}$$

where $\tilde{y}_{t+h}^{(i)}$ is the value simulated using (8.7).

The results in (8.8)–(8.13) are valid for the ETS(A,A$_\text{d}$,A) and ETS(M,A$_\text{d}$,A) models, and any of their special cases in Tables 2.2 and 2.3. Thus, for these models it is clearly not necessary to normalize the seasonal factors if one is only interested in the forecasts and the prediction intervals. It is also possible to apply (8.9) at any time period t to find the normalized components when forecasting with the original models. Observe that it is necessary to adjust the level ℓ_t if its value is to be reported or if one plans to continue the exponential smoothing process with the values of the new components.

Because $\alpha q(\tilde{\boldsymbol{x}}_{t-1})\varepsilon_t + a_t = (\alpha + \gamma/m)q(\tilde{\boldsymbol{x}}_{t-1})\varepsilon_t$, the smoothing parameter for the level in the normalized model (8.7) is $\alpha + \gamma/m$. This model may be simplified slightly by letting $\alpha^* = \alpha + \gamma/m$; that is, only altering the equations for the seasonal components. The values for the components, forecasts, and forecast variances produced using this modification are identical to those from the normalized model in (8.7).

When the trend is multiplicative, a model analogous to (8.7) can be used for normalization. However, in contrast to the case of additive trend, the forecasts will be somewhat altered from the original form of the model (see Exercises 8.2 and 8.3). Nevertheless, we recommend that this model be used whenever normalized components are required.

8.2 Normalizing Multiplicative Seasonal Components

As in the case of additive seasonal components, it may be desirable to report the values of the multiplicative seasonal components. We will continue to denote the seasonal component at time t for the season that corresponds to time period $t-i$ by $s_t^{(i)}$, $i = 0, \ldots, m-1$. The multiplicative seasonal components are said to be *normalized* when the seasonal components from any m consecutive time periods have an average of 1, or equivalently, a sum of m:

$$\sum_{i=0}^{m-1} s_{t-i} = m, \qquad \text{for } t \geq 0.$$

The normalization procedure for multiplicative seasonality was introduced by Archibald and Koehler (2003).

To normalize multiplicative seasonal components, we replace (8.4) by

$$\tilde{s}_t^{(0)} = \left[\tilde{s}_{t-1}^{(m-1)} + \gamma q(\tilde{\boldsymbol{x}}_{t-1})\varepsilon_t\right]/r_t, \tag{8.14a}$$

$$\tilde{s}_t^{(i)} = \tilde{s}_{t-1}^{(i-1)}/r_t, \qquad i = 1, \ldots, m-1, \tag{8.14b}$$

where r_t is a *multiplicative normalizing factor*:

$$r_t = 1 + (\gamma/m)q(\tilde{\boldsymbol{x}}_{t-1})\varepsilon_t,$$

and $q(\tilde{\boldsymbol{x}}_{t-1})$ takes one of the following values:

- Multiplicative error

$$q(\tilde{\boldsymbol{x}}_{t-1}) = \tilde{s}_{t-1}^{(m-1)}. \tag{8.15}$$

- Additive error with no trend

$$q(\tilde{\boldsymbol{x}}_{t-1}) = 1/\tilde{\ell}_{t-1}. \tag{8.16}$$

- Additive error and additive damped trend ($\phi = 1$ for no damping)

$$q(\tilde{\boldsymbol{x}}_{t-1}) = 1/(\tilde{\ell}_{t-1} + \phi\tilde{b}_{t-1}). \tag{8.17}$$

- Additive error and multiplicative damped trend ($\phi = 1$ for no damping)

$$q(\tilde{\boldsymbol{x}}_{t-1}) = 1/(\tilde{\ell}_{t-1}\tilde{b}_{t-1}^{\phi}). \tag{8.18}$$

Assuming that $\sum_{i=0}^{m-1} \tilde{s}_0^{(i)} = m$ for the initial estimates, the sum of seasonal components at any time t is m. This can be shown by noting that if the seasonal components are normalized (i.e., the sum is m) at time period $t-1$, then they are normalized at time period t:

$$\begin{aligned}
\sum_{i=0}^{m-1} \tilde{s}_t^{(i)} &= \frac{1}{r_t}\Big\{\big[\tilde{s}_{t-1}^{(m-1)} + \gamma q(\tilde{\boldsymbol{x}}_{t-1})\varepsilon_t\big] + \sum_{i=0}^{m-2} \tilde{s}_{t-1}^{(i-1)}\Big\} \\
&= \frac{1}{r_t}\Big\{\sum_{i=0}^{m-1} \tilde{s}_{t-1}^{(i)} + \gamma q(\tilde{\boldsymbol{x}}_{t-1})\varepsilon_t\Big\} \\
&= \frac{m + \gamma q(\tilde{\boldsymbol{x}}_{t-1})\varepsilon_t}{1 + (\gamma/m)q(\tilde{\boldsymbol{x}}_{t-1})\varepsilon_t} \\
&= m.
\end{aligned}$$

In the normalized form of the models, we multiply the level and growth by r_t if the trend is additive and multiply only the level equation by r_t if there is no trend or if the trend is multiplicative. For example, the ETS(M,A$_\text{d}$,M) model from Table 2.3 has the form

$$\begin{aligned}
y_t &= (\ell_{t-1} + \phi b_{t-1})s_{t-1}^{(m-1)}(1 + \varepsilon_t), \\
\ell_t &= (\ell_{t-1} + \phi b_{t-1})(1 + \alpha\varepsilon_t), \\
b_t &= \phi b_{t-1} + \beta(\ell_{t-1} + \phi b_{t-1})\varepsilon_t, \\
s_t^{(0)} &= s_{t-1}^{(m-1)}(1 + \gamma\varepsilon_t), \\
s_t^{(i)} &= s_{t-1}^{(i-1)}, \qquad i = 1, \ldots, m-1,
\end{aligned}$$

and the normalized form of the ETS(M,A$_\text{d}$,M) model is given by

$$y_t = (\tilde{\ell}_{t-1} + \phi\tilde{b}_{t-1})\tilde{s}_{t-1}^{(m-1)}(1 + \varepsilon_t), \tag{8.19a}$$

$$\tilde{\ell}_t = \big[(\tilde{\ell}_{t-1} + \phi\tilde{b}_{t-1})(1 + \alpha\varepsilon_t)\big]r_t, \tag{8.19b}$$

$$\tilde{b}_t = \big[\phi\tilde{b}_{t-1} + \beta(\tilde{\ell}_{t-1} + \phi\tilde{b}_{t-1})\varepsilon_t\big]r_t, \tag{8.19c}$$

$$\tilde{s}_t^{(0)} = \big[\tilde{s}_{t-1}^{(m-1)}(1 + \gamma\varepsilon_t)\big]/r_t, \tag{8.19d}$$

$$\tilde{s}_t^{(i)} = \tilde{s}_{t-1}^{(i-1)}/r_t, \qquad i = 1, \ldots, m-1. \tag{8.19e}$$

We now find results for multiplicative seasonality that correspond to those for additive seasonality. Assume that we have observed $y_1, y_2, \dots y_t$, with $\sum_{i=0}^{m-1} \tilde{s}_0^{(i)} = m$. Then the following results are valid (see Exercise 8.4):

- The *cumulative multiplicative normalizing factor* is given by

$$R_t = \frac{1}{m} \sum_{i=0}^{m-1} s_t^{(i)}, \qquad t \geq 0. \tag{8.20}$$

- Recursive formula for the cumulative multiplicative normalizing factor:

$$R_{t+1} = R_t r_{t+1}. \tag{8.21}$$

- Normalized states at time t:

$$\tilde{\ell}_t = \ell_t R_t, \tag{8.22a}$$

$$\tilde{b}_t = \begin{cases} b_t R_t & \text{if the trend is additive} \\ b_t & \text{if the trend is multiplicative,} \end{cases} \tag{8.22b}$$

$$\tilde{s}_t^{(i)} = s_t^{(i)} / R_t \qquad \text{for } i = 0, \dots, m-1. \tag{8.22c}$$

- Point forecasts are equal for all $t \geq 0$ and $h \geq 1$:

$$\tilde{y}_{t+h|t} = \hat{y}_{t+h|t}, \tag{8.23}$$

where $\tilde{y}_{t+h|t}$ is the forecast using (8.19).
- Simulated prediction distributions are the same:

$$\tilde{y}_{t+h}^{(i)} = y_{t+h}^{(i)} \quad \text{for the } i\text{th simulated value at time } t, \tag{8.24}$$

where $\tilde{y}_{t+h}^{(i)}$ is the simulated value using (8.19).
- For Class 3 models in Chap. 6, means and variances are equal:

$$\tilde{\mu}_{t+h|t} = \mu_{t+h|t} \quad \text{see (6.4) and (6.7),} \tag{8.25a}$$

$$\tilde{v}_{t+h|t} = v_{t+h|t} \quad \text{see (6.5) and (6.8).} \tag{8.25b}$$

Because the forecasts are the same with and without normalizing, it is not important to normalize the components unless the values of the components need to be provided. Moreover, one can normalize the components at any time period by using (8.22). Notice that all the components (level, slope and seasonal) must be adjusted if one intends to re-start the exponential smoothing process with the normalized values.

8.3 Application: Canadian Gas Production

To demonstrate the normalization procedure, we will use Canadian gas production data shown in the top panel of Fig. 8.1. Because the seasonal pattern is changing rapidly throughout the series, this series is likely to have a high value of γ in the fitted models, and is therefore likely to have the estimated seasonal component wander away from one.

We fit an ETS(M,N,M) model to these data with $\alpha = 0.2$ and $\gamma = 0.6$. The level and seasonal components are shown in Fig. 8.2. The original seasonal component has wandered some distance away from one, and the level is lower to compensate. After normalization, the seasonal component stays close to one and the level reflects the true level of the original series.

When we divide the original series by the normalized component, we obtain seasonally adjusted data. These are shown in the bottom panel of Fig. 8.1.

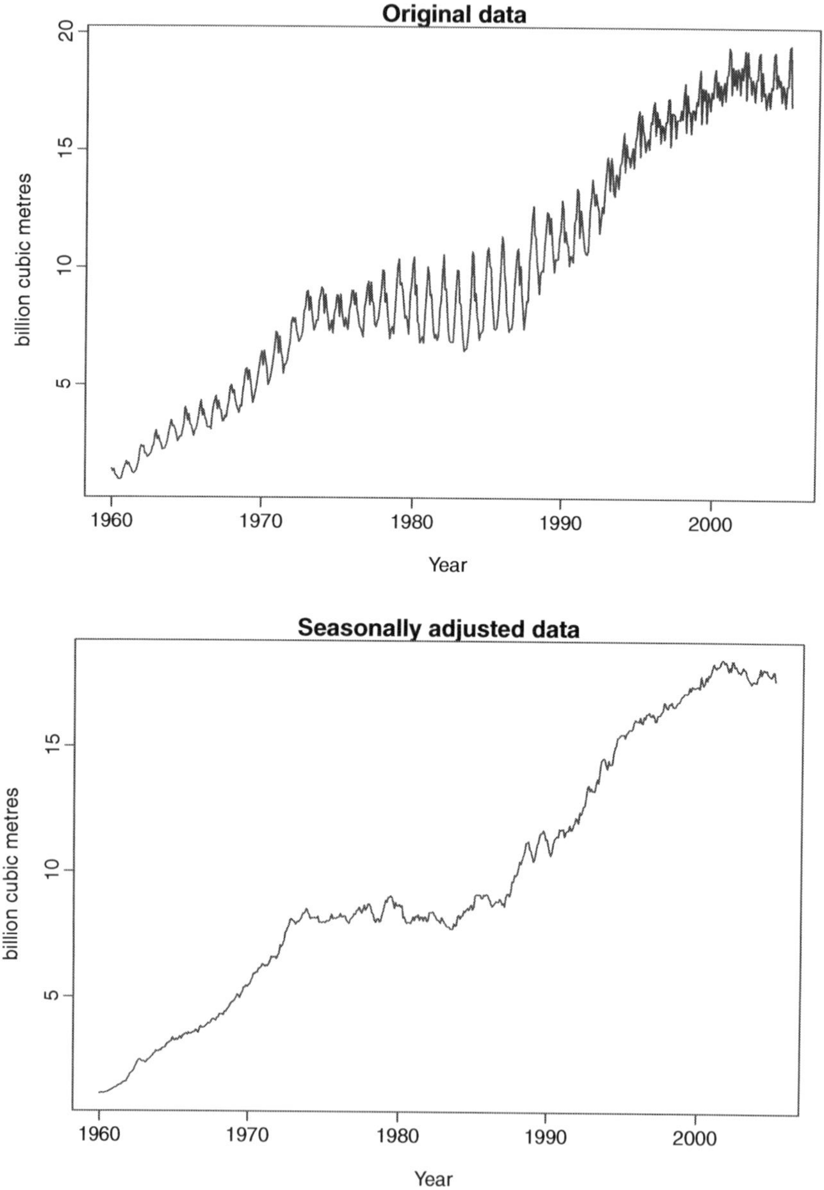

Fig. 8.1. *Top*: Monthly Canadian gas production in billions of cubic meters, January 1960–February 2005. *Bottom*: Seasonally adjusted Canadian gas production.

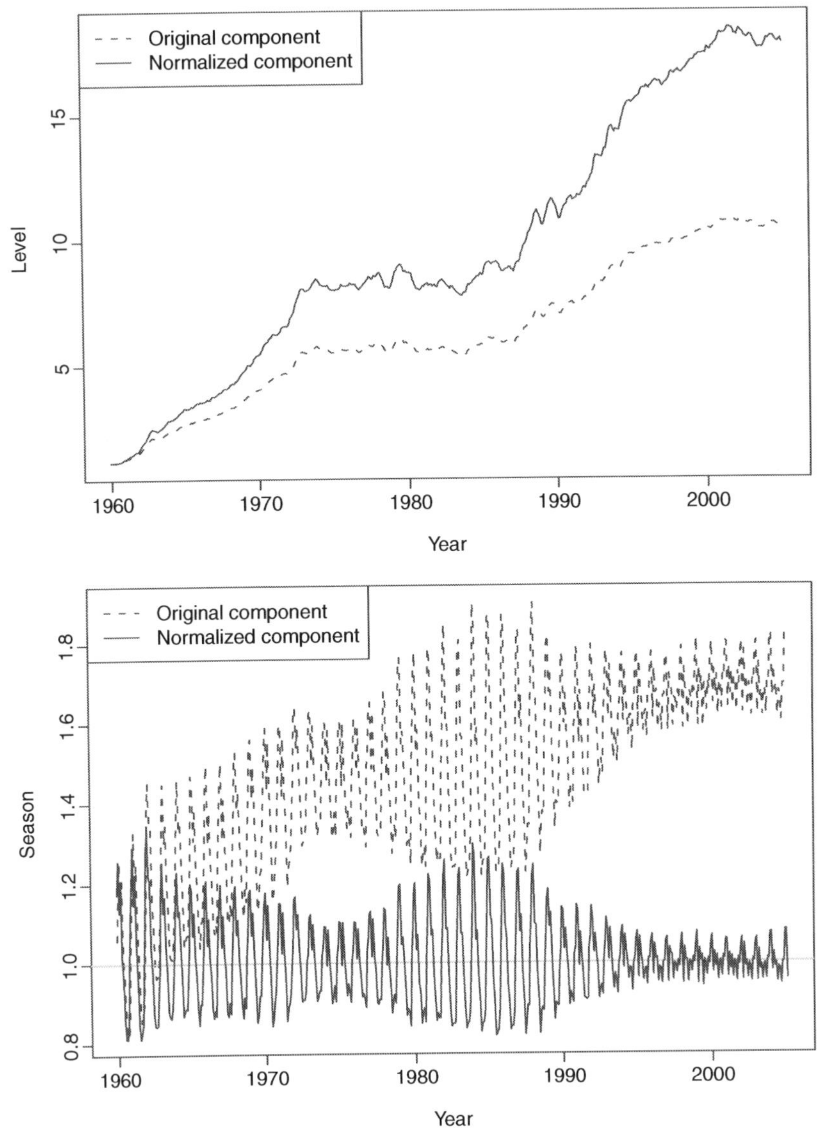

Fig. 8.2. The level and seasonal components for the ETS(M,N,M) model fitted to the monthly Canadian gas production data. The original components are shown as *dashed lines* and the normalized components are shown as *solid lines*.

8.4 Exercises

Exercise 8.1. Use the proofs in Chap. 6 to find $\tilde{v}_{t+h|t}$ for Class 1 and Class 2 models.

Exercise 8.2. Consider the components of the ETS(A,M_d,A) model:

$$y_t = \mu_t + \varepsilon_t, \tag{8.26a}$$
$$\mu_t = \ell_{t-1} b_{t-1}^{\phi} + s_{t-1}^{(m-1)}, \tag{8.26b}$$
$$\ell_t = \ell_{t-1} b_{t-1}^{\phi} + \alpha\varepsilon_t, \tag{8.26c}$$
$$b_t = b_{t-1}^{\phi} + \beta\varepsilon_t/\ell_{t-1}, \tag{8.26d}$$
$$s_t^{(0)} = s_{t-1}^{(m-1)} + \gamma\varepsilon_t, \tag{8.26e}$$
$$s_t^{(i)} = s_{t-1}^{(i-1)}, \qquad i = 1, \dots, m-1. \tag{8.26f}$$

a. Show that the same updating equations apply for the normalized seasonal terms as in (8.6).
b. Show that if the seasonal normalization is applied in isolation, the original and normalized models will produce different forecasts.

Exercise 8.3.

a. Following on from Exercise 8.2, show that for the normalized form to produce the same one-step-ahead forecasts, we require the normalized form of the model represented in (8.26) to be written as:

$$\tilde{\ell}_{t-1} \tilde{b}_{t-1}^{\phi} = \ell_{t-1} b_{t-1}^{\phi} + A_{t-1}$$

so that

$$\tilde{\ell}_t = \ell_t + A_t = \tilde{\ell}_{t-1} \tilde{b}_{t-1}^{\phi} + \alpha\varepsilon_t + a_t.$$

b. Hence show that, in order to have $\tilde{b}_t = b_t$, we must use the recurrence relationship

$$\tilde{b}_t = c_t(\tilde{b}_{t-1}^{\phi} - A_{t-1}/\tilde{\ell}_t + \beta\varepsilon_t/\tilde{\ell}_t),$$

where $c_t = \tilde{\ell}_t/(\tilde{\ell}_t - A_t)$.
c. Show that yet another normalization would be required to maintain the same two-step-ahead forecast.

Exercise 8.4. Derive (8.21), (8.22), (8.23) and (8.24) for the ETS(M,A_d,M), ETS(A,A_d,M), ETS(A,M_d,M) and ETS(M,M_d,M) models.

Appendix: Derivations for Additive Seasonality

The purpose of this appendix is to derive the results in (8.8)–(8.13). We will prove the first three of the six items jointly by mathematical induction. Because we have observed $y_1, y_2, \ldots, y_t$, the values of ε_i for $i = 1, 2, \ldots, t$ are replaced by $\varepsilon_i = (y_t - \hat{y}_{t|t-1})/q(\boldsymbol{x}_{t-1})$ in the non-normalized case and by $\tilde{\varepsilon}_i = (y_t - \tilde{y}_{t|t-1})/q(\tilde{\boldsymbol{x}}_{t-1})$ in the normalized case.

For $t = 1$, result (8.8) is true because

$$\begin{aligned}
A_1 &= \frac{\sum_{i=0}^{m-1} s_1^{(i)}}{m} \\
&= \frac{1}{m}\left\{\left[s_0^{(m-1)} + \gamma q(\boldsymbol{x}_0)\varepsilon_1\right] + \sum_{i=0}^{m-2} s_0^{(i)}\right\} \\
&= 0 + (\gamma/m)q(\boldsymbol{x}_0)\varepsilon_1 \\
&= 0 + (\gamma/m)q(\tilde{\boldsymbol{x}}_0)\varepsilon_1 \\
&= A_0 + a_1
\end{aligned}$$

It is also easily seen that (8.9) and (8.10) hold for $t = 1$.

Assume that (8.8)–(8.10) are true for time t. Observe that $\varepsilon_{t+1} = \tilde{\varepsilon}_{t+1}$ because $\tilde{y}_{t+1|t}$ is assumed to be the same as $\hat{y}_{t+1|t}$. Then, $A_{t+1} = A_t + a_{t+1}$ follows by the same argument as for $t = 1$ and the assumptions for t. The other items can be justified for $t + 1$ as follows:

$$\begin{aligned}
\tilde{\ell}_{t+1} &= \tilde{\ell}_t + \phi\tilde{b}_t + \alpha q(\tilde{\boldsymbol{x}}_t)\tilde{\varepsilon}_{t+1} + a_{t+1} \\
&= \ell_t + A_t + \phi b_t + \alpha q(\boldsymbol{x}_t)\varepsilon_{t+1} + a_{t+1} \\
&= \ell_{t+1} + A_{t+1}, \\
\tilde{b}_{t+1} &= \phi\tilde{b}_t + \beta q(\tilde{\boldsymbol{x}}_t)\varepsilon_{t+1} \\
&= b_{t+1}, \\
\tilde{s}_{t+1}^{(0)} &= \tilde{s}_t^{(m-1)} + \gamma q(\tilde{\boldsymbol{x}}_t)\tilde{\varepsilon}_{t+1} - a_{t+1} \\
&= s_t^{(m-1)} - A_t + \gamma q(\boldsymbol{x}_t)\varepsilon_{t+1} - a_{t+1} \\
&= s_{t+1}^{(0)} - A_{t+1}, \\
\tilde{s}_{t+1}^{(i)} &= \tilde{s}_t^{(i-1)} - a_{t+1} \\
&= s_t^{(i-1)} - A_t - a_{t+1}, \qquad i = 1, \ldots, m-1 \\
&= s_{t+1}^{(i-1)} - A_{t+1}, \qquad i = 1, \ldots, m-1, \\
\tilde{y}_{t+1+h|t+1} &= \tilde{\ell}_{t+1} + \phi_h\tilde{b}_{t+1} + \tilde{s}_{t+1}^{(m-h_m^+)} \\
&= (\ell_{t+1} + A_{t+1}) + \phi_h b_{t+1} + \left[s_{t+1}^{(m-h_m^+)} - A_{t+1}\right] \\
&= \hat{y}_{t+1+h|t+1},
\end{aligned}$$

where $h_m^+ = \left[(h-1) \bmod m\right] + 1$.

To prove (8.11) and (8.12), we use the notation in Chap. 6 to write the normalized Class 1 and Class 2 models as

$$\tilde{y} = \boldsymbol{w}'\tilde{\boldsymbol{x}}_{t-1} + q(\tilde{\boldsymbol{x}}_{t-1})\varepsilon_t, \tag{8.27a}$$

$$\tilde{\boldsymbol{x}}_t = \boldsymbol{F}\tilde{\boldsymbol{x}}_{t-1} + \boldsymbol{g}q(\tilde{\boldsymbol{x}}_{t-1})\varepsilon_t, \tag{8.27b}$$

where $q(\tilde{\boldsymbol{x}}_{t-1}) = 1$ for Class 1 models and $q(\tilde{\boldsymbol{x}}_{t-1}) = \boldsymbol{w}'\tilde{\boldsymbol{x}}_{t-1} = \hat{y}_{t+h|t}$ for Class 2 models. In addition, $\tilde{\boldsymbol{g}} = \boldsymbol{g} + \boldsymbol{\gamma}$, where $\boldsymbol{\gamma} = [\gamma/m, 0, -\gamma/m, (-\gamma/m)\boldsymbol{1}'_{(m-1)}]'$. Recall from Example 6.2 (p. 96) that in the ETS(A,A$_\text{d}$,A) and ETS(M,A$_\text{d}$,A) models

$$\boldsymbol{w}'\boldsymbol{F}^i = [1, \phi_{i+1}, d_{i+1,m}, d_{i+2,m}, \ldots, d_{i+m,m}], \tag{8.28}$$

where $\phi_i = \phi + \phi^2 + \cdots + \phi^i$, and $d_{j,m} = 1$ if $j = 0 (\text{mod } m)$ and $d_{j,m} = 0$ otherwise. It follows that $\boldsymbol{w}'\boldsymbol{F}^i\boldsymbol{\gamma} = 0$.

Because we have shown that $\tilde{y}_{t+h|t} = \hat{y}_{t+h|t}$, the proofs in Chap. 6 can be applied to the two different cases in (8.27) to find $\tilde{v}_{t+h|h}$. The two variances have the forms in (8.11) and (8.12) with $\tilde{c}_j = \boldsymbol{w}'\boldsymbol{F}^{j-1}\tilde{\boldsymbol{g}}$. Because $\boldsymbol{w}'\boldsymbol{F}^i\boldsymbol{\gamma} = 0$,

$$\tilde{c}_j = \boldsymbol{w}'\boldsymbol{F}^{j-1}\tilde{\boldsymbol{g}} = \boldsymbol{w}'\boldsymbol{F}^{j-1}(\boldsymbol{g} + \boldsymbol{\gamma}) = \boldsymbol{w}'\boldsymbol{F}^{j-1}\boldsymbol{g} = c_j,$$

and $\tilde{v}_{t+h|t} = v_{t+h|t}$ for all of the Class 1 and Class 2 models in Tables 2.2 and 2.3.

The verification of (8.13) is now addressed. When we have observed values for $y_1, y_2, \ldots, y_t$, we can use these values and the models in (8.6) and (8.7) to find $\boldsymbol{x}_t$ and $\tilde{\boldsymbol{x}}_t$, respectively. Then starting with $\boldsymbol{x}_t$ and $\tilde{\boldsymbol{x}}_t$, we can use the same models to generate values for $y_{t+1}, y_{t+2}, \ldots, y_{t+h}$ and $\tilde{y}_{t+1}, \tilde{y}_{t+2}, \ldots, \tilde{y}_{t+h}$, respectively, by randomly selecting values $\varepsilon_{t+1}, \varepsilon_{t+2}, \ldots, \varepsilon_{t+h}$ from a probability distribution with mean 0 and standard deviation σ. If we treat the simulated values $y_{t+1}, y_{t+2}, \ldots, y_{t+h}$ as the observed values, we can extend the results in (8.8)–(8.10) to the simulated values for $\boldsymbol{x}_{t+1}, \boldsymbol{x}_{t+2}, \ldots, \boldsymbol{x}_{t+h}$ and $\tilde{\boldsymbol{x}}_{t+1}, \tilde{\boldsymbol{x}}_{t+2}, \ldots, \tilde{\boldsymbol{x}}_{t+h}$. It follows that the simulated prediction distributions using (8.6) and (8.7) are identical because, for the ith simulated value,

$$\begin{aligned}
\tilde{y}_{t+h}^{(i)} &= \tilde{\ell}_{t+h-1}^{(i)} + \phi\tilde{b}_{t+h-1}^{(i)} + \tilde{s}_{t+h-1}^{(m-1)(i)} + q(\tilde{\boldsymbol{x}}_{t+h-1}^{(i)})\varepsilon_{t+h} \\
&= (\ell_{t+h-1}^{(i)} + A_{t+h-1}) + \phi b_{t+h-1}^{(i)} + (s_{t+h-1}^{(m-1)(i)} - A_{t+h-1}) + q(\boldsymbol{x}_{t+h-1}^{(i)})\varepsilon_{t+h} \\
&= y_{t+h}^{(i)}.
\end{aligned}$$

9

Models with Regressor Variables

Up to this point in the book, we have considered models based upon a single series. However, in many applications, additional information may be available in the form of *input* or *regressor* variables; the name may be rather opaque, but we prefer it to the commonly-used but potentially misleading description of *independent* variables. We then refer to the series of interest as the *dependent* series. Regressor series may represent either *explanatory* or *intervention* variables.

An explanatory variable is one that provides the forecaster with additional information. For example, futures prices for petroleum products can foreshadow changes for consumers in prices at the pump. Despite the term "explanatory" we do not require a causal relationship between the input and dependent variables, but rather a series that is available in *timely* fashion to improve the forecasting process. Thus, stock prices or surveys of consumer sentiment are explanatory in this sense, even though they may not have causal underpinnings in their relationship with a dependent variable.

An intervention is often represented by an indicator variable taking values 0 and 1, although more general forms are possible. These variables may represent planned changes (e.g., the introduction of new legislation) or unusual events that are recognized only in retrospect (e.g., extreme weather conditions). Indicator variables may also be used to flag unusual observations or *outliers*; if such values are not identified they can distort the estimates of other parameters in the model.

In the next section we introduce the general linear innovations model and then examine a special case which provides insights into its structure. The model development parallels that of the multiple source of error model (see Harvey 1989, Chap. 7). We illustrate the use of these methods with two examples in Sect. 9.2; the first uses intervention variables to modify a univariate sales series and the second considers a leading indicator model for gasoline prices. We conclude the chapter with a discussion of diagnostic tests based upon the residuals.

9.1 The Linear Innovations Model with Regressors

We start from the standard linear innovations model introduced in Chap. 3:

$$y_t = \boldsymbol{w}'\boldsymbol{x}_{t-1} + \varepsilon_t, \tag{9.1a}$$

$$\boldsymbol{x}_t = \boldsymbol{F}\boldsymbol{x}_{t-1} + \boldsymbol{g}\varepsilon_t. \tag{9.1b}$$

The regressor variables are incorporated into the measurement equation (9.1a) and the model has the general form:

$$y_t = \boldsymbol{w}'\boldsymbol{x}_{t-1} + \boldsymbol{z}_t'\boldsymbol{p} + \varepsilon_t, \tag{9.2a}$$

$$\boldsymbol{x}_t = \boldsymbol{F}\boldsymbol{x}_{t-1} + \boldsymbol{g}\varepsilon_t. \tag{9.2b}$$

The vector $\boldsymbol{p}$, formed from the regression coefficients, consists of unknown quantities that need to be estimated. The vector $\boldsymbol{z}_t$ contains the regressor variables.

Although $\boldsymbol{p}$ is time invariant, it is convenient to provide it with a time subscript and rewrite the model (9.2) as

$$\begin{aligned} y_t &= \boldsymbol{w}'\boldsymbol{x}_{t-1} + \boldsymbol{z}_t'\boldsymbol{p}_{t-1} + \varepsilon_t, \\ \boldsymbol{x}_t &= \boldsymbol{F}\boldsymbol{x}_{t-1} + \boldsymbol{g}\varepsilon_t, \\ \boldsymbol{p}_t &= \boldsymbol{p}_{t-1}. \end{aligned}$$

These equations can be stacked to give

$$y_t = \bar{\boldsymbol{w}}_t'\bar{\boldsymbol{x}}_{t-1} + \varepsilon_t, \tag{9.3a}$$

$$\bar{\boldsymbol{x}}_t = \bar{\boldsymbol{F}}_t\bar{\boldsymbol{x}}_{t-1} + \bar{\boldsymbol{g}}\varepsilon_t, \tag{9.3b}$$

where $\quad \bar{\boldsymbol{x}}_t = \begin{bmatrix} \boldsymbol{x}_t \\ \boldsymbol{p}_t \end{bmatrix}, \quad \bar{\boldsymbol{w}}_t = \begin{bmatrix} \boldsymbol{w}_t \\ \boldsymbol{z}_t \end{bmatrix}, \quad \bar{\boldsymbol{F}}_t = \begin{bmatrix} \boldsymbol{F} & \boldsymbol{0} \\ \boldsymbol{0} & \boldsymbol{I} \end{bmatrix}$ and $\bar{\boldsymbol{g}} = \begin{bmatrix} \boldsymbol{g} \\ 0 \end{bmatrix}.$

Equations (9.3) have the form of a general time invariant innovations state space model.

As an example, consider a local level model where a single intervention occurs at time $t = 10$ that has a transient effect on the series (a spike) of an unknown amount p_1. The measurement equation becomes $y_t = \ell_{t-1} + z_t p_1 + \varepsilon_t$ and the transition equation is simply $\ell_t = \ell_{t-1} + \alpha\varepsilon_t$, where z_t is an indicator variable that is 1 in period 10 and 0 in all other periods. Similarly, if the effect is permanent (a step), we define the regressor variable as $z_t = 1$ if $t \geq 10$ and $z_t = 0$ otherwise.

In either case, the model may be written in the form (9.3) as

$$\begin{aligned} y_t &= \begin{bmatrix} 1 & z_t \end{bmatrix} \begin{bmatrix} \ell_{t-1} \\ p_1 \end{bmatrix} + \varepsilon_t, \\ \begin{bmatrix} \ell_t \\ p_1 \end{bmatrix} &= \begin{bmatrix} 1 & 0 \\ 0 & 1 \end{bmatrix} \begin{bmatrix} \ell_{t-1} \\ p_1 \end{bmatrix} + \begin{bmatrix} \alpha \\ 0 \end{bmatrix} \varepsilon_t. \end{aligned}$$

We make the usual assumption that the error terms are independent and follow Gaussian distributions with zero means and constant variance; that is, $\varepsilon \sim \mathrm{N}(0,\sigma^2)$. The method of estimation in Chap. 5 may be readily adapted to fit a model with the general form (9.3). It treats the seed state as a fixed vector and combines it with the model's parameters for optimization of the likelihood or sum of squared errors functions. Because the regression coefficients form part of the seed state vector, estimates of them are obtained from the optimized value of the seed state vector.

Predictions can be undertaken with a suitable adaptation of the method in Chap. 6. At this stage, when developing prediction distributions we assume that the errors are homoscedastic. It is now necessary to supplement this method with future values of the regressors. If the regressors consist of leading indicator variables, such values will be known up to a certain future point of time. Moreover, if they consist of indicator variables reflecting the effect of known future interventions that have also occurred in the past, then such values are also known. However, when they are unknown, predictions of the future values of the regressors are needed. It is then best to revert to a multivariate time series framework (see Chap. 17).

This approach is easily adapted to accommodate heteroscedastic innovations of the type considered in Chap. 4. A grand model is obtained that has the general form

$$y_t = \bar{\boldsymbol{w}}_t'\bar{\boldsymbol{x}}_{t-1} + r(\bar{\boldsymbol{x}}_{t-1})\varepsilon_t, \tag{9.4a}$$

$$\bar{\boldsymbol{x}}_t = \bar{\boldsymbol{F}}_t\bar{\boldsymbol{x}}_{t-1} + \bar{\boldsymbol{g}}(\bar{\boldsymbol{x}}_{t-1})\varepsilon_t. \tag{9.4b}$$

The model may be fitted using the method from Chap. 5. Forecasts and prediction distributions may then be obtained by methods for heteroscedastic data described in Chap. 6.

9.2 Some Examples

In this section, we assume that a homoscedastic model is adequate and that the errors are independent and follow a Gaussian distribution; that is, $\varepsilon \sim \mathrm{N}(0,\sigma^2)$.

9.2.1 An Example Using Indicator Variables

We now examine a simple example to illustrate the methods developed so far. We consider a series that gives the sales of a product for 62 weeks starting in early 2003. We refer to the series, which was supplied by a company, as "FM Sales." The series is plotted in Fig. 9.1.

We will incorporate three indicator variables as regressors:

- $z_1 = 1$ in weeks 1–12 when product advertising was in a low-profile mode, and $z_1 = 0$ otherwise

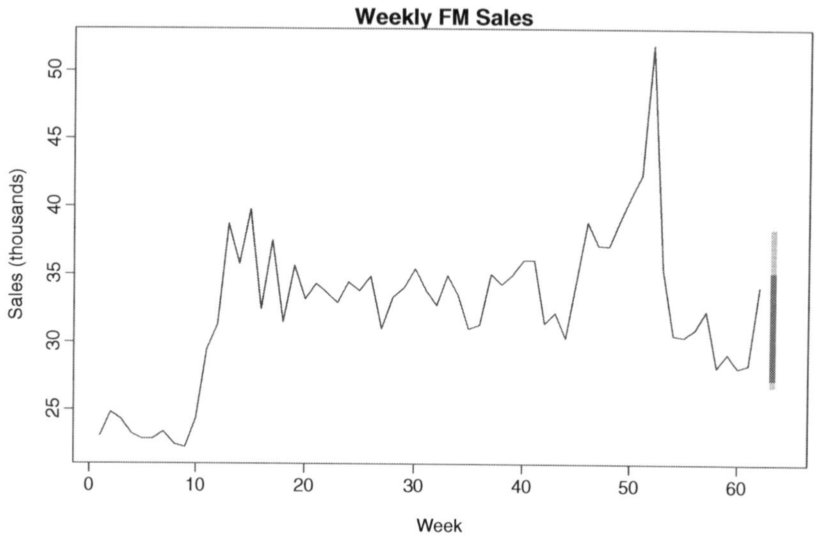

Fig. 9.1. The FM Sales series (thousands of units sold per week). The *shaded regions* show 90% prediction intervals: the *light-shaded region* is from the local level model without regressors; the *dark-shaded region* is from the local level model with regressors.

Table 9.1. Analysis of the FM sales data using the local level model with regressors.

Model	p_1	p_2	p_3	α	ℓ_0	R^2
Regression only	−8.65	5.88	18.73	–	–	0.793
Local level (LL)	–	–	–	0.731	23.47	0.575
LL with regressors	−4.80	4.65	17.29	0.471	28.47	0.808

- $z_2 = 1$ in weeks 13–15 and 48–51 denoting high sales periods before Easter and Christmas, and $z_2 = 0$ otherwise
- $z_3 = 1$ in week 52 to denote the after-Christmas sales peak, and $z_3 = 0$ otherwise

The results are given in Table 9.1; the maximum likelihood estimates were obtained by direct maximization. The contributions of the regressor terms tend to dominate in this case, but the persistence in the series is clearly seen with the value of $\alpha = 0.471$.

The local level with regressors model yields a point forecast for the next period of 31.26. The estimated standard deviation is $\hat{\sigma} = 2.39$, so that a 90% one-step-ahead prediction interval would be $[27.33, 35.19]$. By contrast, the local level model without regressors gives a point forecast of 32.59, $\hat{\sigma} = 3.54$ and the prediction interval $[26.75, 38.43]$.

9.2.2 Use of a Leading Indicator

Time series regression based upon two or more variables requires some care in model development. For example, a researcher may use the transfer function approach of Box et al. (1994, Chaps. 10 and 11). The detailed discussion of such procedures is beyond the scope of this book, because the methodology is similar whether an ARIMA or a state space approach is employed. Accordingly, we use a single explanatory variable to illustrate ideas. Consider two series relating to US gasoline prices:

Y = US retail gas prices (the average price per gallon, in dollars)
X = The spot price of a barrel of West Texas Intermediate (WTI) oil in dollars as traded at Cushing, Oklahoma

The Cushing spot price is widely used in the industry as a "marker" for pricing a number of other crude oil supplies traded in the domestic spot market at Cushing, Oklahoma. The data are monthly and cover the period January 1991 to November 2006.

The two series are plotted in Fig. 9.2 and show both marked nonstationarity and considerably increased variability in the later years. At this point, we will examine the series through the end of 2001 to reduce the effects of

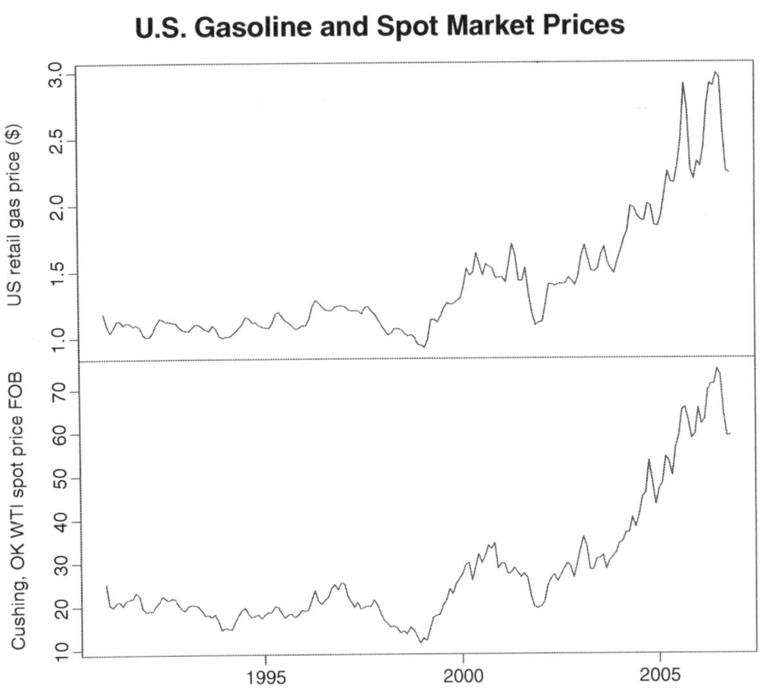

Fig. 9.2. Monthly US gasoline and spot market prices: January 1991–November 2006.

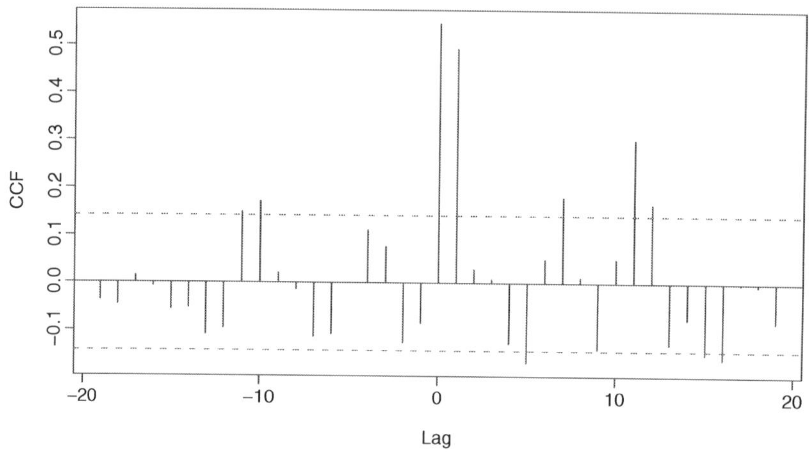

Fig. 9.3. Cross correlation function for gas prices and spot prices (after taking logarithms and differencing).

the increased volatility. Also, we convert to natural logarithms to reduce the variability due to increased prices. We return to the volatility question in Chap. 19. As the spot price should be a long-term leading indicator of the retail price, this relationship could usefully be examined using cointegration methods (see Tsay 2005, pp. 376–390). Given that our present focus is on shorter-term forecasting, we will eschew that approach in favor of a simpler analysis for expository purposes.

Analysis of the logarithms of spot price (Lspot) reveals that the time series may be represented as a random walk ($\hat{\alpha} = 1.17$ for the LL model), so that modeling in terms of first differences is a reasonable place to start; this step is consistent with the pre-whitening approach of Box et al. (1994). The cross correlation function (CCF) for the differenced series of logarithms (DLprice and DLspot) is shown in Fig. 9.3.

The CCF shows that the spot price has strong contemporaneous variation with the retail price, but also that it leads retail prices by one month. Weekly data would show a more refined pattern of leading behavior given the time taken for the price of crude oil to affect pump prices. Because we are interested in forecasting, we restrict our attention to models that use the spot price with a one-month lag, Lspot(1). Models with non-zero slopes did not appear to improve matters, and we also found that a fixed seasonal pattern sufficed.

Accordingly, we consider the following models:

- Local level — ETS(A,N,N)
- Local level with seasonals — ETS(A,N,A)
- Regression on Lspot(1)
- Regression on Lspot(1) and seasonal dummies

Table 9.2. Analysis of the US gas prices data (logarithmic transform).

Model	α	*b(spot)*	R^2	s	AICR
Local level (LL)	1.71	–	0.924	0.0359	−153.3
LL + seasonals	1.63	–	0.938	0.0339	−156.9
Regression on Lspot(1)	–	0.531	0.756	0.0642	−1.2
Regression on Lspot(1) + seasonals	–	0.503	0.795	0.0617	0.0
LL + Lspot(1)	1.52	0.142	0.930	0.0346	−161.8
LL + seasonals + Lspot(1)	1.45	0.151	0.944	0.0325	−167.6
LL + seasonals + Lspot(1) [MSOE]	1.00	0.259	0.937	0.0345	−152.0

b(spot) denotes the slope of Lspot(1)

- Local level with regression on Lspot(1)
- Local level with seasonals and regression on Lspot(1)
- Local level with seasonals and regression on Lspot(1), and $\alpha \leq 1$

The final model is included as it corresponds to the restriction imposed on α in the multiple source of error model.

The results are summarized in Table 9.2. The AIC values are reported relative to the poorest fitting model, and are labeled as AICR. Several features are apparent. First, the regression models perform poorly when the series dynamics are ignored. Second, when the local level effects are included, the coefficient of the spot price variable is much smaller. Third, the seasonal effects are modest but important; the values show a consistent rise in the warmer months when demand is higher. Finally, the estimates of α consistently exceed 1; to understand why this might be, we can rewrite the transition equation as

$$\ell_t = y_t + (\alpha - 1)\varepsilon_t.$$

From this expression, we see that $\alpha > 1$ means that the price is expected to continue to move in the same direction as it did in the previous period. It is worth noting that the MSOE scheme is unable to capture such behavior.

The performance of the models could be improved by identifying outliers and making appropriate adjustments. In order to make such adjustments we now need to identify suitable diagnostic procedures.

9.3 Diagnostics for Regression Models

Hitherto, our approach to model selection has highlighted the use of information criteria, as described in Chap. 7. However, model building with regressors often involves the evaluation of many variables and multiple lags. In principle, all possible models could be evaluated by information criteria, but this approach may be computationally intensive. Also, many researchers prefer a more hands-on developmental process in which a variety of diagnostics may be used. The diagnostics described in the rest of this section are not new, nor do they form an exhaustive list, but they cover the main questions that a researcher may wish to address. For more detailed coverage in a time series context, see Harvey (1989, Chap. 5); much of the discussion below derives from this source. For more general discussions on regression diagnostics, the reader may consult Belsley et al. (1980) particularly on multicollinearity questions, and Cook and Weisberg (1999) on graphical methods, influence measures and outliers.

9.3.1 Goodness-of-Fit Measures

The regression residuals are defined by substituting the parameter estimates into (9.3a); we denote these sample quantities by e_t, the estimates of ε_t. After the model has been fitted, the associated degrees of freedom are $n-q-1$, where q denotes the number of fitted parameters. That is, $q = n_p + n_g + d$, where n_p denotes the number of regression coefficients, n_g the number of parameters in the transition equations, and d the number of free states in the initial state vector $\boldsymbol{x}_0$.

We consider three components that provide information about the goodness of fit. The baseline is the original (or total) sum of squared errors for the dependent variable:

$$SST = \sum_{t=1}^{n} (y_t - \bar{y})^2 .$$

We may then compute the sum of squared errors based upon fitting the innovations model alone:

$$SSE(I) = \sum_{t=1}^{n} e_t^2 .$$

The coefficient of determination is then defined in the usual way as

$$R_I^2 = 1 - SSE(I)/SST. \tag{9.5}$$

We can then incorporate regression elements into the model and generate the sum of squared errors for the complete model [SSE(C)]. Thus, the complete model has a coefficient of determination

$$R_C^2 = 1 - SSE(C)/SST.$$

The quantity $R_C^2 - R_I^2$ represents the improvement due to the inclusion of the regression terms. The efficacy of the regression component may be tested using the ratio:

$$F = \frac{SSE(I) - SSE(C)}{SSE(C)} .$$

Under the usual null hypothesis this measure has an F distribution with $(n_p, n-q-1)$ degrees of freedom.

When the local level model is used, Harvey (1989, p. 268) suggests replacing the denominator of (9.5) by the sum of squares for the first differences

$$SSTD = \sum_{t=1}^{n}(y_t - y_{t-1} - \bar{y}_D)^2,$$

where $\bar{y}_D = (y_n - y_1)/(n-1)$ denotes the mean of the first differences. We could then use the modified coefficient:

$$R_D^2 = 1 - SSE(I)/SSTD.$$

Intuitively, R_D^2 measures the improvement of the model over a simple random walk with drift and may be used to test the value of the innovations model beyond differencing. However, because competing models may imply different orders of differencing in the reduced forms, we prefer to use the measures defined earlier.

9.3.2 Standard Errors

Approximate standard errors for individual coefficients may be determined from the information matrix; details are given in Harvey (1989, pp. 140–143).

9.3.3 Checks of the Residuals

Many diagnostics have been developed that search for structure among the residuals in the series. Again, we mention only some of the standard tests; those seeking a more detailed treatment should consult Harvey (1989, Chap. 5).

Residual Autocorrelation and Cross-Correlations

Plots of the autocorrelation function (ACF) of the residuals and of the cross-correlations (CCF) of the residuals with suitably differenced input series (currently in or outside the model) will help to identify omitted elements. Again, we should note that relying only upon graphical procedures can be both time-consuming and lacking in clarity, so some form of testing or evaluation is preferable. The ACF may be tested using the Box–Ljung–Pierce [BLP] statistic (Ljung and Box 1978). We may define the autocorrelation of order j as

$$r_j = \frac{\sum_{t=j+1}^{n}(e_t - \bar{e})(e_{t-j} - \bar{e})}{\sum_{t=1}^{n}(e_t - \bar{e})^2}.$$

The BLP statistic is then defined as

$$Q(P) = n(n+2)\sum_{j=1}^{P}(n-j)^{-1}r_j^2.$$

Under the null hypothesis of no autocorrelation, Q is asymptotically distributed as chi-squared with $P - n_g$ degrees of freedom, where n_g denotes the number of parameters that are estimated for the transition equations. The choice of P is somewhat arbitrary, and many researchers use several different values in order to gain insight. It is worth noting that, under H_0, $Q(P_2) - Q(P_1)$ is asymptotically chi-squared with $P_2 - P_1$ degrees of freedom, which allows the researcher to check different horizons.

The BLP statistics for the last two models in Table 9.2 based upon the first 12 lags are 47.9 and 46.2 respectively, with 11 degrees of freedom. Both sets of residuals show a significant negative autocorrelation at lag 2.

Tests of Gaussianity

A Gaussian probability plot will often suggest the need for a transformation and help to identify major outliers. A simple test of Gaussianity proposed by Bowman and Shenton (1975) has been popularized in the econometric literature by Jarque and Bera (1980). The test uses the third and fourth moments of the residuals to measure skewness and kurtosis respectively:

$$\sqrt{b_1} = \frac{1}{ns^3} \sum_{t=1}^{n} (e_t - \bar{e})^3,$$

$$b_2 = \frac{1}{ns^4} \sum_{t=1}^{n} (e_t - \bar{e})^4,$$

where $s = \sqrt{\text{MSE}}$. When the error process is Gaussian, these statistics are asymptotically independent. Further, both statistics have sampling distributions that are asymptotically Gaussian, although the approach to the limiting forms is very slow. Thus, a test may be based upon the statistic

$$J = (n/6)b_1 + (n/24)(b_2 - 3)^2,$$

which is asymptotically distributed as chi-squared with 2 degrees of freedom. An improved test of this general form is contained in an unpublished paper by Doornik and Hansen (1994).

Several other tests, such as those of Anderson and Darling (1952) and Lilliefors (1967), are based upon deviations from expectation in the probability plot. These tests have the advantage that the outliers are easily identified. In the present example, a few outliers at the end of the series result in the rejection of the null hypothesis. An extended model that allows for these outliers has only a minor effect upon the estimates, although the forecasts would be more affected as some of the outliers occur at the end of the series. Rather than extend the discussion, we defer further consideration to Chap. 19, where we examine heteroscedastic disturbances.

Heteroscedasticity

A common form of heteroscedasticity arises when the variance at time t is a function of the mean level of the series at that time. Such structures are a major motivation for the multiplicative models introduced in Chap. 4 and discussed at various points throughout this book. If model-building starts from the linear innovations state space form, it is quite likely that the residuals will not indicate a uniform variance over time. Because many series display a positive trend, the variability at the end of the series will often be greater than that at the outset. Based upon this intuition, Harvey (1989, pp. 259–260) suggests dividing the series into three nearly equal parts of length $I = \lfloor (n+1)/3 \rfloor$. The test statistic is defined as

$$H(I) = \sum_{t=n-I+1}^{n} e_t^2 \Big/ \sum_{t=1}^{I} e_t^2.$$

When the error process is Gaussian and the null hypothesis of homoscedasticity holds, the sampling distribution of $H(I)$ is approximately $F(I, I)$.

The *LL + seasonals + Lspot(1)* model has $H = 6.3$ with $I = 44$, which is highly significant and indicates the need to account for heteroscedasticity.

Because the purpose of this section was to illustrate ideas, rather than to discuss model-building in detail (outlier identification, additional variables, etc.), we do not pursue matters further at this stage. However, it is evident that the most critical concern is the increased volatility in the series, and we return to that question in Chap. 19.

9.4 Exercises

Exercise 9.1. Extend the model given in (9.2) to include regressor variables in the transition equations, where the vectors of coefficients $\boldsymbol{p}_1$ and $\boldsymbol{p}_2$ will typically include some zero elements:

$$\begin{aligned} y_t &= \boldsymbol{w}'\boldsymbol{x}_{t-1} + \boldsymbol{z}_t'\boldsymbol{p}_1 + \varepsilon_t, \\ \boldsymbol{x}_t &= \boldsymbol{F}\boldsymbol{x}_{t-1} + \boldsymbol{z}_t'\boldsymbol{p}_2 + \boldsymbol{g}\varepsilon_t. \end{aligned}$$

Express the model in reduced form by eliminating the state variables. Hence show that the regressors in the transition equation only affect the dependent variable after a one-period delay.

Exercise 9.2. Consider the special case of the model defined in Exercise 9.1 corresponding to the ETS(A,N,N) process with a single regressor variable:

$$\begin{aligned} y_t &= x_{t-1} + z_t p_1 + \varepsilon_t, \\ x_t &= x_{t-1} + z_t p_2 + \alpha\varepsilon_t. \end{aligned}$$

Show that the reduced form is:

$$y_t - y_{t-1} = p_1(z_t - z_{t-1}) + p_2 z_{t-1} + \varepsilon_t - (1-\alpha)\varepsilon_{t-1}.$$

Further show that the same reduced form could be obtained by including both z_{t+1} and z_t in the transition equation, but omitting them from the measurement equation.

Exercise 9.3. The residual checks in Sect. 9.3 suggest the need for an autoregressive term at lag 2 in the oil price model. Develop such a model and compare its performance with the results given in Table 9.2.

Exercise 9.4. The events of 11 September 2001 produced a substantial short-term drop in the number of air passengers. Use the monthly series on the number of enplanements on domestic aircraft (data set `enplanements`) to develop an intervention model to describe the series. Use two indicator variables to model the changes: SEPT1 $= 1$ in September 2001 and $= 0$ otherwise; and SEPT2 $= 1$ in and after October 2001 and $= 0$ otherwise. Hence estimate the overall effects upon air travel. [For further discussion, see Ord and Young (2004).]

Exercise 9.5. The US Conference Board carries out a monthly survey on consumer confidence. Although the use of this measure as a true explanation of economic change is debatable, its primary benefit is that it appears before many macroeconomic indices are released. The data set `unemp.cci` contains 100 monthly observations on the consumer confidence index (CCI) and seasonally adjusted civilian unemployment (UNEMP) in the US, covering the period June 1997–September 2005:

a. Develop univariate models for each series and establish that each series is close to a random walk.
b. Develop a state space regression model that uses CCI (lagged one month) and the SEPT2 indicator defined in Exercise 9.4 to predict UNEMP.

10

Some Properties of Linear Models

In this chapter, we discuss some of the mathematical properties of the linear innovations state space models described in Chap. 3. These results are based on Hyndman et al. (2008).

We provide conditions that ensure the model is of minimal dimension (Sect. 10.1) and conditions that guarantee the model is stable (Sect. 10.2). We will see that the non-seasonal models are already of minimal dimension, but that the seasonal models are slightly larger than necessary. The normalized seasonal models, introduced in Chap. 8, are of minimal dimension.

The stability conditions discussed in Sect. 10.2 can be used to derive the associated parameter space. We find that the usual parameter restrictions (requiring all smoothing parameters to lie between 0 and 1) do not always lead to stable models. Exact parameter restrictions are derived for all the linear models.

10.1 Minimal Dimensionality for Linear Models

The linear innovations state space models (defined in Chap. 3) are of the form

$$y_t = \boldsymbol{w}'\boldsymbol{x}_{t-1} + \varepsilon_t, \tag{10.1a}$$

$$\boldsymbol{x}_t = \boldsymbol{F}\boldsymbol{x}_{t-1} + \boldsymbol{g}\varepsilon_t. \tag{10.1b}$$

The model is not unique; for example, an equivalent model can be obtained simply by adding an extra row to the state vector and adding a row containing only zeros to each of $\boldsymbol{w}$, $\boldsymbol{F}$ and $\boldsymbol{g}$. Therefore it is of interest to know when the model has the shortest possible state vector $\boldsymbol{x}_t$, in which case we say it has "minimal dimension."

In particular, we wish to know whether the specific cases of the model given in Table 2.2 on page 21 are of minimal dimension. The coefficient matrices $\boldsymbol{F}$, $\boldsymbol{g}$ and $\boldsymbol{w}$ can easily be determined from Table 2.2, and are given below.

Here $\boldsymbol{I}_k$ denotes the $k \times k$ identity matrix and $\boldsymbol{0}_k$ denotes a zero vector of length k.

$$\text{ETS(A,N,N):} \quad \boldsymbol{w} = 1 \qquad \boldsymbol{F} = 1 \qquad \boldsymbol{g} = \alpha$$

$$\text{ETS(A,A}_\text{d}\text{,N):} \quad \boldsymbol{w} = \begin{bmatrix} 1 \\ 1 \end{bmatrix} \qquad \boldsymbol{F} = \begin{bmatrix} 1 & 1 \\ 0 & \phi \end{bmatrix} \qquad \boldsymbol{g} = \begin{bmatrix} \alpha \\ \beta \end{bmatrix}$$

$$\text{ETS(A,N,A):} \quad \boldsymbol{w} = \begin{bmatrix} 1 \\ \boldsymbol{0}_{m-1} \\ 1 \end{bmatrix} \quad \boldsymbol{F} = \begin{bmatrix} 1 & \boldsymbol{0}'_{m-1} & 0 \\ 0 & \boldsymbol{0}'_{m-1} & 1 \\ \boldsymbol{0}_{m-1} & \boldsymbol{I}_{m-1} & \boldsymbol{0}_{m-1} \end{bmatrix} \qquad \boldsymbol{g} = \begin{bmatrix} \alpha \\ \gamma \\ \boldsymbol{0}_{m-1} \end{bmatrix}$$

$$\text{ETS(A,A}_\text{d}\text{,A):} \quad \boldsymbol{w} = \begin{bmatrix} 1 \\ 1 \\ \boldsymbol{0}_{m-1} \\ 1 \end{bmatrix} \quad \boldsymbol{F} = \begin{bmatrix} 1 & 1 & \boldsymbol{0}'_{m-1} & 0 \\ 0 & \phi & \boldsymbol{0}'_{m-1} & 0 \\ 0 & 0 & \boldsymbol{0}'_{m-1} & 1 \\ \boldsymbol{0}_{m-1} & \boldsymbol{0}_{m-1} & \boldsymbol{I}_{m-1} & \boldsymbol{0}_{m-1} \end{bmatrix} \quad \boldsymbol{g} = \begin{bmatrix} \alpha \\ \beta \\ \gamma \\ \boldsymbol{0}_{m-1} \end{bmatrix}$$

The matrices for ETS(A,A,N) and ETS(A,A,A) are the same as for ETS(A,A$_\text{d}$,N) and ETS(A,A$_\text{d}$,A) respectively, but with $\phi = 1$.

The following definitions are given by Hannan and Deistler (1988, pp. 44–45):

Definition 10.1. *The model (10.1) is said to be* observable *if Rank*$(\mathcal{O}) = p$ *where*

$$\mathcal{O} = [\boldsymbol{w}, \boldsymbol{F}'\boldsymbol{w}, (\boldsymbol{F}')^2\boldsymbol{w}, \ldots, (\boldsymbol{F}')^{p-1}\boldsymbol{w}]$$

and p is the length of the state vector $\boldsymbol{x}_t$.

Definition 10.2. *The model (10.1) is said to be* reachable *if Rank*$(\mathcal{R}) = p$ *where*

$$\mathcal{R} = [\boldsymbol{g}, \boldsymbol{F}\boldsymbol{g}, \boldsymbol{F}^2\boldsymbol{g}, \ldots, \boldsymbol{F}^{p-1}\boldsymbol{g}]$$

and p is the length of the state vector $\boldsymbol{x}_t$.

Reachability and observability are desirable properties of a state space model because of the following result from Hannan and Deistler (1988, p. 48):

Theorem 10.1. *The state space model* (10.1) *is of minimal dimension if and only if it is observable and reachable.*

Example 10.1: ETS(A,A,N)

The observability matrix is

$$\mathcal{O} = [\boldsymbol{w}, \boldsymbol{F}'\boldsymbol{w}] = \begin{bmatrix} 1 & 1 \\ 1 & 2 \end{bmatrix},$$

which has rank 2. The reachability matrix is

$$\mathcal{R} = [\boldsymbol{g}, \boldsymbol{F}\boldsymbol{g}] = \begin{bmatrix} \alpha & \alpha+\beta \\ \beta & \beta \end{bmatrix},$$

which has rank 2 unless $\beta = 0$. Consequently, the model is of minimal dimension provided $\beta \neq 0$.

A similar argument can be used (see Exercise 10.1a) to show that the non-seasonal models ETS(A,N,N) and ETS(A,A$_d$,N) are both reachable and observable, and therefore of minimal dimension.

10.1.1 Seasonal Models

Consider the ETS(A,N,A) model, for which the rank of $\mathcal{O} < p$ and the rank of $\mathcal{R} < p$. This is because, for the ETS(A,N,A) model, $(\boldsymbol{F}')^{p-1} = \boldsymbol{F}^{p-1} = \boldsymbol{I}_p$. Therefore, model ETS(A,N,A) is neither reachable nor observable. A similar argument (Exercise 10.1b) shows that models ETS(A,A,A) and ETS(A,A$_d$,A) are also neither reachable nor observable.

These problems arise because of a redundancy in the model. For example, the ETS(A,N,A) model is given by $y_t = \ell_{t-1} + s_{t-m} + \varepsilon_t$, where the level and seasonal components are given by

$$\ell_t = \ell_{t-1} + \alpha\varepsilon_t \qquad \text{and} \qquad s_t = s_{t-m} + \gamma\varepsilon_t.$$

So both the level and seasonal components have long run features due to unit roots. In other words, both can model the level of the series, and the seasonal component is not constrained to lie anywhere near zero. This is the same problem that led to the use of normalizing in Chap. 8.

Let L denote the lag operator defined by $Ly_t = y_{t-1}$. Then, by expanding $s_t = e_t/(1 - L^m)$, where $e_t = \gamma\varepsilon_t$, it can be seen that s_t can be decomposed into two processes, a level displaying a unit root at the zero frequency and a purely seasonal process, having unit roots at the seasonal frequency:

$$\begin{aligned} s_t &= \ell_t^* + s_t^*, \\ \text{where} \qquad \ell_t^* &= \ell_{t-1}^* + \tfrac{1}{m} e_t, \\ S(L)s_t^* &= \theta(L)e_t, \end{aligned}$$

$S(L) = 1 + L + \cdots + L^{m-1}$ represents the seasonal summation operator and $\theta(L) = m^{-1}\left[(m-1) + (m-2)L + \cdots + 2L^{m-3} + L^{m-2}\right]$. The long run component ℓ_t^* should be part of the level term.

This leads to an alternative model specification where the seasonal equation for models ETS(A,N,A), ETS(A,A,A) and ETS(A,A$_d$,A) is replaced by

$$S(L)s_t = \theta(L)\gamma\varepsilon_t. \tag{10.2}$$

The other equations remain the same, as the additional level term can be absorbed into the original level equation by a simple change of parameters. Noting that $\theta(L)/S(L) = [1 - \frac{1}{m}S(L)]/(1 - L^m)$, we see that (10.2) can be written as

$$s_t = s_{t-m} + \gamma\varepsilon_t - \frac{\gamma}{m}(\varepsilon_t + \varepsilon_{t-1} + \cdots + \varepsilon_{t-m+1}).$$

In other words, the seasonal term is calculated as in the original models, but is then adjusted by subtracting the average of the last m shocks. The effect of this adjustment is equivalent to the normalization procedure outlined in Chap. 8, in which the seasonal terms $s_t, \ldots, s_{t-m+1}$ are adjusted every time period to ensure that they sum to zero. Models using the seasonal component (10.2) will be referred to as "normalized" versions of ETS(A,N,A), ETS(A,A,A) and ETS(A,A$_d$,A). It can be shown (Exercise 10.1c) that the normalized models are of minimal dimension.

10.2 Stability and the Parameter Space

In Chap. 3, we found (p. 36) that, for linear models of the form (10.1), we could write the state vector as

$$\boldsymbol{x}_t = \boldsymbol{D}^t\boldsymbol{x}_0 + \sum_{j=0}^{t-1} \boldsymbol{D}^j\boldsymbol{g}y_{t-j},$$

where $\boldsymbol{D} = \boldsymbol{F} - \boldsymbol{g}\boldsymbol{w}'$ is the discount matrix. So for initial conditions to have a negligible effect on future states, we need $\boldsymbol{D}^t$ to converge to zero. Therefore, we require $\boldsymbol{D}$ to have all eigenvalues inside the unit circle. We call this condition *stability* (following Hannan and Deistler 1988, p. 48).

Definition 10.3. *The model (10.1) is said to be* stable *if all eigenvalues of* $\boldsymbol{D} = \boldsymbol{F} - \boldsymbol{g}\boldsymbol{w}'$ *lie inside the unit circle.*

Stability is a desirable property of a time series model because we want models where the distant past has a negligible effect on the present state.

In Chap. 3, we also found that

$$\hat{y}_{t|t-1} = \boldsymbol{w}'\boldsymbol{x}_t = a_t + \sum_{j=1}^{t-1} c_j y_{t-j},$$

where $a_t = \boldsymbol{w}'\boldsymbol{D}^{t-1}\boldsymbol{x}_0$ and $c_j = \boldsymbol{w}'\boldsymbol{D}^{j-1}\boldsymbol{g}$. Thus, the forecast is a linear function of the past observations and the seed state vector. This result shows that for a model to be stable, we require the weaker *forecastability* condition:

Definition 10.4. *The model* (10.1) *is* forecastable *if*

$$\sum_{j=1}^{\infty} |c_j| < \infty \qquad \text{and} \qquad \lim_{t\to\infty} a_t = a. \tag{10.3}$$

Obviously, any model that is stable is also forecastable.

On the other hand, it is possible for a model to have a unit eigenvalue for $\boldsymbol{D}$, but to satisfy the forecastability condition (10.3). In other words, an unstable model can still produce stable forecasts provided the eigenvalues which cause the instability have no effect on the forecasts. This arises because $\boldsymbol{D}$ may have unit eigenvalues where $\boldsymbol{w}'$ is orthogonal to the eigenvectors corresponding to the unit eigenvalues.

To avoid complications, we will assume that all the eigenvalues are distinct. In this case, we can write the eigendecomposition of $\boldsymbol{D}$ as $\boldsymbol{D} = \boldsymbol{U}\boldsymbol{\Lambda}\boldsymbol{V}$, where the columns of $\boldsymbol{U}$ are the eigenvectors of $\boldsymbol{D}$, $\boldsymbol{\Lambda}$ is a diagonal matrix containing the eigenvalues of $\boldsymbol{D}$, and $\boldsymbol{V} = \boldsymbol{U}^{-1}$. Then

$$c_{j+1} = \boldsymbol{w}'\boldsymbol{D}^j\boldsymbol{g} = \boldsymbol{w}'\boldsymbol{U}\boldsymbol{\Lambda}^j\boldsymbol{V}\boldsymbol{g} = \sum_i \lambda_i^j(\boldsymbol{w}'\boldsymbol{u}_i)(\boldsymbol{v}_i'\boldsymbol{g}),$$

where $\boldsymbol{u}_i$ is a column of $\boldsymbol{U}$ (a right eigenvector) and $\boldsymbol{v}_i$ is a row of $\boldsymbol{V}$ (a left eigenvector). By inspection, we see that the sequence converges to zero provided either $|\lambda_i| < 1$, $\boldsymbol{w}'\boldsymbol{u}_i = 0$ or $\boldsymbol{v}_i'\boldsymbol{g} = 0$, for each i. Further, the sequence only converges under these conditions. Similarly,

$$a_{t+1} = \boldsymbol{w}'\boldsymbol{D}^t\boldsymbol{x}_0 = \sum_i \lambda_i^t(\boldsymbol{w}'\boldsymbol{u}_i)(\boldsymbol{v}_i'\boldsymbol{x}_0).$$

In this case, the sequence converges to a constant if and only if either $|\lambda_i| \leq 1$, $\boldsymbol{w}'\boldsymbol{u}_i = 0$ or $\boldsymbol{v}_i\boldsymbol{x}_0 = 0$, for each i. Thus, we can restate forecastability as follows.

Theorem 10.2. *Let λ_i denote an eigenvalue of $\boldsymbol{D} = \boldsymbol{F} - \boldsymbol{g}\boldsymbol{w}'$, and let $\boldsymbol{u}_i$ be the corresponding right eigenvector and $\boldsymbol{v}_i$ the corresponding left eigenvector. Then the model (10.1) is forecastable if and only if, for each i, at least one of the following four conditions is met:*

1. $|\lambda_i| < 1$
2. $\boldsymbol{w}'\boldsymbol{u}_i = 0$
3. $|\lambda_i| = 1$ *and* $\boldsymbol{v}_i'\boldsymbol{g} = 0$
4. $\boldsymbol{v}_i'\boldsymbol{x}_0 = 0$ *and* $\boldsymbol{v}_i'\boldsymbol{g} = 0$

The concept of forecastability was noted by Sweet (1985) and Lawton (1998) for ETS(A,A,A) (additive Holt-Winters) forecasts, although neither author used a stochastic model as we do here. The phenomenon was also observed by Snyder and Forbes (2003) in connection with the ETS(A,A,A) model. The first general definition of this property was given by Hyndman et al. (2008).

We now establish stability and forecastability conditions for each of the linear models. For the damped models, we assume that ϕ is a fixed damping parameter between 0 and 1, and we consider the values of the other parameters that would lead to a stable model.

The value of D for each model is given below.

$$\text{ETS(A,N,N):} \quad D = 1 - \alpha$$

$$\text{ETS(A,A}_\text{d}\text{,N):} \quad D = \begin{bmatrix} 1-\alpha & 1-\alpha \\ -\beta & \phi-\beta \end{bmatrix}$$

$$\text{ETS(A,N,A):} \quad D = \begin{bmatrix} 1-\alpha & \mathbf{0}'_{m-1} & -\alpha \\ -\gamma & \mathbf{0}'_{m-1} & 1-\gamma \\ \mathbf{0}_{m-1} & \boldsymbol{I}_{m-1} & \mathbf{0}_{m-1} \end{bmatrix}$$

$$\text{ETS(A,A}_\text{d}\text{,A):} \quad D = \begin{bmatrix} 1-\alpha & 1-\alpha & \mathbf{0}'_{m-1} & -\alpha \\ -\beta & \phi-\beta & \mathbf{0}'_{m-1} & -\beta \\ -\gamma & -\gamma & \mathbf{0}'_{m-1} & 1-\gamma \\ \mathbf{0}_{m-1} & \mathbf{0}_{m-1} & \boldsymbol{I}_{m-1} & \mathbf{0}_{m-1} \end{bmatrix}$$

Again, for ETS(A,A,N) and ETS(A,A,A), the corresponding result is obtained from ETS(A,A_d,N) and ETS(A,A_d,A) by setting $\phi = 1$.

Example 10.2: Local level model with drift

The local level model with drift is equivalent to the ETS(A,A,N) model with $\beta = 0$. Thus, the discount matrix for this model is

$$D = \begin{bmatrix} 1-\alpha & 1-\alpha \\ 0 & 1 \end{bmatrix}, \text{ so that } U = \begin{bmatrix} 1 & (1-\alpha)/q \\ 0 & \alpha/q \end{bmatrix} \text{ and } V = \begin{bmatrix} 1 & -(1-\alpha)/\alpha \\ 0 & q/\alpha \end{bmatrix},$$

where $q = \sqrt{1-2\alpha+2\alpha^2}$. The corresponding roots of D are 1 and $1-\alpha$, and corresponding to the unit root we have:

$$w'u = [1,1]\begin{bmatrix} (1-\alpha)/q \\ \alpha/q \end{bmatrix} = 1/q \quad \text{and} \quad v'g = [0, q/\alpha]\begin{bmatrix} \alpha \\ 0 \end{bmatrix} = 0.$$

Thus, the model is not stable as D has a unit root. However, it can be forecastable as the unit root satisfies the third condition of Theorem 10.2.

The other root is $1-\alpha$ with

$$w'u = [1,1]\begin{bmatrix} 1 \\ 0 \end{bmatrix} = 1 \quad \text{and} \quad v'g = [1, -(1-\alpha)/\alpha]\begin{bmatrix} \alpha \\ 0 \end{bmatrix} = \alpha.$$

So the model will be forecastable if $0 < \alpha < 2$ (condition 1 of Theorem 10.2), or if $\alpha = 0$ (condition 3 of Theorem 10.2). If $\alpha = 0$, the model is equivalent to the linear regression model $y_t = \ell_0 + bt + \varepsilon_t$.

The stability conditions for models without seasonality (i.e., ETS(A,N,N), ETS(A,A,N) and ETS(A,A_d,N)) are summarized in Table 10.1. These are given

Table 10.1. Stability conditions for models without seasonality.

ETS(A,N,N):	$0 < \alpha < 2$
ETS(A,A,N):	$0 < \alpha < 2$
	$0 < \beta < 4 - 2\alpha$
ETS(A,A_d,N):	$1 - 1/\phi < \alpha < 1 + 1/\phi$
	$\alpha(\phi - 1) < \beta < (1 + \phi)(2 - \alpha)$
	$0 < \phi \leq 1$

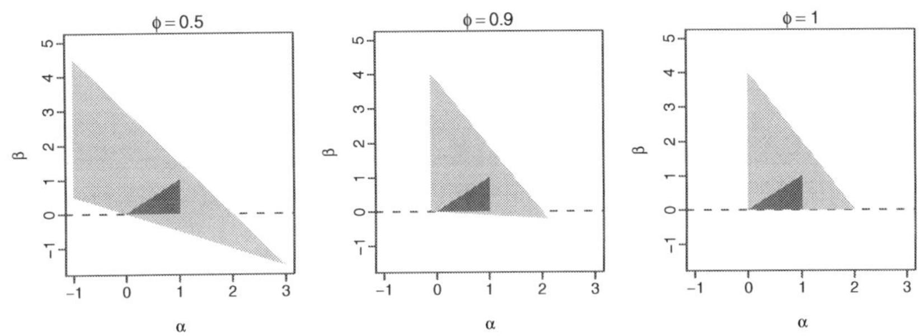

Fig. 10.1. Parameter spaces for model ETS(A,A_d,N). The *right hand graph* shows the region for model ETS(A,A,N) (when $\phi = 1$). In each case, the *light-shaded* regions represent the stability regions; the *dark-shaded* regions are the usual regions constructed by restricting each parameter in the conventional parameterization to lie between 0 and 1.

in McClain and Thomas (1973) for the ETS(A,A,N) model; results for the ETS(A,A_d,N) and ETS(A,N,N) models are obtained in a similar way. To visualize these regions, we have plotted them in Fig. 10.1. The light-shaded regions represent the stability regions; the dark-shaded regions are the usual regions defined by $0 < \beta < \alpha < 1$. Note that the usual parameter region is entirely within the stability region in each case. Therefore non-seasonal models obtained using the usual constraints are always stable (and always forecastable).

10.2.1 Seasonal Models

The characteristic equation for matrix $\boldsymbol{D}$ in the (un-normalized) ETS(A,N,A) model is $f(\lambda) = (1 - \lambda)P(\lambda) = 0$, where

$$P(\lambda) = \lambda^m + \alpha\lambda^{m-1} + \alpha\lambda^{m-2} + \cdots + \alpha\lambda^2 + \alpha\lambda + (\alpha + \gamma - 1). \qquad (10.4)$$

Thus, $\boldsymbol{D}$ has a unit eigenvalue regardless of the values of the model parameters, and so the model is always unstable.

Similarly, the characteristic equation of D for model ETS(A,A$_d$,A) is $f(\lambda) = (1-\lambda)P(\lambda) = 0$, where

$$P(\lambda) = \lambda^{m+1} + (\alpha+\beta-\phi)\lambda^m + (\alpha+\beta-\alpha\phi)\lambda^{m-1} + \cdots + (\alpha+\beta-\alpha\phi)\lambda^2 + (\alpha+\beta-\alpha\phi+\gamma-1)\lambda + \phi(1-\alpha-\gamma). \tag{10.5}$$

Example 10.3: Additive Holt-Winters' model ETS(A,A,A)

In this case,

$$D = \begin{bmatrix} 1-\alpha & 1-\alpha & \mathbf{0}'_{m-1} & -\alpha \\ -\beta & 1-\beta & \mathbf{0}'_{m-1} & -\beta \\ -\gamma & -\gamma & \mathbf{0}'_{m-1} & 1-\gamma \\ \mathbf{0}_{m-1} & \mathbf{0}_{m-1} & \boldsymbol{I}_{m-1} & \mathbf{0}_{m-1} \end{bmatrix}.$$

Solving the equations corresponding to the unit root case (Exercise 10.2) shows that $\boldsymbol{u}$ is proportion to $[-1, 0, 1, \ldots, 1]$. It follows that $\boldsymbol{w}'\boldsymbol{u} = 0$. Consequently, the model is forecastable if the remaining roots are inside the unit circle.

The same argument applies to all of the seasonal models. Thus, the ETS(A,N,A), ETS(A,A,A) and ETS(A,A$_d$,A) models are forecastable if and only if the roots of $P(\lambda)$ lie inside the unit circle. Hyndman et al. (2008) use this result to derive the specific conditions for forecastability; these conditions are summarized in Table 10.2.

The inequalities involving only α and γ provide necessary conditions for forecastability that are easily implemented. The final condition (giving a range for β) is more complicated to use than finding the numerical roots of (10.5). Therefore, we suggest that, in practice, these conditions on α and γ be imposed when estimating the model; the roots of (10.5) can then be calculated and tested.

To visualize these regions, we have plotted them in Figs. 10.2–10.4. The light-shaded regions represent the forecastability regions; the dark-shaded regions are the usual regions given by

$$0 < \alpha < 1, \quad 0 < \beta < \alpha, \quad 0 < \gamma < 1-\alpha, \quad \text{and} \quad 0 < \phi < 1.$$

The forecastable region for α and γ is illustrated in Fig. 10.2. For large values of ϕ, the upper limit of γ is obtained when the upper limit of α equals the lower limit of α. For $\phi = 1$, this simplifies to $\gamma < 2m/(m-1)$, as given by Archibald (1991), but for smaller values of ϕ the upper limit of γ is smaller than this.

The right hand column of Fig. 10.2 shows that the usual parameter region of an ETS(A,N,A) model is entirely within the forecastability region.

Table 10.2. Forecastability conditions for models ETS(A,N,A) and ETS(A,A_d,A).

ETS(A,N,A): $\max(-m\alpha,0) < \gamma < 2-\alpha$ and $\dfrac{-2}{m-1} < \alpha < 2-\gamma$

ETS(A,A_d,A):
$$0 < \phi \le 1$$
$$\max(1-1/\phi-\alpha,0) < \gamma < 1+1/\phi-\alpha$$
$$1-1/\phi-\gamma(1-m+\phi+\phi m)/(2\phi m) < \alpha < (B+C)/(4\phi)$$
$$-(1-\phi)(\gamma/m+\alpha) < \beta < D+(\phi-1)\alpha$$

where

$$B = \phi(4-3\gamma)+\gamma(1-\phi)/m$$

$$C = \sqrt{B^2-8\left[\phi^2(1-\gamma)^2+2(\phi-1)(1-\gamma)-1\right]+8\gamma^2(1-\phi)/m}$$

$$D = \min_\theta\left\{(\phi-\phi\alpha+1)(1-\cos\theta) - \gamma\left[\frac{(1+\phi)(1-\cos\theta-\cos m\theta)+\cos(m-1)\theta+\phi\cos(m+1)\theta}{2(1-\cos m\theta)}\right]\right\}$$

and θ is a solution to

$$\frac{\phi\alpha-\phi+1}{\gamma}+\frac{(\phi-1)(1+\cos\theta-\cos m\theta)+\cos(m-1)\theta-\phi\cos(m+1)\theta}{2(1+\cos\theta)(1-\cos m\theta)}=0$$

Conditions for ETS(A,A,A) can be obtained from ETS(A,A_d,A) by setting $\phi = 1$.

Therefore ETS(A,N,A) models obtained using the usual constraints are always forecastable.

The forecastable region for α and β is depicted in Figs. 10.3 and 10.4 for $m = 4$ and $m = 12$ respectively. For $m = 4$, the usual parameter region is entirely contained within the forecastability region for all values of ϕ and γ, except when both ϕ and γ are relatively small. However, for $m = 12$ (Fig. 10.4), it can be seen that the usual parameter region and the forecastability region intersect for model ETS(A,A_d,A), but neither is contained within the other, even when $\phi = 1$. Therefore, models obtained using the usual constraints may often not be forecastable.

Consequently, we recommend that the usual parameter regions not be used. Instead, when parameters are estimated, the optimization routine should be constrained to return values within the forecastability region. If we constrain the parameters to lie in the intersection of the usual region and the forecastability region, we can retain the interpretation of the model equations as weighted averages. However, such constraints may produce inferior forecasts when the best-fitting model lies outside the more restricted region.

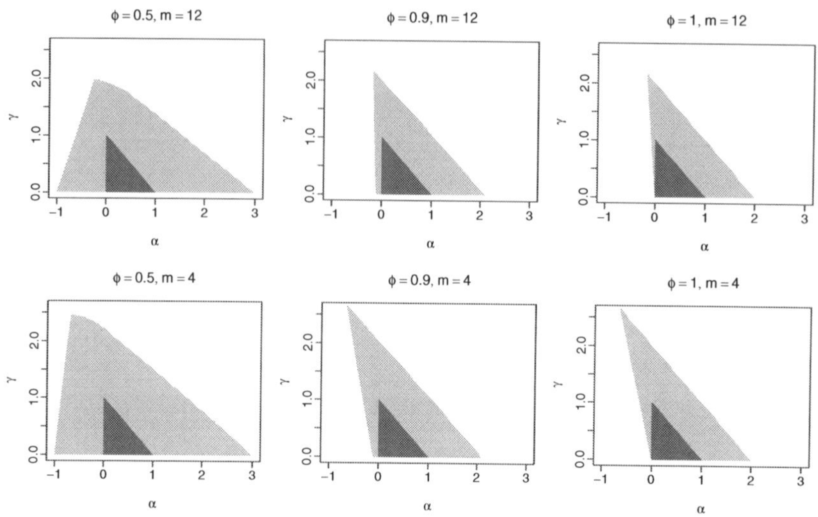

Fig. 10.2. *Light-shaded* region: the forecastable region of α and γ for model ETS(A,A$_d$,A). *Dark-shaded* region: the usual region where $0 < \alpha < 1$ and $0 < \gamma < 1 - \alpha$. The *right column* shows the regions for model ETS(A,A,A) (when $\phi = 1$). These are also the regions for model ETS(A,N,A) as they are independent of β.

10.2.2 Normalized Models

Archibald (1984, 1990) discussed the stable region for the normalized version of ETS(A,A,A), and Archibald (1991) provided some preliminary steps towards finding the stable region for the normalized version of ETS(A,A$_d$,A). Hyndman et al. (2008) extended this analysis and derived the results described below.

In Chap. 8, we showed that the normalized models can be written in state space form with the state vector $\boldsymbol{x}_t = (\ell_t, b_t, s_{1,t}, \ldots, s_{m-1,t})'$, where $s_{i,t}$ is the estimate of the seasonal factor for the ith month ahead made at time t. Note that $s_{m,t} \equiv s_{0,t} = 1 - s_{1,t} - \cdots - s_{m-1,t}$. Following Roberts (1982, Sect. 3), the seasonal updating is defined as follows:

$$\begin{aligned} s_{0,t} &= s_{1,t-1} + \gamma(1 - \tfrac{1}{m})e_t, \\ s_{i,t} &= s_{i+1,t-1} - \tfrac{\gamma}{m}e_t. \end{aligned}$$

The level and trend equations are updated as with the standard model. Then $\boldsymbol{w}' = [1, 1, 1, \boldsymbol{0}'_{m-2}]$,

$$\boldsymbol{F} = \begin{bmatrix} 1 & 1 & 0 & \boldsymbol{0}'_{m-2} \\ 0 & \phi & 0 & \boldsymbol{0}'_{m-2} \\ \boldsymbol{0}_{m-2} & \boldsymbol{0}_{m-2} & \boldsymbol{0}_{m-2} & \boldsymbol{I}_{m-2} \\ 0 & 0 & -1 & -\boldsymbol{1}'_{m-2} \end{bmatrix}, \quad \boldsymbol{g} = \begin{bmatrix} \alpha \\ \beta \\ -(\gamma/m)\boldsymbol{1}_{m-1} \end{bmatrix}$$

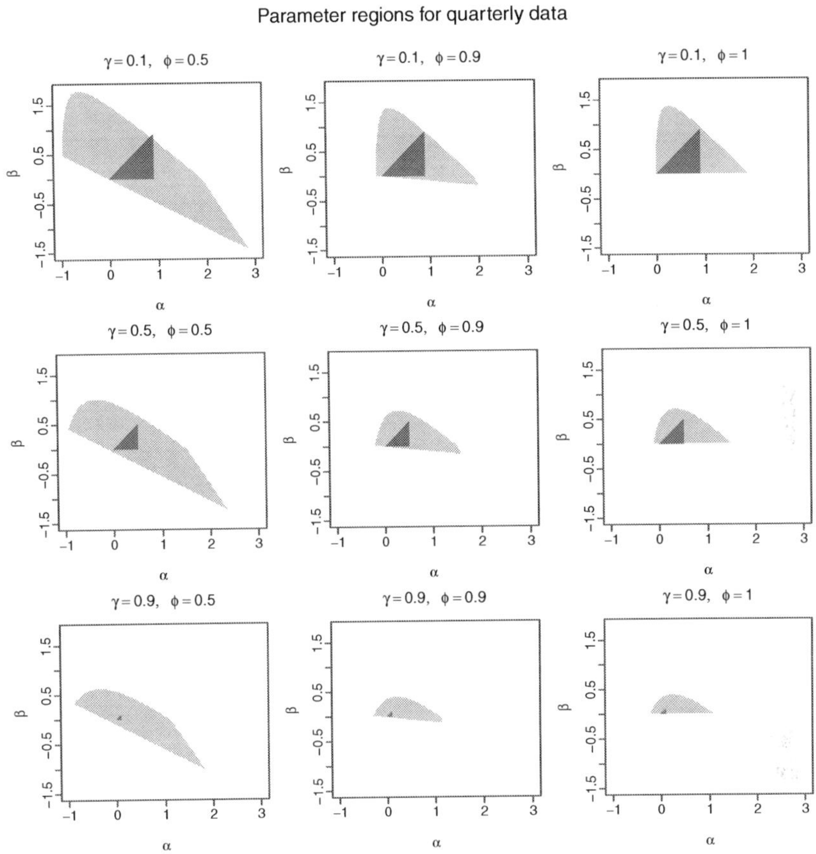

Fig. 10.3. *Light-shaded* region: the forecastable region of α and β for model ETS(A,A$_d$,A) with $m = 4$. *Dark-shaded* region: the usual region where $0 < \alpha < 1 - \gamma$ and $0 < \beta < \alpha$. The *right column* shows the region for model ETS(A,A,A) (when $\phi = 1$).

and

$$\boldsymbol{D} = \begin{bmatrix} 1-\alpha & 1-\alpha & -\alpha & \boldsymbol{0}'_{m-2} \\ -\beta & \phi-\beta & -\beta & \boldsymbol{0}'_{m-2} \\ (\gamma/m)\boldsymbol{1}_{m-2} & (\gamma/m)\boldsymbol{1}_{m-2} & (\gamma/m)\boldsymbol{1}_{m-2} & \boldsymbol{I}_{m-2} \\ \gamma/m & \gamma/m & \gamma/m-1 & -\boldsymbol{1}'_{m-2} \end{bmatrix},$$

where $\boldsymbol{1}_k$ denotes a k-vector of ones. The characteristic equation for $\boldsymbol{D}$ is given by

$$f(\lambda) = \sum_{i=0}^{m+1} \theta_i \lambda^{m+1-i}, \tag{10.6}$$

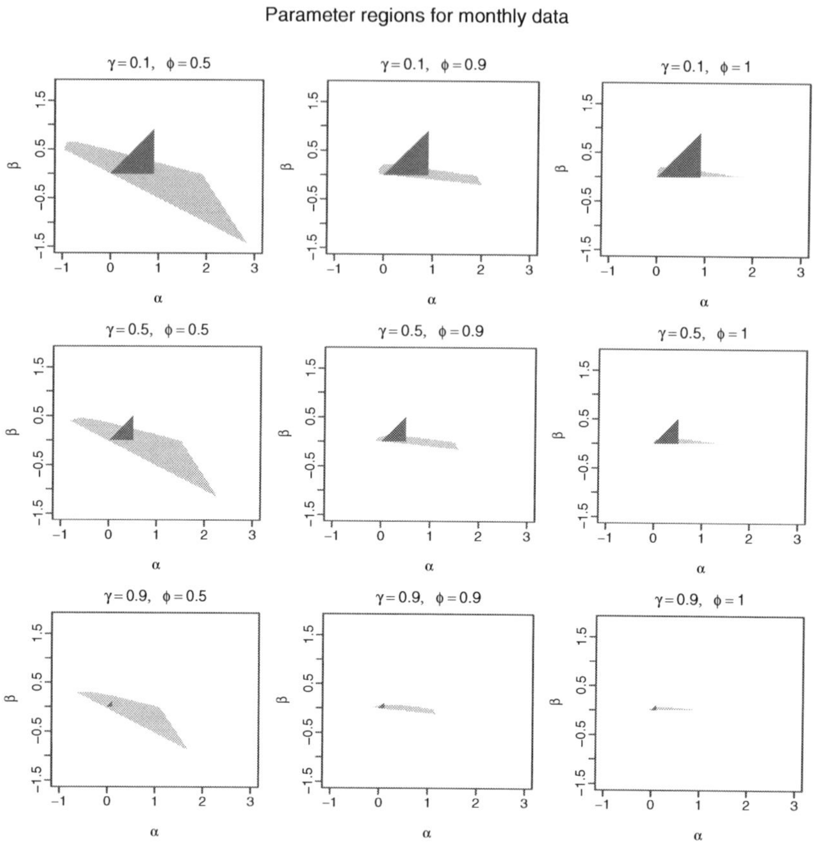

Fig. 10.4. *Light-shaded* region: the forecastable region of α and β for model ETS(A,A$_\text{d}$,A) with $m = 12$. *Dark-shaded* region: the usual region where $0 < \alpha < 1-\gamma$ and $0 < \beta < \alpha$. The *right column* shows the region for model ETS(A,A,A) (when $\phi = 1$).

where

$$\begin{aligned}
\theta_0 &= 1 \\
\theta_1 &= \alpha + \beta - \gamma/m - \phi \\
\theta_i &= \alpha(1-\phi) + \beta - (1-\phi)\gamma/m, \quad i = 2, \ldots, m-1 \\
\theta_m &= \alpha(1-\phi) + \beta + \gamma[1-(1-\phi)/m] - 1
\end{aligned}$$

and

$$\theta_{m+1} = \phi[1 - \gamma(1-1/m) - \alpha].$$

Note that this is equivalent to (10.5) if we reparameterize the model, replacing α in (10.5) by $\alpha - \gamma/m$. Therefore the forecastability conditions for the standard ETS(A,A_d,A) model are the same as the stability conditions for the normalized ETS(A,A_d,A) model, apart from this minor reparameterization. In particular, the normalized models are stable (provided the parameters are within the stability regions).

10.3 Conclusions

With the non-seasonal exponential smoothing models, our results are clear: the models are of minimal dimension and are stable using the usual constraints. In fact, it is possible to allow parameters to take values in a larger space, and still retain a stable model.

With the seasonal exponential smoothing methods, the situation is more complicated. There is a redundancy in the state vector because the seasonal states are not constrained, making the models of larger dimension than necessary. The same redundancy leads to a unit root in the discount matrix, causing all of the linear seasonal models to be unstable for any values of the model parameters. However, we have shown that the model can be made forecastable, and we have provided conditions for the parameters to ensure forecastability.

The normalized model circumvents this problem by requiring the seasonal states to sum to zero, thus removing the inherent redundancy in the seasonal terms. This leads to both a minimal dimension model and a stable model.

10.4 Exercises

Exercise 10.1.

a. Show that the non-seasonal models ETS(A,N,N) and ETS(A,A_d,N) are of minimal dimension.
b. Show that the seasonal models ETS(A,A,A) and ETS(A,A_d,A) are not of minimal dimension.
c. Show that the normalized seasonal models ETS(A,N,A), ETS(A,A,A) and ETS(A,A_d,A) are of minimal dimension.
d. Show that the (unnormalized) seasonal models ETS(A,A,A) and ETS(A,A_d,A) are of minimal dimension if the level component is omitted from the models. (This is an alternative to normalization).

Exercise 10.2. Complete Example 10.3 by showing that $\boldsymbol{u}$ is proportional to $[-1, 0, 1, \ldots, 1]$ for the ETS(A,A,A) model.

Exercise 10.3. The expression $\boldsymbol{x}_t = \boldsymbol{D}\boldsymbol{x}_{t-1} + \boldsymbol{g}y_t$ also applies to some of the nonlinear models discussed in Chap. 4. Use this observation to write down the stability conditions for the relevant nonlinear models.

11

Reduced Forms and Relationships with ARIMA Models

The purpose of this chapter is to examine the links between the (linear) innovations state space models and autoregressive integrated moving average (ARIMA) models, frequently called "Box–Jenkins models" because Box and Jenkins (1970) proposed a complete methodology for identification, estimation and prediction with these models. We will show that when the state variables are eliminated from a linear innovations state space model, an ARIMA model is obtained. This ARIMA form of the state space model is called its *reduced form*.

The process for deriving the reduced form uses the lag operator, defined by $Ly_t = y_{t-1}$, to eliminate the state variables from the state space model. Another procedure that relies on conventional equation solving methods will be explained in Chap. 13. The latter method has the advantage that its algorithm can be implemented relatively easily in a matrix programming language such as **R** or Matlab.

We begin the chapter with a brief summary of ARIMA models and their properties. In Sect. 11.2 we obtain reduced forms for the simple cases of the local level model, ETS(A,N,N), and the local trend model, ETS(A,A,N). Then, in Sect. 11.3 we show how to put a general linear innovations state space model into an ARIMA reduced form. (Causal) stationarity and invertibility conditions for the reduced form model are developed in Sect. 11.4, and we explore the links with causal stationarity and stability of the corresponding innovations state space model.

In the opposite direction, an ARIMA model can also be put in the form of a linear innovations state space model. This reverse procedure is demonstrated in Sect. 11.5.

11.1 ARIMA Models

The general form of an ARMA model is conventionally written as:

$$\phi(L)y_t = \lambda + \theta(L)\varepsilon_t, \tag{11.1}$$

where L is the lag operator defined above, and $\phi(L)$ and $\theta(L)$ are polynomials in L. The random errors, ε_t, are assumed to be independent and identically distributed with zero means and equal variances, σ^2; we write this as $\varepsilon_t \sim \text{IID}(0, \sigma^2)$. The parameter λ represents a constant term.

Several special cases serve to illustrate the general model.

- First order autoregression—AR(1):

$$y_t = \lambda + \phi_1 y_{t-1} + \varepsilon_t. \tag{11.2}$$

- pth order autoregression—AR(p):

$$y_t = \lambda + \phi_1 y_{t-1} + \phi_2 y_{t-2} + \cdots + \phi_p y_{t-p} + \varepsilon_t. \tag{11.3}$$

- First order moving average—MA(1):

$$y_t = \lambda + \varepsilon_t - \theta_1 \varepsilon_{t-1}. \tag{11.4}$$

- qth order moving average—MA(q):

$$y_t = \lambda + \varepsilon_t - \theta_1 \varepsilon_{t-1} - \theta_2 \varepsilon_{t-2} - \cdots - \theta_q \varepsilon_{t-q}. \tag{11.5}$$

- pth order AR, qth order MA—ARMA(p, q):

$$y_t = \lambda + \phi_1 y_{t-1} + \cdots + \phi_p y_{t-p} + \varepsilon_t - \theta_1 \varepsilon_{t-1} - \cdots - \theta_q \varepsilon_{t-q}. \tag{11.6}$$

An important aspect of ARMA modeling is that we assume the series started up in the infinite past, in contrast to the innovations state space models we have considered thus far, where a finite start-up has been employed. Intuitively, if the finite start was a long time ago and the effect of the initial conditions diminishes over time, we might expect that the finite start-up system would converge to the limiting infinite start-up scheme. We now specify the conditions under which this convergence occurs.

11.1.1 Causal Stationarity

The standard assumption made about the autoregressive component is that the roots of the polynomial $\phi(u) = 0$ all lie outside the unit circle. This assumption means that we can rewrite (11.1) as:

$$y_t = \lambda/\phi(1) + [\theta(L)/\phi(L)]\varepsilon_t = \lambda/\phi(1) + \psi(L)\varepsilon_t, \tag{11.7}$$

where

$$\psi(u) = 1 + \psi_1 u + \psi_2 u^2 + \cdots \tag{11.8}$$

is an infinite series which is absolutely convergent for $|u| \leq 1$. An ARMA model satisfying these conditions is said to be *causally stationary*. A univariate process is *causal* if the current value depends only upon current and past values of the error process and past values of the series. Henceforth we assume this to be the case and refer just to stationarity, with the qualifier "causal" always there by implication. Stationarity clearly implies that the coefficients ψ_i converge to zero as we move away from the present time. This condition reduces to the requirement that the roots of the polynomial $\phi(u) = 0$ should lie outside the unit circle. The representation given in (11.7) is known as the Wold representation (or decomposition) of a time series. Thus, any stationary time series may be represented by an infinite order MA scheme. Further, (11.7) shows that such processes may in some cases be represented by finite-order ARMA schemes. By extension, as indicated in Exercise 11.6, state space models always result in finite-order ARIMA reduced-form models. In turn, these conditions imply that the process has an unconditional mean and variance, as illustrated by the following examples.

Example 11.1: Mean and variance for AR(1)

It follows from (11.8) that the AR(1) process is stationary provided $|\phi_1| < 1$. We may then denote the mean by μ and the variance by ω^2. Taking expectations on both sides of (11.2) we obtain:

$$\mathrm{E}(y_t) = \mu = \lambda + \phi_1 \mathrm{E}(y_{t-1}) + \mathrm{E}(\varepsilon_t) = \lambda + \phi_1 \mu$$

so that $\mu = \lambda/(1 - \phi_1)$. In general, the mean of an AR(p) process can be written as $\mu = \lambda/\phi(1)$. Subtracting out the mean, squaring both sides of (11.2) and taking expectations, we arrive at:

$$\mathrm{E}[(y_t - \mu)^2] = \phi_1^2 \mathrm{E}[(y_{t-1} - \mu)^2] + 2\phi_1 \mathrm{E}[\varepsilon_t(y_{t-1} - \mu)] + \mathrm{E}(\varepsilon_t^2).$$

Because ε_t and y_{t-1} are independent, the cross-product term is zero, so this expression reduces to:

$$\mathrm{V}(y_t) = \omega^2 = \sigma^2/(1 - \phi_1^2).$$

Example 11.2: Mean and variance for stationary processes

We can use the ARMA(p, q) form given in (11.6) to arrive at general expressions for the mean and variance, although some simplifications are usually possible for specific models. Because the error terms have zero expectations, we see immediately that $\mu = \lambda/\phi(1)$. Further, because the error terms are uncorrelated and each has variance σ^2, we obtain for the variance:

$$V(y_t) = \omega^2 = \sigma^2 \sum_{i=0}^{\infty} \psi_i^2.$$

How does this definition of stationarity compare with that given in Sect. 3.3.2? The difference lies in the start-up conditions. In Chap. 3 we assumed a finite start-up, whereas here we are assuming that the series started in the infinite past. We can rewrite the AR(1) model as:

$$\begin{aligned} y_t &= \lambda + \phi_1 y_{t-1} + \varepsilon_t, \qquad t = 2, 3, \ldots, \\ y_1 &= \lambda + \phi_1 \ell_0 + \varepsilon_1. \end{aligned}$$

This form corresponds to the damped level model in Sect. 3.5.1 with $\alpha = 1$. This correspondence enables us to see that the coefficients $\{k_j\}$ in (3.8) are equivalent to the $\{\psi_j\}$ in (11.8), and that the constant term is $d_t = \mu + \phi_1^t \ell_0$, which converges to the mean when $|\phi_1| < 1$. Similar equivalences may be established for more general models.

11.1.2 Invertibility

The representation in (11.7) enables us to recast any ARMA(p, q) process as an infinite order MA process. A similar manipulation enables us to rewrite the model as:

$$[\phi(L)/\theta(L)](y_t - \mu) = \varepsilon_t,$$

which may be represented as an infinite order AR process with operator $\pi(L) = \phi(L)/\theta(L)$ provided the series expansion of $\pi(u)$ is absolutely convergent for $|u| \leq 1$ or $\sum_{i=1}^{\infty} |\pi_i| < \infty$. This requirement reduces to the condition that the roots of $\theta(u) = 0$ should lie outside the unit circle. When this condition holds, we can write the model as:

$$\pi(L)y_t = \mu\phi(1)/\theta(1) + \varepsilon_t \quad \text{or} \quad y_t = \lambda/\theta(1) + \pi_1 y_{t-1} + \pi_2 y_{t-2} + \cdots + \varepsilon_t.$$

When this representation is valid, we say the model is *invertible*. We relate this concept to our earlier discussion of forecastability in Sect. 11.4.

11.1.3 ARIMA Models

Our discussion in the previous section focused upon stationary models, yet nearly all of the discussion in earlier chapters has assumed the existence of trends or, at the very least, locally varying mean levels. In our experience, most series in economics and business exhibit such behavior, so that stationary series are relatively rare. Indeed, in many cases where stationarity is observable, the time series has been transformed in some way; for example, a series of stock prices $\{y_t\}$ typically follows a random walk, but the return on the stock, defined as $r_t = (y_t - y_{t-1})/y_{t-1}$ may well be stationary.

We extend the models under consideration to include nonstationary models by specifying the ARIMA class of models, which may be written as

$$\phi(L)(1 - L)^d y_t = \theta(L)\varepsilon_t. \tag{11.9}$$

We drop the constant term in accordance with common usage. Note that we have partitioned the AR operator into two parts: the first term $\phi(L)$ is the standard AR polynomial and the second term $(1 - L)^d$ describes the *differencing*[1] operations. On occasion, it is convenient to represent this product by: $\eta(L) = \phi(L)(1 - L)^d$. Differencing once or twice is sufficient in most applications.

Once the appropriate order of differencing has been performed, the series $z_t = (1 - L)^d y_t$ may be modeled as an ARMA process, as in the previous section. The full model for the original series is then referred to as an ARIMA(p, d, q) process. For full details of ARIMA processes, see Box et al. (1994, Chap. 4).

11.1.4 Seasonal Series

In order to complete the description of ARIMA models, we must also consider the existence of seasonal patterns in the data. If the series was purely seasonal, we could consider a model such as (11.9) but with each "month" relating back only to the same month in previous "years." So, if there are m months, a purely seasonal model could be written as

$$\Phi(L^m)[1 - L^m]^D y_t = \Theta(L^m)\varepsilon_t. \tag{11.10}$$

The operator $1 - L^m$ defines a seasonal difference, whereas $\Phi(L^m)$ is the seasonal autoregressive polynomial and $\Theta(L^m)$ represents the seasonal moving average polynomial. Purely seasonal series may occur from time to time, but a far more common possibility is that there are both regular and seasonal effects to take into account. A natural way to do this is to combine (11.9) and

[1] Note that $(1 - L)y_t = y_t - y_{t-1}$ represents the difference between successive observations.

(11.10) to produce what is termed a "multiplicative" model in the literature (a confusing but now standard term):

$$\phi(L)\Phi(L^m)(1-L)^d[1-L^m]^D y_t = \theta(L)\Theta(L^m)\varepsilon_t. \tag{11.11}$$

As before, we omit the constant term. Differencing may occur either at lag 1, or at lag m, or at both lags. Overall, the seasonal element may comprise P AR terms, Q MA terms and D seasonal differences. The full model is then denoted by ARIMA$(p,d,q)(P,D,Q)_m$.

An advantage of the multiplicative form is that the stationarity and invertibility conditions can be examined separately for the regular and seasonal components, by checking the roots of each polynomial in turn.

Example 11.3: The airline model

The best-known and most widely used model in this extended class is known as the "airline model" used by Box and Jenkins to model a monthly time series of airline passenger counts; see Box et al. (1994, Chap. 9), which is also the standard reference for these seasonal models. The airline model is of the form ARIMA(0,1,1)(0,1,1)$_m$, with $m = 12$ in the example; that is,

$$(1-L)(1-L^m)y_t = (1-\theta_1 L)(1-\theta_m L^m)\varepsilon_t. \tag{11.12}$$

11.2 Reduced Forms for Two Simple Cases

We now consider the relationships between the ARIMA models and the linear innovations state space models considered earlier in the book. The approach we will follow is to eliminate the state variables from the state model, thereby arriving at an ARIMA process, which we refer to as a *reduced form model*. Because the innovations models operate from a finite start-up, but the ARIMA models assume an infinite past, we must restrict our attention to those linear innovations state space models that are *stable* in the sense of Sect. 3.3.1, so that the effect of the start-up conditions can be ignored in sufficiently long series.

Example 11.4: Simple exponential smoothing—ETS(A,N,N) model

In Chap. 3, the ETS(A,N,N) model, which is the innovations state space model for simple exponential smoothing, was defined. Using the lag operator L, this model can be written in a slightly different manner as

$$y_t = \ell_{t-1} + \varepsilon_t, \tag{11.13a}$$
$$(1-L)\ell_t = \alpha\varepsilon_t. \tag{11.13b}$$

To find the ARIMA reduced form of the ETS(A,N,N) model in (11.13), apply the differencing operator $1 - L$ to both sides of the measurement equation (11.13a). The result is the ARIMA(0,1,1) model

$$\begin{aligned}(1-L)y_t &= (1-L)\ell_{t-1} + (1-L)\varepsilon_t \\ &= \alpha\varepsilon_{t-1} + \varepsilon_t - \varepsilon_{t-1} \\ &= (1-\theta_1 L)\varepsilon_t,\end{aligned}$$

where $\theta_1 = 1 - \alpha$. It is well known that the ARIMA(0,1,1) model provides the same point forecasts as simple exponential smoothing. The ETS(A,N,N) model and the ARIMA(0,1,1) model will also produce the same forecast variances and prediction intervals.

The ETS parameter space $\alpha \in (0,2)$ corresponds exactly to the ARIMA parameter space $|\theta_1| < 1$. However, we observe that the finite start-up assumption enables the ETS scheme to handle $\alpha = 0$, corresponding to a constant mean; the ARIMA model does not include this case.

Example 11.5: ETS(A,A,N) model

In a similar manner, we can write the local ETS(A,A,N) model, which is the innovations state space model for Holt's method, as

$$y_t = \ell_{t-1} + b_{t-1} + \varepsilon_t, \tag{11.14a}$$
$$(1-L)\ell_t = b_{t-1} + \alpha\varepsilon_t, \tag{11.14b}$$
$$(1-L)b_t = \beta\varepsilon_t. \tag{11.14c}$$

Applying the differencing operator $(1 - L)^2$ to the measurement equation (11.14a) and using transition equations (11.14b) and (11.14c), we obtain the following ARIMA(0,2,2) model

$$(1-L)^2 y_t = [1 - \theta_1 L - \theta_2 L^2]\varepsilon_t,$$

where $\theta_1 = 2 - \alpha - \beta$ and $\theta_2 = \alpha - 1$. The invertibility conditions for this model are $4 - 2\alpha - \beta > 0$, $\beta > 0$ and $0 < \alpha < 2$, which the reader is asked to verify in Exercise 11.1.

We note that these reduced forms implicitly assume an infinite start-up. If the finite start-up assumption is considered, model (11.14) must be supplemented by the values of the initial states ℓ_0 and b_0.

In the next section we find reduced forms for the general form of the linear innovations state space model.

11.3 Reduced Form for the General Linear Innovations Model

We now consider the reduced form of the general linear innovations state space model. The goal is to transform the state space model into an ARIMA model. As in (11.9), the general ARIMA model may be expressed by

$$\eta(L)y_t = \phi(L)\delta(L)y_t = \theta(L)\varepsilon_t,$$

where $\eta(L)$ and $\theta(L)$ are both polynomials in the lag operator L, and may include powers of L related to the seasonal period m. In addition, $\delta(L)$ contains all the unit roots of the polynomial. If we let $z_t = \delta(L)y_t$,

$$\phi(L)z_t = \theta(L)\varepsilon_t$$

is an autoregressive moving average (ARMA) model.

The general linear innovations state space model $y_t = \boldsymbol{w}'\boldsymbol{x}_{t-1} + \varepsilon_t$ and $\boldsymbol{x}_t = \boldsymbol{F}\boldsymbol{x}_{t-1} + \boldsymbol{g}\varepsilon_t$ can be reduced to an ARIMA model with the help of the lag operator. First, the transition equation can be rewritten as

$$(\boldsymbol{I} - \boldsymbol{F}L)\boldsymbol{x}_t = \boldsymbol{g}\varepsilon_t. \tag{11.15}$$

As it is possible that $\boldsymbol{I} - \boldsymbol{F}L$ may not have an inverse, we multiply both sides of (11.15) by its adjoint, $\text{adj}(\boldsymbol{I} - \boldsymbol{F}L)$, to get

$$\det(\boldsymbol{I} - \boldsymbol{F}L)\boldsymbol{x}_t = \text{adj}(\boldsymbol{I} - \boldsymbol{F}L)\boldsymbol{g}\varepsilon_t. \tag{11.16}$$

Next, apply the operator $\det(\boldsymbol{I} - \boldsymbol{F}L)$ to both sides of the measurement equation to find

$$\det(\boldsymbol{I} - \boldsymbol{F}L)y_t = \boldsymbol{w}'\det(\boldsymbol{I} - \boldsymbol{F}L)\boldsymbol{x}_{t-1} + \det(\boldsymbol{I} - \boldsymbol{F}L)\varepsilon_t.$$

Then, using (11.16), substitute for $\det(\boldsymbol{I} - \boldsymbol{F}L)\boldsymbol{x}_{t-1}$ to obtain the following ARIMA model:

$$\det(\boldsymbol{I} - \boldsymbol{F}L)y_t = \boldsymbol{w}'\text{adj}(\boldsymbol{I} - \boldsymbol{F}L)\boldsymbol{g}\varepsilon_{t-1} + \det(\boldsymbol{I} - \boldsymbol{F}L)\varepsilon_t. \tag{11.17}$$

In this ARIMA model

$$\eta(L) = \det(\boldsymbol{I} - \boldsymbol{F}L)$$

and

$$\theta(L) = \boldsymbol{w}'\text{adj}(\boldsymbol{I} - \boldsymbol{F}L)\boldsymbol{g}L + \det(\boldsymbol{I} - \boldsymbol{F}L).$$

It is possible that the polynomials on both sides of this equation have common factors, in which case they should be canceled, in accordance with the minimal dimension state representation developed in Chap. 10. That is, this reduced form will correspond to the ARIMA model after the elimination of any factors that are common to both $\eta(L)$ and $\theta(L)$. Technically, the elimination of a unit root should lead to the introduction of a constant on the right hand side of the equation. However, such constants are often set to zero when there is at least one unit root remaining on the left hand side. The following example illustrates the process.

Example 11.6: Finding an ARIMA reduced form for the ETS(A,A,A) model

Direct application of (11.17) to the ETS(A,A,A) model in Sect. 3.4.3 yields

$$\begin{aligned} \eta(L) &= (1-L)^2(1-L^m) \\ \text{and} \quad \theta(L) &= L(1-L)(1-L^m)\alpha \\ &\quad + L(1-L^m)\beta + L^m(1-L)^2\gamma + (1-L)^2(1-L^m). \end{aligned}$$

Inspection of the two polynomials reveals the presence of a common factor $1-L$ in both polynomials, indicating that the state space model has been overdifferenced. Elimination of a unit root common to both sides yields the revised expression:

$$\begin{aligned} \eta(L) &= (1-L)(1-L^m) \\ \text{and} \quad \theta(L) &= (1-L)(1-L^m) + L(1-L^m)\alpha \\ &\quad + (L+\cdots+L^m)\beta + L^m(1-L)\gamma. \end{aligned}$$

This model contains $(m+1)$ moving average terms but only three parameters, so it differs from the usual seasonal ARIMA process. When $\beta = 0$, this model is close to the airline model (11.12), differing only by the factor $\alpha\gamma$ in the coefficient of L^{m+1}.

11.4 Stationarity and Invertibility

The conditions for stationarity and invertibility will now be considered. We seek conditions on the matrices in the state space models to indicate when their ARIMA reduced forms have each of these two properties. Recall that a basic identity for the lag operator L and a matrix $\boldsymbol{A}$ is $(\boldsymbol{I}-\boldsymbol{A}L)^{-1} = \sum_{j=0}^{\infty}(\boldsymbol{A}L)^j$.

We stated in Sect. 11.1.1 that an ARMA model of the form $\phi(L)z_t = \theta(L)\varepsilon_t$, where $z_t = y_t - \mu$, is *stationary* provided $\sum_{i=1}^{\infty}|\psi_i| < \infty$. Initially, we will assume that $\det(\boldsymbol{I}-\boldsymbol{F}L) = 0$ has no unit roots. Then we can put the transition equation (11.15) into the form

$$\boldsymbol{x}_t = (\boldsymbol{I}-\boldsymbol{F}L)^{-1}\boldsymbol{g}\varepsilon_t. \tag{11.18}$$

By substituting (11.18) into the measurement equation, we obtain

$$y_t = [1 + \boldsymbol{w}'(\boldsymbol{I} - \boldsymbol{F}L)^{-1}\boldsymbol{g}L]\varepsilon_t = \psi(L)\varepsilon_t. \tag{11.19}$$

Equation (11.19) is the MA form of the state space model, provided the inverse matrix $(\boldsymbol{I} - \boldsymbol{F})^{-1}$ exists. If we write $\boldsymbol{F}^j = \boldsymbol{U}\boldsymbol{\Lambda}^j\boldsymbol{V}$, where $\boldsymbol{\Lambda}$ is the diagonal matrix of eigenvalues and $(\boldsymbol{U}, \boldsymbol{V})$ are the matrices of eigenvectors, then

$$\psi(L) = 1 + \boldsymbol{w}'\boldsymbol{U}\Big(\sum_{j=0}^{\infty} \boldsymbol{\Lambda}^j L^{j+1}\Big)\boldsymbol{V}\boldsymbol{g}.$$

Requiring the eigenvalues of $\boldsymbol{F}$ to lie within the unit circle guarantees that $\sum_{i=1}^{\infty} |\psi_i| < \infty$. Thus, the reduced form ARIMA model is stationary if the eigenvalues of $\boldsymbol{F}$ lie inside the unit circle.

We found in Sect. 3.3.2 that this same condition on the eigenvalues was a sufficient condition for stationarity of the linear innovations state space model. It is also a necessary condition for all of the models we are interested in.

Now we consider the case when $\boldsymbol{F}$ has unit eigenvalues. If the eigenvalues of $\boldsymbol{F}$ do not exceed 1, then $\det(\boldsymbol{I} - \boldsymbol{F}L) = \phi(L)\delta(L)$, where $\delta(L)$ is a polynomial for which the roots are all of the unit eigenvalues of $\boldsymbol{F}$, and so $\phi(L)$ is a polynomial that has no unit roots. Then, (11.17) can be written as the integrated MA model

$$w_t = \delta(L)y_t = \Big(\boldsymbol{w}'\frac{\operatorname{adj}(\boldsymbol{I} - \boldsymbol{F}L)}{\phi(L)}\boldsymbol{g}L + \delta(L)\Big)\varepsilon_t = \psi(L)\varepsilon_t.$$

The process $\{w_t\}$ will be stationary provided $\sum_{i=1}^{\infty} |\psi_i| < \infty$. That is, we have induced stationarity by differencing. If an eigenvalue of $\boldsymbol{F}$ exceeds 1, the process is not stationary and cannot be made stationary by applying difference operators.

Recall that an ARIMA model is *invertible* if $\sum_{i=1}^{\infty} |\pi_i| < \infty$. The AR reduced form of the innovations state space model can be found in a similar manner. In Sect. 3.3.1 we saw that the state vector $\boldsymbol{x}_t$ may be written as $\boldsymbol{x}_t = \boldsymbol{D}\boldsymbol{x}_{t-1} + \boldsymbol{g}y_t$, where $\boldsymbol{D} = \boldsymbol{F} - \boldsymbol{g}\boldsymbol{w}'$. Hence, another form for the transition equation is

$$\boldsymbol{x}_t = (\boldsymbol{I} - \boldsymbol{D}L)^{-1}\boldsymbol{g}y_t. \tag{11.20}$$

Provided all the eigenvalues of $\boldsymbol{D}$ lie inside the unit circle, we may substitute equation (11.20) into the measurement equation to obtain

$$y_t = \boldsymbol{w}'(\boldsymbol{I} - \boldsymbol{D}L)^{-1}\boldsymbol{g}y_{t-1} + \varepsilon_t.$$

Thus, the AR form of the state space model is

$$\Big[1 - \boldsymbol{w}'(\boldsymbol{I} - \boldsymbol{D}L)^{-1}\boldsymbol{g}L\Big]\, y_t = \pi(L)y_t = \varepsilon_t.$$

By employing the same argument that was used for the MA polynomial $\psi(L)$ and the transition matrix $\boldsymbol{F}$, we can see that requiring eigenvalues of $\boldsymbol{D}$ to lie inside the unit circle is equivalent to guaranteeing that the absolute value of the coefficients in the polynomial $\pi(L)$ will converge to zero.

Comparing this with the results in Chap. 10, we see that stability of the linear innovations model implies invertibility of the reduced form ARIMA model.

11.5 ARIMA Models in Innovations State Space Form

In the previous section it was shown how to reduce a linear innovations state space model to an equivalent ARIMA model. It will now be shown that any ARIMA model can be reformulated as an innovations state space model. We start with the general ARIMA model

$$\eta(L)y_t = \theta(L)\varepsilon_t, \tag{11.21}$$

where the polynomials $\eta(L)$ and $\theta(L)$ do not possess any common roots. The polynomial operator $\eta(L)$ contains both the unit root operators and the autoregressive operators. Let $k = \max(r, s)$, where r and s are the degrees of the polynomials $\eta(L)$ and $\theta(L)$, respectively. Then the two polynomials can be written as

$$\eta(L) = 1 - \sum_{i=1}^{k} \eta_i L^i \qquad \text{and} \qquad \theta(L) = 1 - \sum_{i=1}^{k} \theta_i L^i.$$

It follows that (11.21) can be written as

$$y_t = \sum_{i=1}^{k} \eta_i y_{t-i} + \varepsilon_t - \sum_{i=1}^{k} \theta_i \varepsilon_{t-i}.$$

Let $x_{j,t-j}$ be a partial sum that is calculated with information available at period $t - j$ and defined by

$$x_{j,t-j} = \sum_{i=j}^{k} (\eta_i y_{t-i} - \theta_i \varepsilon_{t-i}). \tag{11.22}$$

Note that $x_{j,t} - 0$ when $j > k$, and that

$$y_t = x_{1,t-1} + \varepsilon_t. \tag{11.23}$$

Combining (11.22) and (11.23), we obtain

$$x_{j,t-j} = \sum_{i=j}^{k} (\eta_i x_{1,t-i-1} + (\eta_i - \theta_i)\varepsilon_{t-i}),$$

so that

$$x_{j,t-j} = x_{j+1,t-j-1} + \eta_j x_{1,t-j-1} + (\eta_j - \theta_j)\varepsilon_{t-j}$$

or

$$x_{j,t} = x_{j+1,t-1} + \eta_i x_{1,t-1} + (\eta_j - \theta_j)\varepsilon_t.$$

In summary, the ARIMA model can be rewritten as

$$\begin{aligned} y_t &= x_{1,t-1} + \varepsilon_t, \\ x_{i,t} &= \eta_i x_{1,t-1} + x_{i+1,t-1} + (\eta_i - \theta_i)\varepsilon_t \quad \text{for} \quad i = 1, \ldots, k. \end{aligned}$$

Thus, as shown in Pearlman (1980), the ARIMA process in (11.21) can be represented by the innovations linear state space model where

$$\boldsymbol{w} = \begin{bmatrix} 1 \\ 0 \\ \vdots \\ 0 \end{bmatrix} \qquad \boldsymbol{F} = \begin{bmatrix} \eta_1 & \boldsymbol{I}_{k-1} \\ \vdots & \\ \eta_k & 0 \end{bmatrix} \qquad \text{and} \qquad \boldsymbol{g} = \begin{bmatrix} \eta_1 - \theta_1 \\ \vdots \\ \eta_k - \theta_k \end{bmatrix}.$$

Example 11.7: The innovations state space model for ARIMA(1,1,1)

Consider the following ARIMA model

$$(1 - L)(1 - \phi_1 L)y_t = (1 - \theta_1 L)\varepsilon_t.$$

The polynomial operators for this model are

$$\eta(L) = 1 - (1 + \phi_1)L + \phi_1 L^2 \qquad \text{and} \qquad \theta(L) = 1 - \theta_1 L - 0L^2.$$

Thus, the innovations state space representation would be

$$\boldsymbol{w} = \begin{bmatrix} 1 \\ 0 \end{bmatrix} \qquad \boldsymbol{F} = \begin{bmatrix} 1 + \phi_1 & 1 \\ -\phi_1 & 0 \end{bmatrix} \qquad \text{and} \qquad \boldsymbol{g} = \begin{bmatrix} 1 + \phi_1 - \theta_1 \\ -\phi_1 \end{bmatrix}.$$

Although the result in Example 11.7 does indeed provide a state space representation for this ARIMA model, the form differs considerably from the models described in Chaps. 2 and 3, and the coefficients may be difficult to interpret. We therefore seek linear transformations of the state variables that deliver an appropriate form.

We start with the usual form of the linear innovations model $y_t = \boldsymbol{w}'\boldsymbol{x}_{t-1} + \varepsilon_t$ and $\boldsymbol{x}_t = \boldsymbol{F}\boldsymbol{x}_{t-1} + \boldsymbol{g}\varepsilon_t$ and transform to $y_t = \boldsymbol{w}_0'\boldsymbol{x}^*_{t-1} + \varepsilon_t$ and $\boldsymbol{x}^*_t = \boldsymbol{F}_0\boldsymbol{x}^*_{t-1} + \boldsymbol{g}_0\varepsilon_t$, where

$$\boldsymbol{w} = \boldsymbol{J}'\boldsymbol{w}_0, \quad \boldsymbol{J}\boldsymbol{F} = \boldsymbol{F}_0\boldsymbol{J} \quad \text{or} \quad \boldsymbol{F}_0 = \boldsymbol{J}\boldsymbol{F}\boldsymbol{J}^{-1}, \quad \boldsymbol{g}_0 = \boldsymbol{J}\boldsymbol{g} \quad \text{and} \quad \boldsymbol{x}^*_t = \boldsymbol{J}\boldsymbol{x}^*_{t-1}.$$

The reduced form is unchanged, so the question is whether a suitable matrix $\boldsymbol{J}$ exists. The answer is "sometimes" as the following examples illustrate.

Example 11.8: A modified innovations state space model for ARIMA(1,1,1)

By analogy with the local linear trend model, an appropriate form for the ARIMA(1,1,1) process would have $\boldsymbol{w}_0 = \begin{bmatrix} 1 \\ \phi \end{bmatrix}$ and $\boldsymbol{F}_0 = \begin{bmatrix} 1 & \phi \\ 0 & \phi \end{bmatrix}$, which can be achieved by setting $\boldsymbol{J} = \begin{bmatrix} 0 & -\phi^{-1} \\ \phi^{-1} & \phi^{-2} \end{bmatrix}$; finally, the transformation yields $\boldsymbol{g}_0 = \begin{bmatrix} 1 \\ 1 - \theta\phi^{-1} \end{bmatrix}$. We note in passing that the damped local trend model given in Sect. 2.3.3 may be represented as an ARIMA(1,1,2) model, so the present process is a special case of that process.

The following example shows that such transformations may not always be feasible.

Example 11.9: Innovations state space models for ARIMA(2,0,2)

The ARIMA(2,0,2) model has the form $y_t = \phi_1 y_{t-1} + \phi_2 y_{t-2} + \varepsilon_t - \theta_1 \varepsilon_{t-1} - \theta_2 \varepsilon_{t-2}$, where we ignore the constant term for convenience. If the roots of $\phi(u) = 0$ are real, denote them by (a_1, a_2), where $a_1 + a_2 = \phi_1/\phi_2$ and $-a_1 a_2 = 1/\phi_2$. Then it may be shown that the state space model can be restructured to give the form:

$$\begin{aligned} y_t &= a_1 x_{1,t-1} + a_2 x_{2,t-1} + \varepsilon_t, \\ x_{1,t} &= a_1 x_{1,t-1} + a_2 x_{2,t-1} + g_1 \varepsilon_t, \\ x_{2,t} &= a_2 x_{2,t-1} + g_2 \varepsilon_t, \end{aligned}$$

where $g_1 = 1 - (\theta_2/\phi_2)$ and $g_2 = 1 - \theta_1/a_2 - \theta_2/a_2^2$. There is clearly an element of choice as to which root to use in which equations, but this indeterminacy does not affect the validity of the state space model.

However, when the roots are complex, the representation changes. We can proceed as follows. Denote the AR coefficients by $\phi_1 = 2a$ and $\phi_2 = ac - a^2$. When $c \geq 0$ the roots are real, but when $c < 0$ the roots are complex and we arrive at a state space model of the form:

$$y_t = a x_{1,t-1} + a x_{2,t-1} + \varepsilon_t, \tag{11.24a}$$

$$x_{1,t} = a x_{1,t-1} + a x_{2,t-1} + g_1 \varepsilon_t, \tag{11.24b}$$

$$x_{2,t} = c x_{1,t-1} + a x_{2,t-1} + g_2 \varepsilon_t. \tag{11.24c}$$

The reason for the different form is that the complex roots give rise to cyclical behavior in the forecast function, which cannot be modeled by the exponential smoothing models listed earlier.

11.6 Cyclical Models

The focus of this book is upon the state space models that underlie exponential smoothing. Nevertheless, there are some series that display regular cyclical patterns, such as the famous Wolfer sunspot series (Anderson 1971). As just noted, suitable models for such processes involve complex roots in the reduced form, which cannot be obtained directly from the exponential smoothing formulation. Following Harvey (1989, pp. 38–40), we specify an innovations form of a stationary cyclical model in the following way, rather than using (11.24) above.

$$y_t = \mu + x_{1,t-1} + \varepsilon_t, \tag{11.25a}$$
$$x_{1,t} = \phi x_{1,t-1} \cos \lambda_c + \phi x_{2,t-1} \sin \lambda_c + g_1 \varepsilon_t, \tag{11.25b}$$
$$x_{2,t} = -\phi x_{1,t-1} \sin \lambda_c + \phi x_{2,t-1} \cos \lambda_c + g_2 \varepsilon_t. \tag{11.25c}$$

The parameter ϕ may be viewed as a damping factor, although all we require for stationarity is that $|\phi| < 1$. The parameter λ_c is measured in radians and denotes the cycle frequency. Alternatively, we can say that the time taken for the system to complete a cycle is $2\pi/\lambda_c$. Leaving aside the start-up conditions, this model has a constant mean and four other parameters, as does the ARIMA$(2,0,2)$ scheme. The reader is asked to verify that the state space version reduces to an ARIMA$(2,0,2)$ model in Exercise 11.2.

By way of example, we consider the Wolfer sunspot data, which represents annual sunspot counts for the period 1770–1889. Fitting model (11.25) yields the estimates:

$$\phi = 0.81, \quad \lambda_c = 0.591, \quad \mu = 46.0, \quad g_1 = 2.09, \quad \text{and} \quad g_2 = 0.97.$$

The value of the frequency λ_c corresponds to a cycle of 10.6 years, consistent with other analyses of these data.

In conclusion, we see that an ARIMA model can always be converted into a linear innovations state space model, but that the particular forms introduced in Chap. 3 do not encompass all possible parameter combinations that exist within the ARIMA class. As a practical matter, we can always identify a state space model that corresponds to a particular ARIMA model, but we may not be able to convert it into an exponential smoothing form.

11.7 Exercises

Exercise 11.1. Verify the invertibility conditions for the local linear trend model, given in Sect. 11.2.

Exercise 11.2. Verify that the state space model (11.25) reduces to an ARIMA (2, 0, 2) scheme with complex roots. Find the conditions for this model to reduce to an AR(2) scheme. Verify that the model is stationary provided $|\phi| < 1$.

Exercise 11.3. Show that the parameter spaces for the cyclical AR(2) model given in (11.25) and the real roots AR(2) model defined in Example 11.9 are disjoint. Further, show that their union corresponds exactly to the entire parameter space for the AR(2) model.

Exercise 11.4. Apply the same reasoning as in Sect. 11.3 to obtain a reduced form for the model given in (9.2). To simplify the derivation, assume that $(\boldsymbol{I} - \boldsymbol{F})$ is invertible.

Exercise 11.5. Use the result in Exercise 11.4 to derive explicit results for the local level model with a single regressor variable, and show that the resulting form is the same as that given in Sect. 9.1.

Exercise 11.6. If a state space model has k transition equations, and the maximum lag in equation i is m_i, $i = 1, \ldots, k$, show that the corresponding ARIMA(p, d, q) model has $p + d \leq M$ and $q \leq M$, where $M = m_1 + \cdots + m_k$.

12

Linear Innovations State Space Models with Random Seed States

Exponential smoothing was used in Chap. 5 to generate the one-step-ahead prediction errors needed to evaluate the likelihood function when estimating the parameters of an innovations state space model. It relied on a *fixed* seed state vector to initialize the associated recurrence relationships, something that was rationalized by recourse to a finite start-up assumption. The focus is now changed to stochastic processes that can be taken to have begun prior to the period of the first observed time series value, and which, as a consequence, have a *random* seed state vector. The resulting theory of estimation and prediction is suitable for applications in economics and finance where observations rarely cover the entire history of the generating process.

The Kalman filter (Kalman 1960) can be used in place of exponential smoothing. Like exponential smoothing, it generates one-step-ahead prediction errors, but it works with *random* seed states. It is an enhanced version of exponential smoothing that is used to update the moments of states and associated quantities by conditioning on successive observations of a time series. It will be seen that it was devised for stationary time series and that it cannot be adapted for nonstationary time series without major modifications.

An alternative to the Kalman filter is an information filter, which also conditions on successive observations. However, instead of having a primary focus on the manipulation of moments of associated random quantities, it relies on linear stochastic equations. By using an information filter, the problems encountered with the Kalman filter for nonstationary data conveniently disappear. An information filter can be applied to both stationary and nonstationary time series without modification. The version presented here is an adaptation of the Paige and Saunders (1977) information filter to the linear innovations state space model context.

Section 12.1 discusses the linear innovations state space model when the initial state vector x_0 is random. Section 12.2 is devoted to likelihood functions and their role in estimating the parameters of models of stationary and nonstationary time series. Section 12.3 outlines the information filter used to generate the information needed for the evaluation of a likelihood function. We also illustrate the use of the information filter in fitting a linear trend line. Section 12.4 provides a method for generating prediction distributions. Then the problem of model selection is considered in Sect. 12.5. Section 12.6 examines the problem of smoothing a time series after it has been filtered. Finally, in Sect. 12.7, we consider the Kalman filter for stationary time series and examine its links with exponential smoothing.

12.1 Innovations State Space Models with a Random Seed Vector

The filters in the random seed case will be outlined for the linear innovations state space model considered in Chap. 3. Thus, it will be helpful at this point to examine this model via some examples in order to understand some of the potential problems that can arise when we have a random seed vector. It will be recalled that the linear innovations state space model is defined as

$$y_t = \boldsymbol{w}'\boldsymbol{x}_{t-1} + \varepsilon_t, \tag{12.1a}$$

$$\boldsymbol{x}_t = \boldsymbol{F}\boldsymbol{x}_{t-1} + \boldsymbol{g}\varepsilon_t, \tag{12.1b}$$

where $\varepsilon_t \sim \text{NID}(0, \sigma^2)$, $\boldsymbol{w}$ and $\boldsymbol{g}$ are fixed vectors, and $\boldsymbol{F}$ is a fixed square matrix. In Chap. 3 it was seen that the typical state vector $\boldsymbol{x}_t$ is a linear function of the seed state vector $\boldsymbol{x}_0$. Moreover, because y_t depends on $\boldsymbol{x}_{t-1}$, it also ultimately depends linearly on $\boldsymbol{x}_0$. Thus, all random quantities in the state space model depend on the distribution of the seed state $\boldsymbol{x}_0$.

The distribution of the seed state summarizes the history of a process in the periods preceding the collection of the time series data. Its derivation is illustrated with two examples, the first for a stationary time series and the second for a nonstationary time series.

Example 12.1: Damped local level

The damped local level model was defined in Sect. 3.5.1 in terms of a measurement equation $y_t = \phi\ell_{t-1} + \varepsilon_t$, where the levels ℓ_t are governed by the recurrence relationship $\ell_t = \phi\ell_{t-1} + \alpha\varepsilon_t$ and where $-1 < \phi < 1$. It represents a stationary time series. The recurrence relationship can be solved to give $\ell_t = \alpha\sum_{j=0}^{\infty}\phi^j\varepsilon_{t-j}$, from which it follows that $\text{E}(\ell_t) = 0$ and $\text{V}(\ell_t) = \alpha^2\sigma^2/(1-\phi^2)$. This invariant distribution describes the situation for any period t before incorporating information from the observations. In particular, it represents the only available information about the process at the end of period 0. In other words, $\ell_0 \sim \text{NID}(0, \alpha^2\sigma^2/(1-\phi^2))$.

Example 12.2: Local level model

A local level model is obtained when $\phi = 1$ in Example 12.1. It represents a nonstationary time series. Then the variance of $\ell_0 = \alpha \sum_{j=0}^{\infty} \varepsilon_{-j}$ is infinite, so that the associated density degenerates to 0 over the entire domain of the random variable. In this case there is no effective information from the past.

When the states are governed by a stationary stochastic process, a general method can be established for finding the moments of their steady state distribution. The key is the transition equation (12.1b), which has the solution $\boldsymbol{x}_t = \sum_{j=0}^{\infty} \boldsymbol{F}^j \boldsymbol{g} \varepsilon_{t-j}$. Therefore, $\boldsymbol{x}_t$ has mean $\mathbf{0}$ and variance matrix $\boldsymbol{V}_{\boldsymbol{x}}$, where

$$\boldsymbol{V}_{\boldsymbol{x}} = \sigma^2 \sum_{j=0}^{\infty} \boldsymbol{F}^j \boldsymbol{g} \boldsymbol{g}' \boldsymbol{F}'^j$$

and $\boldsymbol{V}_{\boldsymbol{x}}$ satisfies the linear constraint $\boldsymbol{V}_{\boldsymbol{x}} = \boldsymbol{F} \boldsymbol{V}_{\boldsymbol{x}} \boldsymbol{F}' + \sigma^2 \boldsymbol{g} \boldsymbol{g}'$. When stacking operations are used to convert the latter into a suitable form for solving, we obtain

$$\text{vec}(\boldsymbol{V}_{\boldsymbol{x}}) = \sigma^2 (\boldsymbol{I} - \boldsymbol{F} \otimes \boldsymbol{F})^{-1} \boldsymbol{g} \otimes \boldsymbol{g},$$

where $\otimes$ is the Kronecker product. When the transition matrix $\boldsymbol{F}$ has a triangular structure, it is possible to solve for the variance matrix of the invariant distribution of the states without recourse to these stacking operations. The following example illustrates the basic ideas.

Example 12.3: Double damped local trend model

The invariant distribution of the states of the double damped local trend model

$$\begin{aligned} y_t &= \phi_1 \ell_{t-1} + \phi_2 b_{t-1} + \varepsilon_t, \\ \ell_t &= \phi_1 \ell_{t-1} + \phi_2 b_{t-1} + \alpha \varepsilon_t, \\ b_t &= \phi_2 b_{t-1} + \beta \varepsilon_t \end{aligned}$$

has a variance matrix that satisfies the linear equations

$$\begin{bmatrix} V_{11} & V_{12} \\ V_{21} & V_{22} \end{bmatrix} = \begin{bmatrix} \phi_1 & \phi_2 \\ 0 & \phi_2 \end{bmatrix} \begin{bmatrix} V_{11} & V_{12} \\ V_{21} & V_{22} \end{bmatrix} \begin{bmatrix} \phi_1 & 0 \\ \phi_2 & \phi_2 \end{bmatrix} + \sigma^2 \begin{bmatrix} \alpha^2 & \alpha\beta \\ \beta\alpha & \beta^2 \end{bmatrix}.$$

Unpacking the relationships leads to a triangular set of equations

$$\begin{aligned} V_{22} &= \sigma^2 \beta^2 / (1 - \phi_2^2), \\ V_{21} &= \sigma^2 (\phi_2^2 V_{22} + \beta\alpha) / (1 - \phi_1 \phi_2), \\ V_{11} &= \sigma^2 (2\phi_1 \phi_2 V_{21} + \phi_2^2 V_{22} + \alpha^2) / (1 - \phi_1^2), \end{aligned}$$

which may be solved for the variances and covariances in the order V_{22}, V_{21}, V_{11}.

A nonstationary time series may have both stationary and nonstationary components. The damped trend model in Example 12.4 has a nonstationary level, but a stationary growth rate. It illustrates the point that the moments of a seed vector may be only partially known. In many common cases, the situation is even worse. When all the states follow nonstationary processes, as in Example 12.2, the moments of the seed vector are completely unknown.

Example 12.4: Damped trend model

The damped trend model ETS(A,A$_d$,N) is

$$\begin{aligned} y_t &= \ell_{t-1} + \phi b_{t-1} + \varepsilon_t, \\ \ell_t &= \ell_{t-1} + \phi b_{t-1} + \alpha\varepsilon_t, \\ b_t &= \phi b_{t-1} + \beta\varepsilon_t. \end{aligned}$$

By iterating the last two equations, we find that

$$b_t = \beta \sum_{j=0}^{\infty} \phi^j \varepsilon_{t-j}$$

and

$$\ell_t = \sum_{j=1}^{\infty} \left[\alpha + \beta(1-\phi^j)(1-\phi)^{-1} + \phi^{j-1}\right] \varepsilon_{t-j} + \alpha\varepsilon_t,$$

from which it follows that

$$\mathrm{V}\begin{bmatrix}\ell_t \\ b_t\end{bmatrix} = \frac{\beta\sigma^2}{1-\phi^2}\begin{bmatrix} \infty & \frac{\phi\beta}{1-\phi} + (1+\phi)\alpha \\ \frac{\phi\beta}{1-\phi} + (1+\phi)\alpha & 1 \end{bmatrix}.$$

The levels follow a random walk and so have infinite variance. The growth rates are governed by an AR(1) process with a damping factor satisfying the stationarity condition $|\phi| < 1$. The variance of the seed growth rate is therefore finite. The covariance between the seed level and seed growth rate is also finite.

12.2 Estimation

It is rare for all of the elements in the transition matrix $\boldsymbol{F}$ and persistence vector $\boldsymbol{g}$ of the innovations state space model to be known a priori, and so they must be estimated. Apart from the common scaling factor σ^2, the unknown parameters of $\boldsymbol{F}$ and $\boldsymbol{g}$ can be collected together into a vector $\boldsymbol{\theta}$ which must be estimated. The option to be examined here is estimation by maximizing the likelihood.

Under the finite start-up assumption, the likelihood was based on the probability density function $p(\boldsymbol{y}|\sigma^2, \boldsymbol{\theta}, \boldsymbol{x}_0)$, where $\boldsymbol{x}_0$ is a fixed vector of seed states. Under the infinite start-up assumption, the seed state vector $\boldsymbol{x}_0$ is random and must be integrated out of the density function to give the new density $p(\boldsymbol{y}|\sigma^2, \boldsymbol{\theta})$. The latter forms the basis of a new form of likelihood function when the seed vector is random.

The likelihood function, based on the joint distribution of the sample, is not particularly simple to evaluate directly. It requires knowledge of the variance matrix of the sample, which is often difficult to derive. Moreover, given that it has the dimensions of the sample size, the required inversion of the variance matrix is no trivial matter. Fortunately, the likelihood can be decomposed into a product of one-step-ahead prediction distributions. It can then be more simply evaluated using the moments of these univariate distributions, which are obtained from either the information filter in Sect. 12.3 or the Kalman filter in Sect. 12.7

The prediction form of the likelihood is examined for both stationary and nonstationary time series. In the stationary case, the seed vector has a known distribution, and so the number of well-defined predictions corresponds to the number of observations. In the nonstationary case, it will be seen that the number of well-defined predictions falls short of the number of observations. The problem of defining the likelihood in this case is also addressed.

12.2.1 Stationary Time Series

The likelihood function for a stationary time series is based on the density $p(\boldsymbol{y}|\boldsymbol{\theta}, \sigma^2)$. Let $y_{t|t-1}$ designate y_t conditioned on $\boldsymbol{y}_{1:t-1}$, $\boldsymbol{\theta}$ and σ^2, where $\boldsymbol{y}_{s:t} = [y_s, y_{s+1}, \ldots, y_t]'$ denotes a vector containing a subseries of observations and $\boldsymbol{y}_{1:0}$ is interpreted as a series with no elements. Then we can write

$$p(\boldsymbol{y}|\boldsymbol{\theta}, \sigma^2) = \prod_{t=1}^{n} p(y_{t|t-1})$$

by successive conditioning. The typical component of this product is given by

$$p(y_{t|t-1}) = (2\pi v_{t|t-1})^{-1/2} \exp\left(-\frac{(y_t - \mu_{t|t-1})^2}{2v_{t|t-1}}\right),$$

where $\mu_{t|t-1}$ and $v_{t|t-1}$ ($t \geq 2$) are the mean and variance of the one-step-ahead predictions, and $\mu_{1|0}$ and $v_{1|0}$ are the corresponding quantities for the invariant distribution of the series. The likelihood can therefore be written as

$$\mathcal{L}(\sigma^2, \boldsymbol{\theta} \mid \boldsymbol{y}) = (2\pi\sigma^2)^{-n/2} \left(\prod_{t=1}^{n} \bar{v}_{t|t-1}\right)^{-1/2} \exp\left(-\sum_{t=1}^{n} \frac{(y_t - \mu_{t|t-1})^2}{2\bar{v}_{t|t-1}\sigma^2}\right),$$

where $\bar{v}_{t|t-1} = v_{t|t-1}/\sigma^2$ is the standardized one-step-ahead variance. This likelihood may be compared with (5.1) for the likelihood obtained under the finite start-up assumption in Sect. 5.1. Because $y_t - \mu_{t|t-1}$ is the one-step-ahead prediction error, this is called the *prediction error decomposition of the likelihood function* (Schweppe 1965). It will be seen that a filter, when applied to each successive element of $\boldsymbol{y}$, yields the mean $\mu_{t|t-1}$ and the standardized variance $\bar{v}_{t|t-1}$ of the one-step-ahead prediction distribution, the mean being the point prediction of y_t using the past observations $\boldsymbol{y}_{1:t-1}$. The maximum likelihood estimate of σ^2 is

$$\hat{\sigma}^2 = \frac{1}{n}\sum_{t=1}^{n}(y_t - \mu_{t|t-1})^2/\bar{v}_{t|t-1}.$$

We use this to concentrate σ^2 out of the likelihood. Then, deleting nuisance constants and inverting the resultant function, we obtain the criterion

$$S(\boldsymbol{\theta}) = \left(\prod_{t=1}^{n}\bar{v}_{t|t-1}\right)^{1/n}\sum_{t=1}^{n}(y_t - \mu_{t|t-1})^2/\bar{v}_{t|t-1}.$$

This is the *augmented sum of squared errors*, the random seed state analogue of (5.4).

The parameter vector $\boldsymbol{\theta}$ that minimizes the augmented sum of squared errors also maximizes the likelihood function. For computational purposes, greater numerical stability is achieved by minimizing the log augmented sum of squared errors

$$\log S(\boldsymbol{\theta}) = \frac{1}{n}\sum_{t=1}^{n}\log(\bar{v}_{t|t-1}) + \log\left(\sum_{t=1}^{n}(y_t - \mu_{t|t-1})^2/\bar{v}_{t|t-1}\right).$$

12.2.2 Nonstationary Time Series

In nonstationary cases, the formation of a likelihood is more problematic. It will be seen that when a filter is applied to a state space model with r nonstationary states, usually the first r predictions are arbitrary and the corresponding one-step-ahead prediction distributions are zero across the entire real line. To obtain a properly defined distribution, these must be discarded. It is standard practice to base the likelihood on the well-defined one-step-ahead prediction distributions from period $r+1$ onwards. The prediction error decomposition of the likelihood becomes

$$\mathcal{L}(\sigma^2, \boldsymbol{\theta} \mid \boldsymbol{y}) = (2\pi\sigma^2)^{-(n-r)/2}\left(\prod_{t=r+1}^{n}\bar{v}_{t|t-1}\right)^{-1/2}\exp\left(-\sum_{t=r+1}^{n}\frac{(y_t - \mu_{t|t-1})^2}{2\bar{v}_{t|t-1}\sigma^2}\right).$$

The maximum likelihood estimate of σ^2 is now

$$\hat{\sigma}^2 = \frac{1}{n-r}\sum_{t=r+1}^{n}(y_t - \mu_{t|t-1})^2/\bar{v}_{t|t-1}.$$

Using this to concentrate σ^2 out of the likelihood, and simplifying again, we obtain the augmented sum of squares criterion

$$S(\theta) = \left(\prod_{t=r+1}^{n} \bar{v}_{t|t-1}\right)^{1/(n-r)} \sum_{t=r+1}^{n} (y_t - \mu_{t|t-1})^2 / \bar{v}_{t|t-1}.$$

12.3 Information Filter

12.3.1 Background

The information filter is an algorithm capable of generating the moments of the one-step-ahead prediction distributions needed for the evaluation of the likelihood function. It is a recursive procedure which passes forward, processing each observation in turn. Each stage of this procedure has a prediction step and a revision step. It relies on triangular factorizations of inverse variance matrices, which means that it may be viewed as a square-root filter.

Before describing the information filter, we first provide some necessary background information on triangular stochastic equations. The basic idea is that if $\boldsymbol{x}$ designates a generic random vector with mean $\boldsymbol{m_x}$ and variance matrix $\boldsymbol{V_x}$, then the inverse variance matrix $\boldsymbol{V_x}^{-1}$ can be decomposed using Gaussian elimination to give the factorization

$$\boldsymbol{V_x}^{-1} = \boldsymbol{R}'\boldsymbol{V}^{-1}\boldsymbol{R}, \tag{12.2}$$

where $\boldsymbol{R}$ is a unit upper triangular matrix[1] and $\boldsymbol{V}$ is a diagonal matrix.

A new random vector $\boldsymbol{c}$ can be defined by the equation $\boldsymbol{c} = \boldsymbol{Rx}$. It is easy to see that $\boldsymbol{c}$ has mean vector $\boldsymbol{m_c} = \boldsymbol{Rm}_x$ and variance matrix $\boldsymbol{V_c} = \boldsymbol{V}$. In effect, the transformation $\boldsymbol{R}$ converts a random vector $\boldsymbol{x}$, which may have correlated elements, to a random vector $\boldsymbol{c}$ with uncorrelated elements. The matrix $\boldsymbol{R}$ is an alternative to covariances for summarizing all the available information on interdependencies between the elements of $\boldsymbol{x}$. The elements of $\boldsymbol{c}$ are fundamental components that encapsulate information about central tendency and dispersion without being confounded by information about interdependencies. It follows that all the information about a random vector $\boldsymbol{x}$ is encapsulated by the triangular stochastic equation

$$\boldsymbol{Rx} = \boldsymbol{c}, \tag{12.3}$$

where $\boldsymbol{V_c}$ is a diagonal matrix. The moments of a random vector $\boldsymbol{x}$ can always be constructed from the triangular equations $\boldsymbol{Rm_x} = \boldsymbol{m_c}$ and $\boldsymbol{RV_xR}' = \boldsymbol{V_c}$.

It will be seen that the filter is built on stochastic equations of the general form

$$\boldsymbol{Ax} = \boldsymbol{b}, \tag{12.4}$$

[1] A *unit upper triangular matrix* has 1s on the diagonal and 0s below the diagonal.

where $\boldsymbol{A}$ is a fixed, but not necessarily triangular, matrix and $\boldsymbol{b}$ is a random vector with uncorrelated elements. When $\boldsymbol{A}$ is non-singular, it is always possible to derive a triangular stochastic equation like (12.3) that is equivalent. The key to this derivation is a QR decomposition (Golub and Van Loan 1996) of the matrix $\boldsymbol{A}$ as

$$\boldsymbol{A} = \boldsymbol{Q}\boldsymbol{R}, \tag{12.5}$$

where $\boldsymbol{R}$ is a unit upper triangular matrix. Unlike a traditional QR decomposition where $\boldsymbol{Q}$ is an orthogonal matrix, the matrix $\boldsymbol{Q}$ is constructed so that the product $\boldsymbol{\Lambda} = \boldsymbol{Q}'\boldsymbol{V}_b^{-1}\boldsymbol{Q}$ is a diagonal matrix. Such a decomposition can always be constructed using fast Givens transformations (Stirling 1981), the theory of which is presented in Appendix "Triangularization of stochastic equations."

On obtaining this decomposition, the triangular equations may be derived as follows. Substitute (12.5) into (12.4) to give $\boldsymbol{Q}\boldsymbol{R}\boldsymbol{x} = \boldsymbol{b}$. Premultiply by $\boldsymbol{Q}'\boldsymbol{V}_b^{-1}$ to give $\boldsymbol{Q}'\boldsymbol{V}_b^{-1}\boldsymbol{Q}\boldsymbol{R}\boldsymbol{x} = \boldsymbol{Q}'\boldsymbol{V}_b^{-1}\boldsymbol{b}$, and replace the term $\boldsymbol{Q}'\boldsymbol{V}_b^{-1}\boldsymbol{Q}$ by the diagonal matrix $\boldsymbol{\Lambda}$ to give $\boldsymbol{\Lambda}\boldsymbol{R}\boldsymbol{x} = \boldsymbol{Q}'\boldsymbol{V}_b^{-1}\boldsymbol{b}$. Thus, $\boldsymbol{R}\boldsymbol{x} = \boldsymbol{\Lambda}^{-1}\boldsymbol{Q}'\boldsymbol{V}_b^{-1}\boldsymbol{b}$. Define the random vector $\boldsymbol{c}$ by $\boldsymbol{c} = \boldsymbol{\Lambda}^{-1}\boldsymbol{Q}'\boldsymbol{V}_b^{-1}\boldsymbol{b}$. Our triangular equation then takes the form (12.3). Then it follows that the moments of $\boldsymbol{c}$ are $\boldsymbol{m}_c = \boldsymbol{\Lambda}^{-1}\boldsymbol{Q}'\boldsymbol{V}_b^{-1}\boldsymbol{m}_b$ and $\boldsymbol{V}_c = \boldsymbol{\Lambda}^{-1}$. In particular, the elements of the constructed $\boldsymbol{c}$ are uncorrelated, as required. It is convenient to write the generalized orthogonality condition for $\boldsymbol{Q}$ as

$$\boldsymbol{Q}'\boldsymbol{V}_b^{-1}\boldsymbol{Q} = \boldsymbol{V}_c^{-1}. \tag{12.6}$$

An information filter begins with a triangular equation representation for the seed state $\boldsymbol{x}_0$. It then processes each observation in turn, each stage terminating with the triangular equation representing what is known about the random state vector $\boldsymbol{x}_t$ conditional on the sub-series $\boldsymbol{y}_{1:t}$.

The innovations variance σ^2 can be omitted from many calculations if we work with standardized variances such as $\bar{\boldsymbol{V}}_x = \boldsymbol{V}_x/\sigma^2$ and $\bar{v}_{t|s} = v_{t|s}/\sigma^2$, where $\boldsymbol{V}_x = \mathrm{V}(\boldsymbol{x})$ and $v_{t|s} = \mathrm{V}(y_t \mid \boldsymbol{y}_{1:s})$. Using standardized variances allows us to avoid the problem of needing to know σ^2 when computing predictions from the model.

12.3.2 Initialization

For stationary time series, the seed state $\boldsymbol{x}_0$ is assigned the invariant distribution of the process. The inverse standardized variance matrix of the invariant distribution is factorized according to (12.2). The following examples illustrate how this is done.

Example 12.5: Damped level model

In Example 12.1, the damped local level model had $\mathrm{V}(\ell_0) = \alpha^2\sigma^2/(1-\phi^2)$. Here $\boldsymbol{R} = 1$ and $\bar{\boldsymbol{V}}_c = \alpha^2/(1-\phi^2)$.

Example 12.6: Damped trend model

The invariant distribution of the states of the double damped local trend model

$$\begin{aligned} y_t &= 0.9\ell_{t-1} + 0.8b_{t-1} + \varepsilon_t, \\ \ell_t &= 0.9\ell_{t-1} + 0.8b_{t-1} + 0.5\varepsilon_t, \\ b_t &= 0.8b_{t-1} + 0.1\varepsilon_t \end{aligned}$$

has, using the formulae from Example 12.3, a standardized variance matrix $\begin{bmatrix} 3.2439 & 0.2421 \\ 0.2421 & 0.0278 \end{bmatrix}$ with corresponding inverse $\begin{bmatrix} 0.8806 & -7.6692 \\ -7.6692 & 102.7591 \end{bmatrix}$. The triangular factorization of this inverse matrix has components

$$\boldsymbol{R} = \begin{bmatrix} 1.0000 & -8.7086 \\ 0 & 1.0000 \end{bmatrix} \quad \text{and} \quad \bar{\boldsymbol{V}}_{\boldsymbol{c}} = \begin{bmatrix} 1.1356 & 0 \\ 0 & 0.0278 \end{bmatrix}.$$

Example 12.7: Local level model

The local level model in Example 12.2 represents a nonstationary time series. The seed local level has an infinite variance. Here $\boldsymbol{R} = 1$ and $\bar{\boldsymbol{V}}_{\boldsymbol{c}} = \infty$. The infinite diagonal matrix means that no prior information is being provided to the filter.

Example 12.8: Damped trend model

The matrix $\begin{bmatrix} a & b \\ b & c \end{bmatrix}$ has an inverse $(ac - b^2)^{-1} \begin{bmatrix} c & -b \\ -b & a \end{bmatrix}$ which, as $a \to \infty$, converges to $\begin{bmatrix} 0 & 0 \\ 0 & c^{-1} \end{bmatrix}$. Therefore, the inverse standardized variance matrix associated with the invariant distribution for the damped trend model in Example 12.4 is $\begin{bmatrix} 0 & 0 \\ 0 & (1-\phi^2)/\beta \end{bmatrix}$. Here

$$\boldsymbol{R} = \begin{bmatrix} 1 & 0 \\ 0 & 1 \end{bmatrix} \quad \text{and} \quad \bar{\boldsymbol{V}}_{\boldsymbol{c}} = \begin{bmatrix} \infty & 0 \\ 0 & \beta/(1-\phi^2) \end{bmatrix}.$$

The first diagonal element of ∞ in the diagonal matrix means that the first row of $\boldsymbol{R}$ provides no information to the filter. It is interesting that the covariances can safely be ignored because, by taking the limit, the finite covariances have disappeared from the inverse variance matrix.

12.3.3 Prediction Step

We now step forward in time to the start of period t, before the random variable y_t has been observed. The aim is to find the prediction distribution of y_t informed by (or conditioned on) past observations $\boldsymbol{y}_{1:t-1}$. A summary of past information is conveyed by the triangular stochastic equation

$$\boldsymbol{R}_{t-1}\boldsymbol{x}_{t-1|t-1} = \boldsymbol{c}_{t-1|t-1}, \tag{12.7}$$

obtained by previous applications of the filter. The exception occurs at the start of period 1, where the triangular equation representation of the invariant state distribution is used. Here, we use the notation $\boldsymbol{x}_{t-1|t-1}$ to mean that $\boldsymbol{x}_{t-1}$ is conditioned on $\boldsymbol{y}_{1:t-1}$.

To specify the prediction step of the information filter, it is convenient to eliminate ε_t from the transition equation (12.1b) and rewrite the invariant innovations state space model (12.1) as

$$y_t = \boldsymbol{w}'\boldsymbol{x}_{t-1} + \varepsilon_t, \tag{12.8a}$$
$$\boldsymbol{x}_t = \boldsymbol{D}\boldsymbol{x}_{t-1} + \boldsymbol{g}y_t, \tag{12.8b}$$

where $\varepsilon_t \sim \text{NID}(0, \sigma^2)$ and $\boldsymbol{D} = \boldsymbol{F} - \boldsymbol{g}\boldsymbol{w}'$. The information from the past, represented by (12.7), is combined with information from the present, represented by the state space equations (12.8a) and (12.8b), to give

$$\begin{bmatrix} \boldsymbol{R}_{t-1} & \boldsymbol{0} & \boldsymbol{0} \\ -\boldsymbol{w}' & \boldsymbol{0} & 1 \\ -\boldsymbol{D} & \boldsymbol{I} & -\boldsymbol{g} \end{bmatrix} \begin{bmatrix} \boldsymbol{x}_{t-1|t-1} \\ \boldsymbol{x}_{t|t-1} \\ y_{t|t-1} \end{bmatrix} = \begin{bmatrix} \boldsymbol{c}_{t-1|t-1} \\ \varepsilon_t \\ 0 \end{bmatrix}.$$

Because this stochastic equation has the general form $\boldsymbol{A}\boldsymbol{x} = \boldsymbol{b}$, the triangularization algorithm from Appendix "Triangularization of stochastic equations" may be applied to give

$$\begin{bmatrix} * & * & * \\ 0 & \boldsymbol{R}_t & \boldsymbol{r}_t \\ 0 & 0 & 1 \end{bmatrix} \begin{bmatrix} \boldsymbol{x}_{t-1|t-1} \\ \boldsymbol{x}_{t|t-1} \\ y_{t|t-1} \end{bmatrix} = \begin{bmatrix} * \\ \boldsymbol{c}_{t|t-1} \\ d_{t|t-1} \end{bmatrix},$$

where an asterisk ($*$) designates quantities that are of no direct interest. These equations can be unpacked to give

$$\boldsymbol{R}_t\boldsymbol{x}_{t|t-1} + \boldsymbol{r}_t y_{t|t-1} = \boldsymbol{c}_{t|t-1} \tag{12.9a}$$

and

$$y_{t|t-1} = d_{t|t-1}. \tag{12.9b}$$

Thus, the mean and variance of $d_{t|t-1}$ are the one-step-ahead mean and variance of y_t. As we saw in Sect. 12.2, these are needed for the evaluation of the likelihood function.

12.3.4 Revision Step

We now step forward to the end of period t, at which point it is assumed that y_t has been observed. The now fixed value of y_t may be substituted into (12.9a) to give

$$\boldsymbol{R}_t \boldsymbol{x}_{t|t} = \boldsymbol{c}_{t|t}, \tag{12.10}$$

where

$$\boldsymbol{c}_{t|t} = \boldsymbol{c}_{t|t-1} - \boldsymbol{r}_t y_t.$$

It is (12.10) that is carried forward to the next period. It represents the most current information about the past behavior of the process.

This completes the specification of the information filter. Unlike its more common counterpart, the Kalman filter, it applies to both stationary and nonstationary time series without major change. In the nonstationary case, one must simply be aware that some of the row variances of the triangular stochastic equations for the seed states are infinite. In the case where no information is available from the past, all of these row variances are set to infinity. As is shown in Appendix "Triangularization of stochastic equations," the recommended triangular factorization algorithm used in the information filter has a limiting form in the presence of infinite variances. It follows, as asserted, that the information filter can then be applied to nonstationary time series without modification.

The above revision step involves conditioning on the new observation y_t. It is remarkable how simple it becomes to conduct the conditioning operation when it is recast like this in terms of triangular stochastic equations.

12.3.5 Linear Trend Example

The information filter will now be illustrated by fitting a global linear trend to the time series $\{5, 10, 12, 15, 20\}$. It may be considered a special case of fitting a local trend model where $\alpha = \beta = 0$. The state space form of a linear trend line is

$$\begin{aligned} y_t &= \ell_{t-1} + b_{t-1} + \varepsilon_t, \\ \ell_t &= \ell_{t-1} + b_{t-1}, \\ b_t &= b_{t-1}, \end{aligned}$$

where ℓ_t designates the level in period t and b_t designates the growth rate.

It is convenient to express the stochastic equations in tableau form. The first columns of a tableau are used to store the left hand matrix of a stochastic equation; the next column contains the means of the right hand side; and the final column lists the standardized variances of the right hand side.

Example 12.9: Tableau form of stochastic equations

The stochastic equation

$$\begin{bmatrix} 2 & 1 \\ 1 & 3 \end{bmatrix} \begin{bmatrix} x_1 \\ x_2 \end{bmatrix} = b,$$

where

$$b \sim \text{NID}\left(\begin{bmatrix} 3 \\ 5 \end{bmatrix}, 81 \begin{bmatrix} 0.50 & 0 \\ 0 & 0.25 \end{bmatrix} \right),$$

has a tableau

x_1	x_2	m	$\bar{v}$
2	1	3	0.50
1	3	5	0.25 .

An overview of the calculations is shown in Table 12.1. Each row lists some results from a stage of the algorithm. The second column contains the stochastic equations for the one-step-ahead prediction distributions that emerge after the application of the prediction step. The third column has the triangular equations that carry information about the states from one stage to the next. To save space, redundant columns of zeros have been dropped from the equations.

The following inferences can be drawn from these calculations:

1. The predictions following stages 1 and 2 appear to be 0. However, the variances are infinite in both cases, suggesting that the predictions are really arbitrary.
2. The triangular equations system following stage 1 has an arbitrary solution for the mean growth rate. The mean level, however, is equal to 5. This corresponds to the situation where there are an infinite number of possible trend lines passing through the point $(1, 5)$, a situation depicted in Fig. 12.1 left.
3. At the end of stage 2, the variances in the triangular equations are both finite. Back-solving gives the solution $m^b_{2|2} = 5$ and $m^\ell_{2|2} = 10$. There is now enough information to define a unique trend line. Passing through the points $(1, 5)$ and $(2, 10)$, it is graphed in Fig. 12.1 right.

After processing all the observations, the distribution of the level and slope for period 5 is governed by the equations shown at the bottom right of the table. Back-solving them gives

$$m_{5|5} = \begin{bmatrix} 19.4 \\ 3.5 \end{bmatrix} \quad \text{and} \quad \bar{V}_{5|5} = \begin{bmatrix} 0.6 & 0.2 \\ 0.2 & 0.1 \end{bmatrix}.$$

Table 12.1. Trend calculations summary.

Sample	$y_{t\|t-1} = d_{t\|t-1}$	$\boldsymbol{R}_t \boldsymbol{x}_{t\|t} = \boldsymbol{c}_{t\|t}$
Initialization		$\begin{array}{cc\|cc} \ell_{0\|0} & b_{0\|0} & \boldsymbol{m}_{0\|0} & \bar{\boldsymbol{v}}_{0\|0} \\ \hline 1 & 0 & 0 & \infty \\ 0 & 1 & 0 & \infty \end{array}$
$y_1 = 5$	$\begin{array}{c\|cc} y_{1\|0} & \mu_{1\|0} & \bar{v}_{1\|0} \\ \hline 1 & 0 & \infty \end{array}$	$\begin{array}{cc\|cc} \ell_{1\|1} & b_{1\|1} & \boldsymbol{m}_{1\|1} & \bar{\boldsymbol{v}}_{1\|1} \\ \hline 1 & 0 & 5 & 1 \\ 0 & 1 & 0 & \infty \end{array}$
$y_2 = 10$	$\begin{array}{c\|cc} y_{2\|1} & \mu_{2\|1} & \bar{v}_{2\|1} \\ \hline 1 & 0 & \infty \end{array}$	$\begin{array}{cc\|cc} \ell_{2\|2} & b_{2\|2} & \boldsymbol{m}_{2\|2} & \bar{\boldsymbol{v}}_{2\|2} \\ \hline 1 & -0.5 & 7.5 & 0.5 \\ 0 & 1 & 5 & 2 \end{array}$
$y_3 = 12$	$\begin{array}{c\|cc} y_{3\|2} & \mu_{3\|2} & \bar{v}_{3\|2} \\ \hline 1 & 15 & 6 \end{array}$	$\begin{array}{cc\|cc} \ell_{3\|3} & b_{3\|3} & \boldsymbol{m}_{3\|3} & \bar{\boldsymbol{v}}_{3\|3} \\ \hline 1 & -1 & 9 & 0.333 \\ 0 & 1 & 3.5 & 0.5 \end{array}$
$y_4 = 15$	$\begin{array}{c\|cc} y_{4\|3} & \mu_{4\|3} & \bar{v}_{4\|3} \\ \hline 1 & 16 & 3.33 \end{array}$	$\begin{array}{cc\|cc} \ell_{4\|4} & b_{4\|4} & \boldsymbol{m}_{4\|4} & \bar{\boldsymbol{v}}_{4\|4} \\ \hline 1 & -1.5 & 10.5 & 0.25 \\ 0 & 1 & 3.2 & 0.2 \end{array}$
$y_5 = 20$	$\begin{array}{c\|cc} y_{5\|4} & \mu_{5\|4} & \bar{v}_{5\|4} \\ \hline 1 & 18.5 & 2.5 \end{array}$	$\begin{array}{cc\|cc} \ell_{5\|5} & b_{5\|5} & \boldsymbol{m}_{5\|5} & \bar{\boldsymbol{v}}_{5\|5} \\ \hline 1 & -2 & 12.4 & 0.2 \\ 0 & 1 & 3.5 & 0.1 \end{array}$

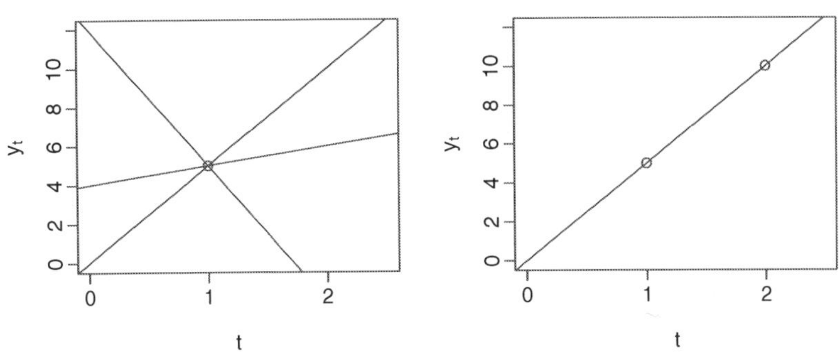

Fig. 12.1. Examples of trend lines based on one and two observations.

This solution provides the standardized variance matrix instead of the variance matrix. To get the variance matrix, it is necessary to scale the standardized variance matrix by an estimate of σ^2. It was seen in Sect. 12.2 that the maximum likelihood estimate of σ^2 is given by

$$\hat{\sigma}^2 = \mathrm{SSE}/(n-r), \tag{12.11}$$

where SSE designates the sum of squared errors and r is the number of free non-informative states. In this trend example $r = 2$.

The sum of squared errors can be calculated in two different ways. One of them is to use the final results to obtain retrospective predictions of the series and calculate the associated errors. More specifically, the retrospective predictions are based on the means $\mu_{t|5} = \mathrm{E}(y_t|y_{1:5})$, quantities that are more commonly referred to as the *smoothed observations*. These may be calculated using the formula

$$\mu_{t|5} = 19.4 - 3.5(5 - t).$$

Table 12.2 shows the smoothed observations, the errors and the sum of squared errors.

A second approach to calculating the sum of squared errors works directly with the one-step-ahead prediction errors from the information filter. The one-step-ahead predictions, together with the associated mean squared errors, are obtained from the last row of the tableaux of the second column of Table 12.1. They have been recorded in columns 2 and 6 of Table 12.3. In the first two periods, the one-step-ahead prediction distributions have arbitrary moments, and so their cells are left blank. Unlike their smoothed counterparts, the one-step-ahead prediction errors do not have the same variance. The squared errors must therefore be adjusted, dividing them by the standardized variances, as shown in the final column of Table 12.3. The column sum also turns out to be 2.7. It is this second approach that is normally used with filters to calculate the sum of squared errors.

Table 12.2. Sum of squared error calculations from smoothed observations.

Period	Smoothed values	Actual values	Errors	Squared errors
1	5.4	5	−0.4	0.16
2	8.9	10	1.1	1.21
3	12.4	12	−0.4	0.16
4	15.9	15	−0.9	0.81
5	19.4	20	0.6	0.36
			SSE	2.70

Table 12.3. Sum of squared error calculations from predictions.

Period	Predicted values	Actual values	Errors	Squared errors	Standardized variances	Ratio
1	Arbitrary	5				
2	Arbitrary	10				
3	15.0	12	−3.0	9.00	6.00	1.50
4	16.0	15	−1.0	1.00	3.33	0.30
5	18.5	20	1.5	2.25	2.50	0.90
					SSE	2.70

The variance $\hat{\sigma}^2$ is now calculated using (12.11) to give 0.90. Therefore, the variance matrix is

$$V_{n|n} = 0.90\bar{V}_{n|n} = \begin{bmatrix} 0.54 & 0.18 \\ 0.18 & 0.09 \end{bmatrix}.$$

12.4 Prediction

Future values of the time series are unknown and must be treated as random variables. When the parameters and seed states are set to their maximum likelihood values, their distributions are Gaussian. So, the prediction problem is to unravel their means and variances. Two methods for doing this are presented here for any linear innovations state space model.

12.4.1 Direct Method

The information filter ends in period n with the stochastic equation $\boldsymbol{R}_n \boldsymbol{x}_{n|n} = \boldsymbol{c}_{n|n}$ that indirectly describes the conditional distribution of the state vector $\boldsymbol{x}_n$. It may be solved for the mean $\boldsymbol{m}_{n|n}$ and variance $\boldsymbol{V}_{n|n}$. These moments will be used to seed a recursive procedure for finding the moments of the future observations.

To derive the recursive procedure, we augment the state vector $\boldsymbol{x}_t$ in future period $t = n + h$ by y_t and so obtain the first-order recurrence relationship

$$\begin{bmatrix} \boldsymbol{x}_t \\ y_t \end{bmatrix} = \begin{bmatrix} \boldsymbol{F} & \boldsymbol{0} \\ \boldsymbol{w}' & 0 \end{bmatrix} \begin{bmatrix} \boldsymbol{x}_{t-1} \\ y_{t-1} \end{bmatrix} \begin{bmatrix} \boldsymbol{g} \\ 1 \end{bmatrix} \varepsilon_t.$$

This has the general form

$$z_t = \boldsymbol{A} z_{t-1} + \boldsymbol{b}\varepsilon_t, \tag{12.12}$$

and thus the mean and variance of the z_t can be computed recursively using the equations

$$\begin{aligned} \boldsymbol{m}_t^z &= \boldsymbol{A}\boldsymbol{m}_{t-1}^z, \\ \boldsymbol{V}_t^z &= \boldsymbol{A}\boldsymbol{V}_{t-1}^z\boldsymbol{A}' + \sigma^2 \boldsymbol{b}\boldsymbol{b}'. \end{aligned}$$

The mean and variance of $y_{t|t-1}$ are the last elements in $\boldsymbol{m}_t^z$ and $\boldsymbol{V}_t^z$.

12.4.2 Indirect Method

By working with the stochastic equations representations of random vectors, the moments of the future observations can be obtained without direct recourse to the means and variances of the future states. The algorithm is based on the recurrence relationship (12.12). The distribution of the state z_t summarizes the past at the beginning of the first future period $t = n+1$. It is represented by its triangular representation $S_n z_{n|n} = c_{n|n}$ where S_n is a unit upper triangular matrix. The following steps are repeated for $t = n+1, \ldots, n+h$.

Step 1 Form the tableau

$z_{t-1\|n}$	ε_t	$z_{t\|n}$	Mean	Variance
S_{t-1}	0	0	$m^c_{t-1\|n}$	$v^c_{t-1\|n}$
0	1	0	0	σ^2
A	b	$-I$	0	0

Step 2 Apply the appropriate linear transformations to obtain the equivalent unit upper triangular stochastic equation system

$z_{t-1\|n}$	ε_t	$z_{t\|n}$	Mean	Variance
*	*	*	*	*
0	1	*	*	*
0	0	S_t	$m^c_{t\|n}$	$v^c_{t\|n}$

where the asterisks are again used to represent quantities of only secondary interest. The bottom elements in m^c_t and v^c_t are the required mean and variance of the distribution of $y_{t|n}$.

12.5 Model Selection

When there are a number of candidate models for a time series, it is necessary to seek that one which is most likely to yield the best forecasts. This issue was considered in Chap. 7 in the context of the finite start-up assumption where we conditioned on the seed states. We now revisit the issue under the infinite start-up assumption where the seed state is now treated as a random rather than a fixed vector.

It was argued in Chap. 7 that the maximized likelihood should not be used as a model selection criterion because it favors models with a large number of parameters. One possible solution to this problem is to use a penalized likelihood to discourage the selection of models with large numbers of parameters. The Akaike information criterion (AIC) was suggested as a possibility.

Now that x_0 is to be treated as a random vector, the AIC is redefined to be

$$\text{AIC} = -2\log \mathcal{L}(\hat{\theta} \mid y) + 2q,$$

where $\hat{\theta}$ designates the maximum likelihood value of the parameter vector θ and q is the number of parameters (excluding σ^2). The key difference is that the seed state $\boldsymbol{x}_0$ is now integrated out of the likelihood function. Moreover, the parameter count no longer includes the number of free seed states.

Evidence was provided in Chap. 7 that the AIC is quite good at model selection on real time series. When the seed state is treated as a random vector, however, things become problematic. The predicament is best illustrated by the behavior of the AIC in the case of the AR(1) process $y_t = \phi y_{t-1} + \varepsilon_t$. The likelihood function, needed for the calculation of the AIC, may be obtained using its prediction error decomposition. This begins with the marginal distribution of y_1, which in Example 11.1 (with $\mu = 0$), was shown to be $\mathrm{N}(0, \sigma^2/(1-\phi^2))$. The one-step-ahead prediction distribution of y_t is $N(\phi y_{t-1}, \sigma^2)$, and so the likelihood function is

$$\mathcal{L}(\phi, \sigma^2 \mid \boldsymbol{y}) = \frac{(1-\phi^2)^{1/2}}{(2\pi\sigma^2)^{n/2}} \exp\left(-\frac{1}{2\sigma^2} \left[(1-\phi^2) y_1^2 + \sum_{t=2}^{n} (y_t - \phi y_{t-1})^2 \right] \right).$$

Suppose we wish to use the AIC to compare an AR(1) process where $\hat{\phi} = 0.99$ with a random walk $y_t = y_{t-1} + \varepsilon_t$ where the constraint $\phi = 1$ effectively applies. The optimal value of the likelihood function is finite for the AR(1) process but 0 for the random walk. In other words, the AIC will be finite in the first case but infinite in the second. This problem with the random walk model is typically circumvented by conditioning on the first sample value y_1 and basing the likelihood on the conditional joint density $p(y_2, \ldots y_n | y_1, \phi, \sigma^2)$. It can be shown that this is equivalent to a likelihood function based on the first-differences of the series. This now gives a finite value for the likelihood of the random walk, but it is unrelated to the likelihood of an AR(1) process: their likelihoods are non-comparable. This, problem carries over to the AIC and it is for this reason that the level of differencing in an ARIMA approach is typically determined with unit root tests rather than information criteria.

In our view, the avoidance of this problem provides a strong incentive to condition on the seed state vector (as was done in Chap. 7). The cost might be a small loss of statistical efficiency in the estimates, and the creation of slight inconsistencies between the state space and standard ARIMA approaches. The gain, however, is a coherent, viable approach to model selection which avoids the complexities of unit root tests. This argument means that exponential smoothing is to be preferred to a Kalman or information filter in the context of the innovations form of the state space model.

12.6 Smoothing Time Series

A filter involves a forward pass through a time series, directly or indirectly computing the moments of the conditional distributions of random states $\boldsymbol{x}_{t|t-1}$ and the one-step-ahead predictions $y_{t|t-1}$. It finishes with information, direct or indirect, about the moments of $\boldsymbol{x}_{n|n}$. The last state is the only one conditioned on the entire sample. All other states are conditioned on a partial sample representing the history at the points of time to which they refer.

We now consider the problem of conditioning the states on the entire sample. We seek an algorithm that finds the moments of $\boldsymbol{x}_{t|n}$ rather than the moments of $\boldsymbol{x}_{t|t-1}$ or $\boldsymbol{x}_{t|t}$. We shall see that it involves a backward pass through the data following the forward pass with the information filter.

The key is the prediction step of the information filter which resulted in the triangular equation

$$\begin{bmatrix} * & * & * \\ 0 & \boldsymbol{R}_t & \boldsymbol{r}_t \\ 0 & 0 & 1 \end{bmatrix} \begin{bmatrix} \boldsymbol{x}_{t-1|t-1} \\ \boldsymbol{x}_{t|t-1} \\ y_{t|t-1} \end{bmatrix} = \begin{bmatrix} * \\ \boldsymbol{c}_{t|t-1} \\ d_{t|t-1} \end{bmatrix}.$$

It is now necessary to replace the asterisks by symbols because the top component turns out to be the key to smoothing. It is rewritten as:

$$\begin{bmatrix} \boldsymbol{T}_t & \boldsymbol{L}_t & \boldsymbol{q}_t \\ 0 & \boldsymbol{R}_t & \boldsymbol{r}_t \\ 0 & 0 & 1 \end{bmatrix} \begin{bmatrix} \boldsymbol{x}_{t-1|t-1} \\ \boldsymbol{x}_{t|t-1} \\ y_{t|t-1} \end{bmatrix} = \begin{bmatrix} \boldsymbol{a}_{t|t-1} \\ \boldsymbol{c}_{t|t-1} \\ d_{t|t-1} \end{bmatrix},$$

where $\boldsymbol{T}_t$ is a unit upper triangular matrix.

After the revision step, the top two sub-equations reduce to

$$\begin{bmatrix} \boldsymbol{T}_t & \boldsymbol{L}_t \\ 0 & \boldsymbol{R}_t \end{bmatrix} \begin{bmatrix} \boldsymbol{x}_{t-1|t} \\ \boldsymbol{x}_{t|t} \end{bmatrix} = \begin{bmatrix} \boldsymbol{a}_{t|t} \\ \boldsymbol{c}_{t|t} \end{bmatrix},$$

where

$$\boldsymbol{a}_{t|t} = \boldsymbol{a}_{t|t-1} - \boldsymbol{q}_t y_t,$$
$$\boldsymbol{c}_{t|t} = \boldsymbol{c}_{t|t-1} - \boldsymbol{r}_t y_t.$$

The top equation is

$$\boldsymbol{T}_t \boldsymbol{x}_{t-1|t} = \boldsymbol{a}_{t|t-1} - \boldsymbol{L}_t \boldsymbol{x}_{t|t}.$$

This provides a mechanism for calculating a preceding state from its successor. It is therefore suitable for the backward pass. Thus, during the forward pass with the information filter, these equations are stored for use during the backward pass.

The backward pass begins by solving for the moments of $\boldsymbol{x}_{n-1|n}$ from the moments of $\boldsymbol{x}_{n|n}$. Then each preceding equation is used to unravel the moments of preceding states, terminating with the moments of the conditioned seed state $\boldsymbol{x}_{0|n}$.

As we show in Sect. 13.2.2, $\boldsymbol{x}_{t|t}$ converges stochastically to $\boldsymbol{x}_t$, so that the additional information provided by $\boldsymbol{x}_{t|n}$ becomes negligible. This result has sometimes been viewed as implying that a two-sided filter cannot be applied within the innovations framework. However, this interpretation is not correct. The reason for the confusion lies in the specification of the state variables. In the innovations model, we can express $\boldsymbol{x}_t$ in terms of $\{y_t, \ldots, y_1, \boldsymbol{x}_0\}$ as demonstrated in (3.5), which clearly demonstrates the one-sided nature of the structure of the state variable. An appropriate framework for two-sided filters is developed in Sect. 13.5.

12.7 Kalman Filter

The Kalman filter is an alternative mechanism for progressively revising the moments of the distributions of states and current unobserved series values with information gleaned from past observed values of a series. As part of this process, it produces the point predictions and the conditional variances needed to evaluate the likelihood function. It is widely recommended and used for this purpose (Harvey 1989), so for completeness, its logic will be outlined here. However, its role in this respect should be supplanted by the information filter. Unlike the information filter, the Kalman filter has no limiting form in the presence of infinite state variances, and cannot be applied without further modification to nonstationary time series.

The Kalman filter is presented here in terms of variances. Where required, its equations can be easily adapted to involve standardized variances instead of variances. Either way, each stage of the Kalman filter, like the information filter, has a prediction step and revision step.

12.7.1 Prediction Step

We jump to the situation that prevails at the *beginning* of period t. The moments of the conditional state vector $\boldsymbol{x}_{t-1|t-1}$ are known from the application of the filter to the past series values $\boldsymbol{y}_{1:t-1}$. The exception is period 1, where the moments of $\boldsymbol{x}_{1|0}$ are set to the moments of the invariant distribution[2] of the states, obtained using the method described in Sect. 12.1.

The prediction step, at the start of period t, is geared to finding the moments of $y_{t|t-1}$ and $\boldsymbol{x}_{t|t-1}$. The covariance between them is also needed. It is therefore best to combine $y_{t|t-1}$ and $\boldsymbol{x}_{t|t-1}$ into a single random vector

$$z_{t|t-1} = \begin{bmatrix} y_{t|t-1} \\ \boldsymbol{x}_{t|t-1} \end{bmatrix}. \tag{12.13}$$

Equations (12.1a) and (12.1b), after conditioning on $\boldsymbol{y}_{1:t-1}$, can be stacked to give

$$z_{t|t-1} = \boldsymbol{A}\boldsymbol{x}_{t-1|t-1} + \boldsymbol{b}\varepsilon_t, \tag{12.14}$$

[2] This is where the stationarity assumption is critical. Nonstationary time series do not possess an invariant distribution.

where $\boldsymbol{A} = \begin{bmatrix} \boldsymbol{w}' \\ \boldsymbol{F} \end{bmatrix}$ and $\boldsymbol{b} = \begin{bmatrix} 1 \\ \boldsymbol{g} \end{bmatrix}$. Using standard rules for calculating the moments of linear combinations of uncorrelated random vectors, the moments of $z_{t|t-1}$ can then be derived with these equations:

$$\boldsymbol{m}^z_{t|t-1} = \boldsymbol{A}\boldsymbol{m}_{t-1|t-1},$$
$$\boldsymbol{V}^z_{t|t-1} = \boldsymbol{A}\boldsymbol{V}_{t-1|t-1}\boldsymbol{A}' + \sigma^2\boldsymbol{b}\boldsymbol{b}'.$$

The moments which emerge after the application of these formulae can be partitioned in conformity with $y_{t|t-1}$ and $\boldsymbol{x}_{t|t-1}$ as follows

$$\boldsymbol{m}^z_{t|t-1} = \begin{bmatrix} \mu_{t|t-1} \\ \boldsymbol{m}_{t|t-1} \end{bmatrix}, \tag{12.15}$$

$$\boldsymbol{V}^z_{t|t-1} = \begin{bmatrix} v_{t|t-1} & \zeta'_{t|t-1} \\ \zeta_{t|t-1} & \boldsymbol{V}_{t|t-1} \end{bmatrix}. \tag{12.16}$$

The quantity $\mu_{t|t-1} = \boldsymbol{w}'\boldsymbol{m}_{t-1|t-1}$ is the point prediction of y_t made at time $t-1$ utilizing the information from the past series values $\boldsymbol{y}_{1:t-1}$. In other words, it is the one-step-ahead prediction. The quantity $v_{t|t-1} = \boldsymbol{w}'\boldsymbol{V}_{t-1|t-1}\boldsymbol{w} + \sigma^2$ is a measure of the risk associated with this prediction, sometimes referred to as the mean squared error. Other information includes the prediction $\boldsymbol{m}_{t|t-1}$ of $\boldsymbol{x}_t$, the associated mean squared error $\boldsymbol{V}_{t|t-1}$, and the covariance $\zeta_{t|t-1}$.

12.7.2 Revision Step

Time now advances to the *end* of period t, at which point y_t will have been observed; the latter changes from a random to a fixed quantity. The goal of the revision step at this point in time is to find the moments of $\boldsymbol{x}_{t|t}$. This step relies on the following general rules for conditioning a generic random vector $\boldsymbol{x}$ on a generic random variable y (see, for example, Anderson (1958)):

$$\mathrm{E}(\boldsymbol{x}|y) = \boldsymbol{m} + \zeta v^{-1}(y - \mu_y)$$

and

$$\mathrm{V}(\boldsymbol{x}|y) = \boldsymbol{V} - \zeta\zeta' v^{-1},$$

where $\mathrm{V}(y) = v$, $\mathrm{E}(\boldsymbol{x}) = \boldsymbol{m}$, $\mathrm{V}(\boldsymbol{x}) = \boldsymbol{V}$ and $\mathrm{Cov}(\boldsymbol{x}, y) = \zeta$.

The various components required to find the moments of $\boldsymbol{x}_{t|t}$ are available from the output of the prediction step, and the equations for the revision step in the Kalman filter are

$$\boldsymbol{m}_{t|t} = \boldsymbol{m}_{t|t-1} + \zeta_{t|t-1}v^{-1}_{t|t-1}(y_t - \mu_{t|t-1})$$

and

$$\boldsymbol{V}_{t|t} = \boldsymbol{V}_{t|t-1} - \boldsymbol{\zeta}_{t|t-1}\boldsymbol{\zeta}'_{t|t-1}v^{-1}_{t|t-1}.$$

A more common form of these equations is

$$\boldsymbol{m}_{t|t} = \boldsymbol{F}\boldsymbol{m}_{t-1|t-1} + \boldsymbol{k}_t(y_t - \mu_{t|t-1}) \tag{12.17}$$

and

$$\boldsymbol{V}_{t|t} = \boldsymbol{V}_{t|t-1} - v_{t|t-1}\boldsymbol{k}_t\boldsymbol{k}'_t, \tag{12.18}$$

where

$$\boldsymbol{k}_t = \boldsymbol{\zeta}_{t|t-1}v^{-1}_{t|t-1} = \frac{\boldsymbol{F}\boldsymbol{V}_{t-1|t-1}\boldsymbol{w} + \boldsymbol{g}\sigma^2}{\boldsymbol{w}'\boldsymbol{V}_{t-1|t-1}\boldsymbol{w} + \sigma^2}. \tag{12.19}$$

This expression follows directly from (12.16). The vector $\boldsymbol{k}_t$ is called the *Kalman gain.*

In summary, the revision step involves the calculation of the Kalman gain $\boldsymbol{k}_t$ with (12.19), and the use of (12.17) and (12.18) to update the mean and variance of the state. Equation (12.17) is of particular interest. It is reminiscent of (3.3c) for exponential smoothing. The only difference is that the vector $\boldsymbol{k}_t$, which determines the impact of the error on the new state vector, changes over time.

Example 12.10: Simple average

A simple average

$$\bar{y}_n = \frac{1}{n}\sum_{t=1}^{n} y_t$$

can be calculated recursively with

$$\bar{y}_t = \bar{y}_{t-1} + (y_t - \bar{y}_{t-1})/t.$$

Here the gain $k_t = 1/t$. This recurrence relationship can be seeded with an arbitrary value for $\bar{y}_0$. After observing y_1, the recurrence relationship yields $\bar{y}_1 = y_1$. The effect of the arbitrary seed value disappears after one period.

It can be shown, under quite general conditions, that $\boldsymbol{k}_t \rightarrow \boldsymbol{k}$ when a time series is governed by an invariant linear innovations state space model (12.1). Moreover, $v_{t|t-1} \rightarrow \sigma^2$. In other words, the Kalman filter converges to exponential smoothing. It should be appreciated that exponential smoothing is used with fixed seed states, and the Kalman filter with random seed states. In long time series subject to structural change, the seed state typically has little effect on all but a few subsequent states. After that there is little effective difference between the Kalman filter and exponential smoothing.

Example 12.11: Simple average

The Kalman filter in the previous example has a gain given by $k_t = 1/t$. Hence $k_t \to 0$ as $t \to \infty$. This means that eventually new observations will contribute no additional information to the estimate of the mean, reflecting the fact that the state of the system is constant over time. New observations only add value in large samples if the underlying state changes randomly over time, such as in a local level model with $\alpha > 0$.

In many applications, exponential smoothing can be used as a convenient approximation for the Kalman filter. The effect is to ignore knowledge about the process prior to period 1. In other words, a process with an infinite start-up is approximated by one with a finite start-up.

In Sect. 12.1 we saw that when the model contains nonstationary states, some or all of the variances are infinite. The formulae of the Kalman filter have no limiting form in this situation and can no longer be used. Special adaptations of the Kalman filter (Ansley and Kohn 1985; de Jong 1991a; Snyder and Forbes 2003) are available for this important situation; they are, however, more complex, and their logic is more opaque as a consequence. The simpler option is to use the information filter.

12.8 Exercises

Exercise 12.1. The moments of a random vector $\boldsymbol{x}$ satisfying the triangular equation (12.3) are readily derived using the equations $\boldsymbol{R}\boldsymbol{m}_x = \boldsymbol{m}_c$ and $\boldsymbol{R}\boldsymbol{V}_x\boldsymbol{R}' = \boldsymbol{V}_c$. To reverse this process and derive the triangular representation of a random vector $\boldsymbol{x}$ from its moments, we need the triangular factorization of the variance matrix. Suppose the variance matrix is

$$\boldsymbol{V}_x = \begin{bmatrix} 10 & 2 \\ 2 & 4 \end{bmatrix}. \tag{12.20}$$

Expand the expression $\boldsymbol{R}\boldsymbol{V}_x\boldsymbol{R}' = \boldsymbol{V}_c$ for this particular matrix to yield equations that must be satisfied by the unknown elements of $\boldsymbol{R}$ and $\boldsymbol{V}_c$. By solving these equations, establish that

$$\boldsymbol{R}_x = \begin{bmatrix} 1 & -0.5 \\ 0 & 1 \end{bmatrix} \quad \text{and} \quad \boldsymbol{V}_c = \begin{bmatrix} 9 & 0 \\ 0 & 4 \end{bmatrix}.$$

Exercise 12.2. Suppose we have a time series of only two values, represented by a random vector $(y_1, y_2)'$ with mean $(0,0)'$ and variance matrix given by (12.20). The problem is to find the one-step-ahead prediction distribution $y_2|y_1$. It is hypothesized that the mean of $y_2|y_1$ is a linear predictor $m_{2|1} = ay_1$ where a is an unknown constant. Define a new random variable $e_2 = y_2 - ay_1$. Derive an expression for the covariance of e_2 and y_1

and demonstrate that when $a = 0.2$, the covariance is 0 and the variance of e_2 is 3.6. The relationship between y_2 and y_1 can now be summarized as $y_2 = 0.2y_1 + e_2$ where y_1 and e_2 are uncorrelated. Now suppose that it is the beginning of period 2 and y_1 has been observed to be 20. Show that $y_2|y_1$ is normally distributed with a mean of 4 and a variance of 3.6.

Exercise 12.3. Generalize the method in Exercise 12.2 to the random vector $(y_1, y_2)'$ where y_i has mean μ_i and variance σ_i, and $\text{Cov}(y_1, y_2) = \sigma_{12}$. Thus demonstrate that $y_2|y_1$ has mean $\mu_2 + a(y_1 - \mu_1)$ and variance $\sigma_2 - a^2\sigma_1$, where $a = \sigma_{12}/\sigma_1$.

Exercise 12.4. Using the result from Exercise 12.1, express the information on the relationship between y_1 and y_2 as a stochastic equation. Use it to derive the moments of $y_2|y_1$.

Exercise 12.5. Consider the following stochastic equations (in tableau form).

y_2	y_1	m_c	v_c
1	−0.2	0	3.6
0	1	0	10

Show that these equations imply that the mean and variance of the vector $(y_1, y_2)'$ are, after rearranging rows and columns, the same as those in Exercise 12.2. Also confirm that the first stochastic equation corresponds to the linear representation of y_2 obtained in Exercise 12.2.

Exercise 12.6. The aim of this exercise is to illustrate the use of the information filter to estimate the mean μ in the white noise process $y_t = \mu + \varepsilon_t$, where $\varepsilon_t \sim \text{NID}(0, \sigma^2)$. We would anticipate that the information filter yields a simple average as the estimate of μ. At the beginning of period 1 there is no information about the series, so μ must be thought of as random variable with an infinite variance. Assume that y_1 turns out to be 5, so that at the *beginning* of period 2 we have a stochastic equation $\mu = 5 - \varepsilon_1$ which represents the available information from the past. The stochastic equation $\mu - y_2 = -\varepsilon_2$ is the only information on the process in period 2. The situation can summarized in tableau from as follows:

μ	y_2	m_c	v_c
1	0	5	1
1	−1	0	1

a. Applying fast Givens transformations,[3] show that this stochastic equation system can be converted to the following equivalent *upper* triangular form.

μ	y_2	m_c	v_c
1	−1/2	5/2	1/2
0	1	5	2

[3] See Appendix " Triangularization of Stochastic Equations".

b. Using the second row in this tableau, determine the moments of the one-step-ahead prediction distribution of y_2. Is this result consistent with the use of a simple average for prediction?
c. Suppose y_2 is observed to be 10. Use the first equation in the above tableau to revise the estimate of μ. Is your result consistent with the use of a simple average?

Appendix: Triangularization of Stochastic Equations

Stochastic equations take the general form $\boldsymbol{Ax} = \boldsymbol{b}$, where $\boldsymbol{A}$ is a square matrix and both $\boldsymbol{x}$ and $\boldsymbol{b}$ are random vectors. Moreover, the elements of $\boldsymbol{b}$ are mutually uncorrelated. The solution to conventional simultaneous linear equations is typically obtained by Gaussian elimination; a variation of this strategy applies when the equations contain random variables.

Gaussian Elimination

We begin by reviewing the operations associated with Gaussian elimination, examining their application to the deterministic simultaneous equations

$$2x_1 + x_2 = 6, \tag{12.21a}$$

$$3x_1 + 4x_2 = 14. \tag{12.21b}$$

The aim is to obtain an equivalent unit upper triangular equation system and to use back-substitution to solve for the variables x_1 and x_2.

Step 1 Do a simple pivot on x_1 in (12.21a): divide by the pivot element 2 to give

$$x_1 + 0.5x_2 = 3. \tag{12.22a}$$

Step 2 Do a conventional pivot with the aim of eliminating x_1 from (12.21b): multiply the revised pivot equation (12.22a) by 3 and subtract from (12.21b) to give

$$2.5x_2 = 5. \tag{12.23}$$

Step 3 Do a simple pivot on (12.23): divide by 2.5, to give $x_2 = 2$.

The resulting triangular equation system is

$$x_1 + 0.5x_2 = 3, \tag{12.24a}$$

$$x_2 = 2. \tag{12.24b}$$

Triangularity is important because it admits back-substitution as a solution method. Equation (12.24b) immediately provides the solution for x_2. When $x_2 = 2$ is back-substituted into (12.24a), the solution $x_1 = 2$ is obtained.

Stochastic Triangularization

To see how triangularization works for stochastic equations, it is now assumed that $b_1 \sim \mathrm{N}(6,8)$ and $b_2 \sim \mathrm{N}(14,12)$, so that (12.21a) and (12.21b) become, in tableau form:

x_1	x_2	μ	v
2	1	6	8
3	4	14	12

Again Gaussian elimination is employed, but the quirk is to add rules to ensure that the new variances created by row operations are computed correctly, and that the random right hand side variables are uncorrelated.

Step 1 As before, a *simple pivot* is done on the first equation by dividing the first row by 2 to give

$$x_1 + 0.5x_2 = b_1^+, \tag{12.25}$$

where $b_1^+ = b_1/2$. Thus, $\mu_1^+ = 6/2 = 3$ and $v_1^+ = 8/(2)^2 = 2$; the square of the pivot element is divided into the original variance. The revised equation system following the simple pivot is

x_1	x_2	μ	v
1	0.5	3	2
3	4	14	12

Step 2 A *conventional pivot* aimed at eliminating x_1 from the second equation is undertaken next. The first equation above is multiplied by 3 and subtracted from the second equation to give

$$2.5x_2 = b_2^+,$$

where $b_2^+ = b_2 - 3b_1$. Thus, $\mu_2^+ = \mu_2 - 3\mu_1 = 14 - 3 \times 3 = 5$ and $v_2^+ = v_2 + (-3)^2 v_1 = 12 + (-3)^2 \times 2 = 30$. The right hand side b_2^+ created by the elimination step, however, is correlated with b_1. An additional sub-step is required to preserve the diagonal structure of the variance matrix of the right hand side vector $\boldsymbol{b}$ after it is transformed. The pivot equation (12.25) is replaced by the new pivot equation formed from a linear combination of itself and the second equation. When w_1 and w_2 designate the (as yet unknown) coefficients of this linear combination, the pivot equation is replaced by

$$(w_1 + 3w_2)x_1 + (0.5w_1 + 4w_2)x_2 = (w_1 b_1 + w_2 b_2). \tag{12.26}$$

The right hand side of the new first equation is $b_1^+ = w_1 b_1 + w_2 b_2$. Its variance must be given by $v_1^+ = w_1^2 v_1 + w_2^2 v_2$. The unknown w_1 and w_2 are selected so that the coefficient of x_1 in the new equation remains equal to 1, something that requires that $w_1 + 3w_2 = 1$. They are also selected so that the new b_1^+ and b_2^+ are uncorrelated. It is convenient when focussing on covariances to work with a random disturbance vector defined by $\boldsymbol{u} = \boldsymbol{b} - \boldsymbol{m}$, where $\boldsymbol{m}$ is the vector of means. In these terms, we require that $\mathrm{E}(u_1^+ u_2^+) = 0$. Substituting the formulae for the new disturbances gives $\mathrm{E}(w_1 u_1 + w_2 u_2)(u_2 - 3u_1) = 0$. Expanding, and applying the expectation operator to the individual terms of the resulting expression, the equation $-6w_1 + 12w_2 = 0$ is obtained, which simplifies to $-w_1 + 2w_2 = 0$. Taken together, the unit diagonal condition and the zero correlation condition imply that w_1 and w_2 satisfy the simultaneous equations

$$w_1 + 3w_2 = 1,$$
$$-w_1 + 2w_2 = 0.$$

Their solution is $w_1 = 0.4$ and $w_2 = 0.2$. Equation (12.26) becomes $x_1 + x_2 = b_1^+$, where $\mu_1^+ = w_1\mu_1 + w_2\mu_2 = 0.4 \times 3 + 0.2 \times 14 = 4$ and $v_1^+ = 0.4^2 \times 2 + 0.2^2 \times 12 = 0.8$. The new tableau is

x_1	x_2	μ	v
1	1	4	0.8
0	2.5	5	30

Step 3 A simple pivot is now performed on the second row to convert the coefficient of x_2 to 1. The tableau, after dividing the second row by 2.5, is

x_1	x_2	μ	v
1	1	4	0.8
0	1	2	4.8

The required unit upper triangular form has been obtained. If required, it can be back-solved to give the means and variances of x_1 and x_2, together with their mutual covariance.

It has been seen that the existence of random right hand side vectors in linear equations leads to a situation where the simple and conventional pivot operations associated with Gaussian elimination must be supplemented by other row operations which ensure that the new right hand sides created by them remain uncorrelated. All of these operations are essentially linear, and the implicit matrices used at each step in transforming one system of stochastic equations into another are orthogonal. The particular method of constructing these implicit transformation matrices and applying them to the stochastic equations is referred to as the *fast Givens transformation method* (Golub and Van Loan 1996).

The operations undertaken in the example can be generalized so that they may be applied to any linear stochastic equation system. The resulting fast Givens transformation algorithm involves the repetitive application of augmented conventional and simple pivots to each successive row. Stage i begins with an echelon structure in the sub-matrix formed from the first $i-1$ rows. The diagonal of this rectangular sub-matrix consists of cells $(1,1),(2,2),\ldots,(i-1,i-1)$ which contain only ones. All elements below the diagonal equal zero. The elements above the diagonal are typically non-zero. The aim at stage i is to extend the special structure to row i so that the new sub-matrix formed from rows 1 to i has an echelon structure. This may be done by eliminating any non-zero elements before the cell (i,i) in row i using conventional augmented pivots. Then, when necessary, an augmented simple pivot can be used to convert the element a_{ii} to 1.

Equation Objects

Fast Givens transformations are typically applied to a collection of stochastic equations given by $\boldsymbol{A}\boldsymbol{x} = \boldsymbol{b}$. To specify the rules governing fast Givens transformations in general terms, it is convenient to introduce another object that represents a stochastic equation in this collection. The typical equation, written as $\boldsymbol{a}'\boldsymbol{x} = b$, can be conveniently summarized by the triple $(\boldsymbol{a}, \mu, v)$, where $\boldsymbol{a}'$ is a row vector, and μ and v are the mean and variance of b. The random vector $\boldsymbol{x}$ is not shown explicitly because it is common to all equations and its moments are typically unknown.

Let $q = (\boldsymbol{a}, \mu, v)$ designate this *equation object*. Operations can be defined on equation objects as follows:

Addition If q_1, q_2 are stochastic equations with uncorrelated right hand sides, $q_1 + q_2 = (\boldsymbol{a}_1 + \boldsymbol{a}_2, \mu_1 + \mu_2, v_1 + v_2)$.

Subtraction If q_1, q_2 are moments equations with uncorrelated right hand sides, $q_1 - q_2 = (\boldsymbol{a}_1 - \boldsymbol{a}_2, \mu_1 - \mu_2, v_1 + v_2)$.

Multiplication If c is a non-zero scalar and q is a moments equation, then $cq = (c\boldsymbol{a}, c\mu, c^2 v)$.

The interpretation of the equation index is now changed to be the position of the equation in the system $\boldsymbol{A}\boldsymbol{x} = \boldsymbol{b}$. The augmented simple pivot involves dividing the ith equation by the pivot element a_{ii}. It can be summarized by the single statement

$$q_i^+ = a_{ii}^{-1} q_i.$$

The augmented conventional pivot involves using a pivot row q_p to eliminate the element a_{ip} in the row q_i where $i > p$. The Gaussian elimination part can be described by the statement

$$q_i^+ = q_i - a_{ip} q_p.$$

The pivot row is also revised, but by using the formula

$$q_p^+ = w_p q_p + w_i q_i,$$

where

$$w_p = v_i / v_i^+,$$
$$w_i = a_{ip} v_p / v_i^+.$$

It can be shown that when w_p and w_i are chosen using these formulae, then:

a. The pivot element after the transformation still equals 1
b. The new right hand sides b_q^+ and b_i^+ are uncorrelated

It is conceivable, when used in the context of nonstationary time series, that some of the variances are infinite. This possibility has so far been ignored. Fortunately, the formulae associated with fast Givens transformations converge to well-defined limits when a variance tends to infinity.

It may be established that the revised variance for the ith row is

$$v_i^+ = v_i + a_{ip}^2 v_p,$$

so that

$$w_p = \frac{v_i}{v_i + a_{ip}^2 v_p} \quad \text{and} \quad w_i = \frac{a_{ip} v_p}{v_i + a_{ip}^2 v_p}.$$

If $v_p \to \infty$ then $w_p \to 0$ and $w_i \to a_{ip}^{-1}$. Hence, when $v_p = \infty$, the conventional augmented pivot is replaced by the simple augmented pivot $\boldsymbol{a}_p^+ = \boldsymbol{a}_i / a_{ip}$ and $v_i^+ = \infty$.

The introduction of the simple augmented pivot for the infinite variance case opens up the possibility of a more refined transformation strategy than is typically used in computer implementations. It begins with the original stochastic equations $\boldsymbol{Ax} = \boldsymbol{b}$ and a triangular stochastic equation $\boldsymbol{Rx} = \boldsymbol{c}$ with the variances of $\boldsymbol{c}$ all set to infinity. This means, in the beginning, that the triangular equation system contains no information. This is done even for stationary time series. In this case, prior information about the process is not lost; it is embedded in the equation system $\boldsymbol{Ax} = \boldsymbol{b}$.

In this refined version of the elimination strategy, each row of the original stochastic equations $\boldsymbol{Ax} = \boldsymbol{b}$ is processed in turn. We consider the situation that prevails when it is the turn of the ith row, written as $\boldsymbol{a}'\boldsymbol{x} = b$, to be processed. Using the rows of the triangular equation system as pivot rows, we can eliminate successive non-zero values in $\boldsymbol{a}'\boldsymbol{x} = b$ preceding the element a_{ii}. On pivoting on a row of the triangular equations with an infinite variance, the simple augmented pivot is undertaken, and the process is then terminated for the row $\boldsymbol{a}'\boldsymbol{x} = b$. We then proceed to process the $(i+1)$st row of the equation system.

By using equation objects to define fast Givens transformations, we can define the transformations more succinctly. Furthermore, equation objects can be implemented in object oriented computer programming languages, and the common arithmetic operations can be "overloaded" using the above definitions. This provides an elegant, practical way to implement fast Givens transformations.

A more formal treatment of fast Givens transformations is provided in Golub and Van Loan (1996). The transformations are normally used for fitting homoscedastic or heteroscedastic regressions (Gentleman 1973; Stirling 1981). They are an alternative to other orthogonalization methods such as Gram-Schmidt and Householder transformations. An advantage of using fast Givens transformations is that they can exploit the sparsity of matrices to reduce computational loads; the elimination operation is skipped for any zero elements before the element a_{ii} in row i. The other advantage, as seen above, is that they have a well-defined limiting form in the presence of infinite variances. It should also be noted that the variances can be replaced by standardized variances in the fast Givens transformation equations when it is used with algorithms which rely on standardized variances like the information filter in Sect. 12.3.

13

Conventional State Space Models

The primary purpose of this book is to demonstrate that the innovations form of the state space model provides a simple but flexible approach to forecasting time series. However, for reasons that are not completely clear, the innovations form has been largely over-shadowed in the literature by another version of the state space model that has multiple sources of randomness. We refer to this version as the multi-disturbance or multiple source of error (MSOE) model. The two approaches are compared and contrasted in this chapter. When we are comparing the two frameworks directly, both the finite and infinite start-up assumptions are valid; however, when the two are compared via their ARIMA reduced forms, the infinite start-up assumption will be used. The emphasis will be almost exclusively upon linear state space models, because, as we shall see in Sect. 13.4, the MSOE formulation becomes difficult to manage in the nonlinear case.

In Chap. 2, we introduced the local level and local trend models, together with their seasonal extensions. It will be seen that these innovations, or single source of error (SSOE), models all have their counterparts within a multiple source of error framework. It is often thought that the MSOE provides a better modeling framework than the SSOE because the multiple sources of error appear to allow greater generality. We will show that any MSOE model has an innovations representation, so that this viewpoint cannot be correct.

A general definition of the state space framework is presented in Sect. 13.1. It is seen to encompass both the innovations and the multiple disturbance forms of the state space model. Several important special cases of the MSOE are also given. A general approach to estimation is given in Sect. 13.2. Reduced forms of the MSOE models are examined in Sect. 13.3. The SSOE and MSOE approaches are then compared in Sect. 13.4.

13.1 State Space Models

The overall state of a system in period t is represented by a random vector z_t, which incorporates both the observations and the unobservable states. The elements in z_t are arranged so that $z_t = (y_t, \boldsymbol{x}_t)'$, where y_t denotes the observation at time t, which will be recorded over the periods 1 to n, and $\boldsymbol{x}_t$ is the random vector of k unobservable states.

The evolution of the state of the system is governed by the first-order recurrence relationship

$$z_t = \boldsymbol{A} z_{t-1} + \boldsymbol{u}_t, \tag{13.1a}$$

$$\text{where} \quad \boldsymbol{u}_t \sim \text{NID}(\boldsymbol{0}, \boldsymbol{V}_u), \tag{13.1b}$$

$\boldsymbol{A}$ is a fixed matrix and $\boldsymbol{V}_u$ is a positive semi-definite variance matrix. This general format is particularly useful when we consider parameter estimation in Sect. 13.2. When expressed in terms of the observable and unobservable states, (13.1) may be written as

$$y_t = \boldsymbol{w}'\boldsymbol{x}_{t-1} + \varepsilon_t, \tag{13.2a}$$

$$\boldsymbol{x}_t = \boldsymbol{F}\boldsymbol{x}_{t-1} + \boldsymbol{\eta}_t, \tag{13.2b}$$

$$\begin{bmatrix} \varepsilon_t \\ \boldsymbol{\eta}_t \end{bmatrix} \sim \text{NID}\left(\begin{bmatrix} 0 \\ \boldsymbol{0} \end{bmatrix}, \begin{bmatrix} V_\varepsilon & \boldsymbol{V}_{\varepsilon\eta} \\ \boldsymbol{V}_{\eta\varepsilon} & \boldsymbol{V}_\eta \end{bmatrix} \right), \tag{13.2c}$$

where $\boldsymbol{w}$ is a fixed vector and $\boldsymbol{F}$ is a fixed matrix. That is, $z_t = \begin{bmatrix} y_t \\ \boldsymbol{x}_t \end{bmatrix}$, $\boldsymbol{A} = \begin{bmatrix} 0 & \boldsymbol{w}' \\ \boldsymbol{0} & \boldsymbol{F} \end{bmatrix}$, $\boldsymbol{u}_t = \begin{bmatrix} \varepsilon_t \\ \boldsymbol{\eta}_t \end{bmatrix}$, and $\boldsymbol{V}_{\boldsymbol{u}} = \begin{bmatrix} V_\varepsilon & \boldsymbol{V}_{\varepsilon\eta} \\ \boldsymbol{V}_{\eta\varepsilon} & \boldsymbol{V}_\eta \end{bmatrix}$.

As in earlier chapters, (13.2a) and (13.2b) are called the measurement and transition equations respectively. Further, we again assume that the observation y_t depends only on the unobserved states $\boldsymbol{x}_{t-1}$ as they prevailed at the beginning of period t (at time $t-1$).

When $\boldsymbol{\eta}_t = \boldsymbol{g}\varepsilon_t$ (where $\boldsymbol{g}$ is a fixed vector of persistence parameters), the state space model becomes

$$y_t = \boldsymbol{w}'\boldsymbol{x}_{t-1} + \varepsilon_t, \tag{13.3a}$$

$$\boldsymbol{x}_t = \boldsymbol{F}\boldsymbol{x}_{t-1} + \boldsymbol{g}\varepsilon_t, \tag{13.3b}$$

$$\varepsilon_t \sim \text{NID}(0, V_\varepsilon). \tag{13.3c}$$

Equations (13.3) describe the vector form of the innovations model, which was introduced in Sect. 2.5.2. Another form of state space model assumes that $\boldsymbol{V}_{\varepsilon\eta} = \boldsymbol{0}$ and that $\boldsymbol{V}_\eta$ is diagonal. We refer to this model as the *multi-disturbance* or *MSOE* state space model. Both possibilities involve restrictions, but the second form places independence assumptions on the disturbances. When there are k states, this formulation includes $k+1$ unknown variances as parameters, just as the innovations model includes $k+1$ parameters: a single variance and k persistence parameters. These choices represent the maximum number of parameters that can be built into the models that retain the *estimability* (or *identifiability*) of all parameters.

At first sight the MSOE model appears to be more general than the innovations form because it involves more random disturbances. However, as we will show in Sect. 13.4, any MSOE model may be represented in innovations form so that there is only a need for one primary source of randomness for each observable state. This conclusion, it should be noted, is derived under the assumption that the disturbances have a Gaussian distribution; it may not be true for non-Gaussian state space models. Nevertheless, because most applications rely upon the mean and variance structures of the models, the practical implication is that little, if anything, will be lost by using the SSOE approach. Furthermore, as we will see later in this chapter, the innovations model approach provides several benefits.

In earlier chapters, we have examined the use of the innovations form of the state space framework to model evolving common features such as trends and seasonal patterns. Particularly important cases included the local level, local trend and damped trend, and their seasonal extensions. It will now be shown that each case has a multi-disturbance analogue.

The multi-disturbance versions form what has been called a *structural approach* to time series (Harvey 1989), one that has been widely used in economic studies. The following table shows corresponding standard structural models from the two approaches. We note that although the common symbols ℓ, b and ε are used to represent the level, slope and innovation respectively, their values and meaning differ between the two frameworks. The multiple disturbance versions presented here differ slightly from those of Harvey (1989); a point we explore in the next subsection.

Model	Conventional models	Innovations models
Level	$y_t = \ell_{t-1} + \varepsilon_t$	$y_t = \ell_{t-1} + \varepsilon_t$
	$\ell_t = \ell_{t-1} + \eta_t$	$\ell_t = \ell_{t-1} + \alpha\varepsilon_t$
Trend	$y_t = \ell_{t-1} + b_{t-1} + \varepsilon_t$	$y_t = \ell_{t-1} + b_{t-1} + \varepsilon_t$
	$\ell_t = \ell_{t-1} + b_{t-1} + \eta_t$	$\ell_t = \ell_{t-1} + b_{t-1} + \alpha\varepsilon_t$
	$b_t = b_{t-1} + \xi_t$	$b_t = b_{t-1} + \beta\varepsilon_t$
Seasonal	$y_t = \ell_{t-1} + b_{t-1} + s_{t-m} + \varepsilon_t$	$y_t = \ell_{t-1} + b_{t-1} + s_{t-m} + \varepsilon_t$
	$\ell_t = \ell_{t-1} + b_{t-1} + \eta_t$	$\ell_t = \ell_{t-1} + b_{t-1} + \alpha\varepsilon_t$
	$b_t = b_{t-1} + \xi_t$	$b_t = b_{t-1} + \beta\varepsilon_t$
	$s_t = s_{t-m} + \omega_t$	$s_t = s_{t-m} + \gamma\varepsilon_t$

13.1.1 Canonical Forms

The MSOE scheme assumes that the various error processes are independent. Thus, in (13.2) we would set $\boldsymbol{V}_{\varepsilon\eta} = \mathbf{0}$. However, most MSOE formulations (e.g., Harvey 1989; West and Harrison 1997) specify the measurement equation as

$$y_t = \boldsymbol{w}'\boldsymbol{x}_t + \varepsilon_t^*. \tag{13.4}$$

If we substitute the transition equation (13.2b) into this expression we arrive at

$$y_t = \boldsymbol{w}'\boldsymbol{F}\boldsymbol{x}_{t-1} + \boldsymbol{w}'\eta_t + \varepsilon_t^*,$$

so that the errors in the measurement and transition equations are now correlated. In order to make the independence assumption operational, we must choose a specific model, termed the *canonical model* by West and Harrison (1997, Chap. 5). Further, we must recognize that any transformation of the state variables may result in previously uncorrelated errors becoming correlated. The innovations approach provides a simple way out of this dilemma. Because the errors are perfectly correlated, any linear transformation leaves them perfectly correlated. The details are provided in Exercise 13.2. It may be shown that the different forms of the model have no effect on predictions, but the choices mean that individual components such as the local level may have different values; see Exercise 13.3.

13.1.2 Other State Space Models

A number of other formulations have appeared in the literature over the years. Akaike (1974) proposed an innovations model that maps directly into an ARMA$(k, k-1)$ scheme. The details are given in Exercise 13.1. Aoki (1987) also presents an innovations form, but we do not pursue these alternatives further in this book.

13.2 Estimation

The unknown parameters of both innovations and multi-disturbance state space models must be estimated. Because they both conform to the structure described in (13.1), a theory of estimation encompassing both cases is developed in terms of the more general framework. The seed state vector $\boldsymbol{z}_0$ is assumed to be random rather than fixed. Two points need to be made at this stage. The first is that virtually all the current literature on the multi-disturbance model relies upon the Kalman filter to develop the estimates. The second is that this reliance is not necessary as the information filter that is described in Sect. 12.3 is applicable in both frameworks. We proceed by first providing a general framework and then adapting the Kalman filter from Sect. 12.7 to its familiar MSOE form.

Using arguments similar to those in Chap. 12, it may be argued that the likelihood function can be rewritten in a prediction error form. For the moment, the focus is restricted to the case where all the states, both observable and unobservable, are stationary. The observations are represented by the n-vector $\boldsymbol{y}$. The unknown parameters are collected together into a vector $\boldsymbol{\theta}$. The prediction error decomposition of the likelihood function is (see Schweppe 1965)

$$\mathcal{L}(\boldsymbol{\theta} \mid \boldsymbol{y}) = \prod_{t=1}^{n} (v_{t|t-1})^{-1/2} \exp\Big(-\tfrac{1}{2}(y_t - \mu_{t|t-1})^2 \big/ v_{t|t-1}\Big),$$

where $\mu_{t|t-1}$ and $v_{t|t-1}$ are the mean and variance of the one-step-ahead prediction distribution. We use $\boldsymbol{\theta}$ to denote the variances of the different error terms in the general model (13.1). The aim is to maximize the likelihood with respect to $\boldsymbol{\theta}$. To implement the maximum likelihood approach, we need a mechanism to generate the updated values $\mu_{t|t-1}$ and $v_{t|t-1}$.

13.2.1 Kalman Filter

The Kalman filter was considered in Sect. 12.7 for the innovations state space model. The version presented here is based on the more general model (13.1) and so encompasses both the innovations model and the MSOE model as special cases. The argument follows along the same lines as that in Sect. 12.7. We let $\boldsymbol{y}_{1:s} = y_1, \ldots, y_s$ and define

$$\begin{aligned}
\mu_{t|s} &= \mathrm{E}(y_t \mid y_{1:s}),\\
v_{t|s} &= \mathrm{V}(y_t \mid y_{1:s}),\\
\boldsymbol{m}_{t|s} &= \mathrm{E}(\boldsymbol{x} \mid y_{1:s}),\\
\boldsymbol{V}_{t|s} &= \mathrm{V}(\boldsymbol{x} \mid y_{1:s}),\\
\zeta_{t|t-1} &= \mathrm{Cov}(\boldsymbol{x}_t, y_t \mid \boldsymbol{y}_{1:t-1}),\\
\boldsymbol{m}^z_{t|s} &= \mathrm{E}(\boldsymbol{z} \mid y_{1:s}),
\end{aligned}$$

$$\text{and} \qquad \boldsymbol{V}^z_{t|s} = \mathrm{V}(\boldsymbol{z} \mid y_{1:s}) = \begin{bmatrix} v_{t|s} & \zeta'_{t|s} \\ \zeta_{t|s} & \boldsymbol{V}_{t|s} \end{bmatrix}.$$

Then, using the notation of (13.1), the Kalman filter is given in part by the equations

$$z_{t|t-1} = \boldsymbol{A} z_{t-1|t-1} + \boldsymbol{u}_t, \tag{13.5a}$$

$$\boldsymbol{m}^z_{t|t-1} = \boldsymbol{A}\boldsymbol{m}^z_{t-1|t-1}, \tag{13.5b}$$

$$\boldsymbol{V}^z_{t|t-1} = \boldsymbol{A}\boldsymbol{V}^z_{t-1|t-1}\boldsymbol{A}' + \boldsymbol{V}_u. \tag{13.5c}$$

It is assumed that the distribution of $z_{t-1|t-1}$ is available from the preceding iteration of the filter after processing $t-1$ observations. The exception is period $t = 1$ where $z_{0|0}$ is described by the steady state distribution. Equations (13.5) are obtained from the general model (13.1). These equations form the *prediction step*, whose application yields the quantities $\mu_{t|t-1}$ and $v_{t|t-1}$

from the top part of (13.5). The remaining part of the Kalman filter is the *revision step*. By an argument similar to that employed in Sect. 12.7.2, we arrive at the relationships:

$$\begin{aligned} \boldsymbol{m}_{t|t} &= \boldsymbol{m}_{t|t-1} + \boldsymbol{k}_t(y_t - \mu_{t|t-1}) \\ \text{and} \quad \boldsymbol{V}_{t|t} &= \boldsymbol{V}_{t|t-1} - v_{t|t-1}\boldsymbol{k}_t\boldsymbol{k}_t', \\ \text{where} \quad \boldsymbol{k}_t &= \boldsymbol{\zeta}_{t|t-1}/v_{t|t-1}. \end{aligned}$$

These expressions provide all the information needed to evaluate the likelihood function for given values of the parameters.

As soon as the assumption of stationarity is dropped, the variances of nonstationary components are infinite and the Kalman filter formulae have no proper limiting form (Ansley and Kohn 1985). Moreover, the likelihood, as defined for stationary time series, is 0 everywhere. The traditional escape from this dilemma is to condition on the first p values of the time series, where p is the number of free nonstationary unobservable components. The density upon which the likelihood is based is then given by

$$p(y_{p+1}, y_{p+2}, \ldots, y_n | \boldsymbol{\theta}, y_1, y_2, \ldots, y_p) = \prod_{t=p+1}^{n} (v_{t|t-1})^{-1/2} \exp\left(-\tfrac{1}{2}(y_t - \mu_{t|t-1})^2 \big/ v_{t|t-1}\right).$$

One approach (Harvey 1989) is based on the assumption that all the unobserved states are nonstationary, so that $p = k$. A set of simultaneous equations is formed by stacking the model equations for the first k periods. The number of unknown unobservable state variables then exactly matches the number of equations and may normally be solved for the unobserved components including the moments of $\boldsymbol{x}_{p|p}$. The Kalman filter is then seeded with the distribution of $\boldsymbol{x}_{p|p}$ in period $p+1$ and used to generate the predictions and associated variance matrices for periods $p+1, p+2, \ldots, n$ needed to evaluate the likelihood function. This approach works in most circumstances, but must be adapted to handle potential complexities such as linear dependencies in the equations, missing values or partially known starting conditions when there is a mix of stationary and nonstationary unobserved state variables. A modern recursive version that allows for these potential complications may be found in de Jong (1991a, b). His algorithm is referred to as an *augmented Kalman filter*.

13.2.2 Convergence of Estimates

As the length of the series t increases, the variance matrix for $\boldsymbol{x}_{t|t}$, defined as $\boldsymbol{V}_{t|t}$ (see (12.18) for this expression in the innovations case) converges to a limiting value, say $\boldsymbol{V}_0$; for a proof, see Anderson and Moore (1979) and Harrison (1997). Harrison's proof applies to the MSOE scheme and does not require an assumption of Gaussian errors. His approach can be extended using the general form of the Kalman filter outlined in Sect. 12.7. For the MSOE model, this matrix is non-null, but for the innovations model, the limiting value is $\mathbf{0}$ as shown by Leeds (2000, pp. 78–79). This latter proof is a correction to errors in both the original proof by Caines and Mayne (1970) and a revised proof by the same authors (1971). Thus, it has been shown that, as t increases, the estimates of the state variables in the innovations model will converge in probability to the true values of the unobserved state variables at time t.

Many computer implementations ignore the distinction between states and their estimates. This result suggests that, in sufficiently long series, this practice is justifiable in the innovations framework.

13.3 Reduced Forms

13.3.1 Multi-Disturbance Models

Unobserved components are very useful in the sense that they enable us to specify plausible candidate state space models for the patterns that one may observe in a time series. However, from a strict mathematical perspective, their role is largely redundant. If a time series is stationary, its behavior is essentially determined by its autocorrelation function (ACF). Two state space models may appear to have a different structure because they are based on different states. However, if they yield the same mean, variance and ACF, they are equivalent from a forecasting perspective. Matters are more complicated for nonstationary time series because the unconditional mean and variances are not defined. An appropriate level of differencing may yield a stationary series. In this case, if the same transformations are applied to two models to induce stationarity, and both transformed models have the same mean, variance and ACF, they have the same properties. In the terminology of Chap. 10, the two models have the same minimal state representation.

The Wold representation theorem states that any linear stationary time series can be expressed as a moving average process and that this representation is unique. These moving average representations may involve infinite series and a more parsimonious structure is often achieved by converting to autoregressive moving average (ARMA) processes (Box et al. 1994).

The ARIMA representation is the reduced form corresponding to the *minimal dimension* representation of the state space model.

Common multi-disturbance state space models and their reduced forms are shown in Table 13.1. The right hand sides of the reduced forms are multi-disturbance moving average processes. However, the Granger–Newbold theorem (Granger and Newbold 1986) asserts that

(a) The sum of *uncorrelated* moving average processes is itself a moving average process
(b) The covariance function of the sum is the sum of the component covariance functions

Table 13.1. Reduced forms of multi-disturbance state space models.

Multi-disturbance model	Reduced form
Level	
$y_t = \ell_{t-1} + \varepsilon_t$	$\Delta y_t = \Delta\varepsilon_t + \eta_{t-1}$
$\ell_t = \ell_{t-1} + \eta_t$	
Trend	
$y_t = \ell_{t-1} + b_{t-1} + \varepsilon_t$	$\Delta^2 y_t = \zeta_{t-1} + \Delta\eta_{t-1} + \Delta^2\varepsilon_t$
$\ell_t = \ell_{t-1} + b_{t-1} + \eta_t$	
$b_t = b_{t-1} + \zeta_t$	
Seasonal	
$y_t = \ell_{t-1} + b_{t-1} + s_{t-m} + \varepsilon_t$	$\Delta^2\Delta_m y_t = \Delta_m\zeta_{t-1} + \Delta\Delta_m\eta_{t-1} + \Delta^2\omega_{t-m}$
$\ell_t = \ell_{t-1} + b_{t-1} + \eta_t$	$+ \Delta^2\Delta_m\varepsilon_t$
$b_t = b_{t-1} + \zeta_t$	
$s_t = s_{t-m} + \omega_t$	

In other words, any multiple disturbance moving average process has an equivalent traditional innovations moving average representation. The reduced forms, in terms of the multiple error terms, are shown in Table 13.1. Note that in the table, the difference operators are defined as $\Delta x_t = x_t - x_{t-1}$ and $\Delta_m x_t = x_t - x_{t-m}$. The reduced forms for the innovations models follow immediately when we replace the error terms from the transition equations by the appropriate linear functions of the single source of error.

The reduced forms may be obtained as an equation solving exercise. In any period, the model consists of $k+1$ equations. Stacking the model $k+1$ equations over the k periods $(t-1), \dots, (t-k)$ gives $k(k+1)$ equations involving the $k+1$ state vectors $\boldsymbol{x}_{t-1}, \dots, \boldsymbol{x}_{t-k-1}$; these equations also involve $y_{t-1}, \dots, y_{t-k-1}$ and the disturbances. Because each state vector contains k elements, the number of *state* variables exactly matches the number of equations. Ignoring the possibility of linear dependence for the moment, the stacked equations can be solved for the state variables in terms of the lagged y values and the disturbances. The solution for $\boldsymbol{x}_{t-1}$ found in this manner can be substituted into the measurement equation to yield an expression that no longer depends on the state variables. It is the required reduced form.

In deriving the reduced form, ultimately only the solution for $\boldsymbol{x}_{t-1}$ is required. It is possible to adapt the above procedure to avoid finding the solutions for $\boldsymbol{x}_{t-2}, \dots, \boldsymbol{x}_{t-k-1}$. The procedure can be understood by placing the stacked equations in tableau form; the tableau is then supplemented by the equation for y_t, which is placed in the final row. We then apply Gaussian elimination to eliminate the state variables from the measurement equation. The approach is illustrated in Example 13.1 for a local level model ($k = 1$).

Example 13.1: Local level model

The relevant equations for the local level may be stacked in detached form as follows:

y_t	y_{t-1}	ℓ_{t-1}	ℓ_{t-2}	ε_t	ε_{t-1}	η_{t-1}
0	0	−1	1	0	0	1
0	1	0	1	0	1	0
1	0	1	0	1	0	0

The aim is to eliminate all the state variables from the final row. The process begins by eliminating the state ℓ_{t-1} by adding the first row to the final row to give

y_t	y_{t-1}	ℓ_{t-1}	ℓ_{t-2}	ε_t	ε_{t-1}	η_{t-1}
0	0	−1	1	0	0	1
0	1	0	1	0	1	0
1	0	0	1	1	0	1

Now ℓ_{t-2} appears in the final row. It is eliminated by subtracting the second row to give:

y_t	y_{t-1}	ℓ_{t-1}	ℓ_{t-2}	ε_t	ε_{t-1}	η_{t-1}
0	0	−1	1	0	0	1
0	1	0	1	0	1	0
1	−1	0	0	1	−1	1

The reduced form is shown in the final row of the third tableau. It was obtained without explicitly solving for the local levels.

It may be observed that only the bottom row was changed by the transformation process in this example. This is not true in general. The rows before the final row may not always have a triangular structure in the columns corresponding to the state variables. Then below-diagonal elements must be eliminated using conventional pivoting operations associated with Gaussian elimination. It is also sensible to undertake simple pivots to ensure that all diagonal elements equal one. Only then should the elements in the final row of the tableau corresponding to state variables be eliminated, as illustrated in the example, to yield the reduced form.

The tableaux associated with this method explode in size when models with more states are considered, and so the derivation of the reduced form is more difficult to illustrate in the available space. Nevertheless, the method is readily coded and is quite tractable when implemented on a computer.

When a zero pivot is encountered, the elements in the pivot column below the pivot are searched for a non-zero value. The row containing this non-zero value is swapped with the pivot row so that resulting new pivot element in the new pivot row is non-zero. Then the reduction algorithm continues as normal.

When no non-zero element lies below a zero pivot element, the required row swap is not possible. In this case, there is not a unique solution for some of the state vectors in terms of the observations, and so the states are not identifiable. This occurs when the model is not of minimal dimension (see Chap. 10).

Identifiability of the state variables is not always necessary in order to derive a unique reduced form. In the ETS(A,A,A) model, the level and seasonal indexes are not identified, yet the reduced form is unique because the linear combination of the state vectors in the measurement equation is unique. The above reduction method can be adapted to handle such cases. As shown in Chap. 10, the solution typically involves the elimination of common factors from the two sides of the reduced model to achieve a canonical form.

In general, the reduced form may be written as

$$y_t = \sum_{i=1}^{k} \phi_i y_{t-i} - \sum_{j=1}^{k} \theta_j \varepsilon_{t-j} + \varepsilon_t, \tag{13.6}$$

where ε_t is formed from all the disturbances associated with period t via the Granger–Newbold theorem. The autoregressive elements may involve unit roots, which can be separated out as in Chap. 10. As is evident from Table 13.1, the moving average component can be expressed as the sum of uncorrelated moving average components:

$$y_t = \sum_{i=1}^{k} \phi_i y_{t-i} + \sum_{j=0}^{k} \xi_j' \eta_{t-j}, \tag{13.7}$$

where η_t represents the independent errors in the state equations, as in (13.2b) with coefficients ξ_j. The individual moving average components have autocovariance functions that can be combined to provide the autocovariance function for ε_t.

13.3.2 Innovations Models

The triangularization method described in the previous section for finding the reduced form of a multi-disturbance state space model is readily adapted to the innovations state space model. The independence assumption of the disturbances is nowhere used in the algorithm, and so it applies in exactly the same way when the disturbances are correlated. In the particular case where the disturbances are perfectly correlated, the reduced form of an innovations model can be obtained from the reduced form of a multi-disturbance model using the substitution $\eta_t = g\varepsilon_t$. For example, in the case of a local trend model, the substitutions $\eta_t = \alpha\varepsilon_t$ and $\xi_t = \beta\varepsilon_t$ yield the innovations reduced form $\Delta y_t = -\theta_2\varepsilon_{t-2} - \theta_1\varepsilon_{t-1} + \varepsilon_t$ where $\theta_1 = 2 - \alpha - \beta$ and $\theta_2 = \alpha - 1$.

13.4 Comparison of State Space Models

Multi-disturbance state space models encompass two special cases: the MSOE model where the disturbances are uncorrelated and the innovations form where they are perfectly correlated. It is often thought that the first assumption is less restrictive than the second; the argument is that the MSOE model has many sources of randomness, and should therefore be more flexible than the innovations form.

Paradoxically, the opposite is true. Anderson and Moore (1979) appear to be the first to have asserted, for discrete time contexts at least, that *any* multi-disturbance linear state space model has an equivalent innovations form. Their claim was remarkably general: it encompassed non-invariant as well as invariant state space models. They provided strong evidence that this has to be true by recognizing that the Kalman filter for any multi-disturbance state space model is always expressed in terms of the one-step-ahead prediction error, and that this implies the existence of an innovations model with the same gains. Their proof is opaque and possibly incomplete, so we do not pursue it further.

Hannan and Deistler (1988) proved the conjecture for stationary time series. They relied on transfer functions (i.e., polynomial functions of the lag operator) for their proof. They did not cover nonstationary time series. However, for those nonstationary time series that can be differenced to create a stationary time series, the same basic theory may be applied.

The general result can be stated as follows and the proof is derived by identifying various results presented earlier in the book.

Theorem 13.1. *The following statements hold for linear time series with invariant coefficients and Gaussian disturbances:*

A. Any MSOE model may be represented as an ARIMA model
B. Any innovations model may be represented as an ARIMA model
C. Any ARIMA model may be represented as an innovations model
D. Not all ARIMA models are representable by an MSOE model

Proof. This proof is somewhat informal and proceeds by drawing together results presented earlier in the book:

A. This property was discussed in Chap. 11 and in Sect. 13.3. The property holds provided that (13.7) corresponds to a minimal state space model. That is, the AR component $\phi(L) = 0$ has roots on or outside the unit circle, the MA component is invertible and any common factors in the two polynomials have been eliminated.

B. This property was demonstrated in Sect. 11.3. The same requirements on the polynomials apply.

C. This property was demonstrated in Sect. 11.5, but we should recall that the innovations model will correspond specifically to an exponential smoothing form only when the polynomial $\theta(L) = 0$ has real roots.

D. This negative statement can be demonstrated by means of counter-examples. Two models are equivalent in this framework if their (differenced) reduced forms have the same autocorrelation function (ACF). We define the autocorrelations by $\rho_j = \gamma_j/\gamma_0$, where $j = 1, 2, \ldots$ and $\gamma_j = \text{Cov}(y_t, y_{t-j})$.

   The ACF of the MSOE reduced form depends on the system variances; for the innovations model it is determined by the persistence parameters.

   The ACFs for the local level and local trend models are given in Table 13.2; all autocovariances not listed in the table are zero. Examination of the expressions in the table reveals that for the local level model $-0.5 < \rho_1 \leq 0$ for MSOE with the limiting value corresponding to $\sigma_\eta^2 = 0$. The ARIMA scheme has $-0.5 < \rho_1 < 0.5$, with the limits corresponding to $|\theta_1| = 1$. Thus, an ARIMA(0,1,1) model with $\theta_1 < 0$ does not have an MSOE counterpart.

   Likewise, for the local trend model, we have $-0.667 < \rho_1 \leq 0$ for the MSOE, but $-0.707 < \rho_1 < 0.707$ for the ARIMA scheme, so that some ARIMA(0,2,2) models do not possess an MSOE form. The derivation is left as an exercise.

   A counter-claim to these examples could be that the parameter spaces may be extended by allowing correlation among the disturbances. We explore this conjecture below. □

We may use the entries in Table 13.2 to explore the relationships between the MSOE and innovations models. The general point may be illustrated using the local level model. The first-order autocorrelation is always negative for the MSOE version. It may be either positive or negative in the innovations model. When the autocorrelation is negative, it is always possible to find a

Table 13.2. Reduced forms of common state space models.

Model	Multiple error	Innovations
Local Level	$\gamma_0 = \sigma_\eta^2 + 2\sigma_\varepsilon^2$	$\gamma_0 = [(\alpha-1)^2 + 1]\sigma_\varepsilon^2$
	$\gamma_1 = -\sigma_\varepsilon^2$	$\gamma_1 = (\alpha-1)\sigma_\varepsilon^2$
Local Trend	$\gamma_0 = (\sigma_\xi^2 + 2\sigma_\eta^2 + 6\sigma_\varepsilon^2)$	$\gamma_0 = \left[(\alpha+\beta-2)^2 + (1-\alpha)^2 + 1\right]\sigma_\varepsilon^2$
	$\gamma_1 = -(\sigma_\eta^2 + 4\sigma_\varepsilon^2)$	$\gamma_1 = -(2-\alpha-\beta)(2-\alpha)\sigma_\varepsilon^2$
	$\gamma_2 = \sigma_\varepsilon^2$	$\gamma_2 = (1-\alpha)\sigma_\varepsilon^2$

corresponding value for α by equating the two expressions and solving the quadratic in α to obtain

$$\alpha = -\frac{q}{2} + \sqrt{\left(1 + \frac{q}{2}\right)^2 - 1}, \tag{13.8}$$

where q is the so-called signal-to-noise ratio defined by $q = \sigma_\eta^2/\sigma_\varepsilon^2$.

This analysis serves to illustrate a further point, relating to the relative ease of use of the MSOE and innovations models. As we saw in Sect. 11.3, the relationships between the parameters in the ARIMA and innovations models are linear. Those between the MSOE model and the other two are quadratic, making it more difficult to establish relationships between the sets of parameters.

13.4.1 Size of the Parameter Space

In order to explore the size of the parameter space under different assumptions about the correlations among the error terms in the measurement and transition equations, we revert to a consideration of the general form given in (13.2). The general argument is due to Leeds (2000, pp. 50–56), and the details are given in the Appendix. The argument given there shows that when there are J transition equations, we must consider 2^J possible solutions, and only one of these solutions will satisfy the forecastability conditions. To be specific, we demonstrate the argument for the local trend model, although it applies quite generally.

We first put the problem into an appropriate framework that enables us to apply the linear fractional programming approach described in the Appendix. The notation we now use is specific to this subsection and purely for convenience in the present discussion. The local trend model has three error terms in the general case, and we may write the variance matrix for $(\varepsilon_t, \eta_t, \xi_t)$ as

$$\begin{bmatrix} v_0^2 & \rho_1 v_0 v_1 & \rho_2 v_0 v_2 \\ \rho_1 v_0 v_1 & v_1^2 & \rho_3 v_1 v_2 \\ \rho_2 v_0 v_2 & \rho_3 v_1 v_2 & v_2^2 \end{bmatrix}.$$

Extending the result in Table 13.2, the general form of the lag one autocorrelation for the twice-differenced series is

$$\frac{-(4v_0^2 + 4\rho_1 v_0 v_1 + 2\rho_2 v_0 v_2 + v_1^2 + \rho_3 v_1 v_2)}{(6v_0^2 + 6\rho_1 v_0 v_1 + 2\rho_2 v_0 v_2 + 2v_1^2 + 2\rho_3 v_1 v_2 + v_2^2)}.$$

A comparable expression may be obtained for the autocorrelation at lag two; see Exercise 13.4. All higher-order autocorrelations are zero. We may determine the maximum size of the parameter space by finding the smallest and largest possible values for each autocorrelation, provided the extremes are achieved for the same choices of the correlations.

If we fix the value of $\boldsymbol{v} = (v_0, v_1, v_2)'$, the numerator and denominator of the autocorrelation are linear in the correlations, and we may maximize (minimize) the value of the expression using linear fractional programming. The details are given in the Appendix. We find that the same extreme solutions apply whatever the value of $\boldsymbol{v}$, and so we conclude that the size of the parameter space is maximized when the errors are perfectly correlated. However, an innovations model with J transition equations still has 2^J possible solutions, and we now proceed to select a unique solution from this set.

The local trend model has $J = 2$, and the transition equations have the error terms $(g_1\varepsilon_t, g_2\varepsilon_t)$. From Table 13.1 we may write the right hand side of the reduced form equation for the local trend model as

$$\varepsilon_t - (2 - g_1 - g_2)\varepsilon_{t-1} - (g_1 - 1)\varepsilon_{t-2} \equiv (1 - \theta_1 L - \theta_2 L^2)\varepsilon_t.$$

The forecastability conditions may be written as:

$$|\theta_2| < 1, \quad 1 - \theta_1 - \theta_2 > 0, \quad 1 + \theta_1 - \theta_2 > 0.$$

These conditions reduce to the requirements that $(g_1 > 0, g_2 > 0)$, which establishes the uniqueness of the solution. The reader is asked to verify these conditions in Exercise 13.5.

13.4.2 Seasonal Models

In order to compare the seasonal models we make use of the autocovariance generating function (ACGF) for the differenced series. Consider an ARIMA model written in moving average form with the error variance equal to 1 (without loss of generality) and the auxiliary polynomial

$$\theta(z) = 1 - \theta_1 z - \theta_2 z^2 - \cdots - \theta_q z^q - \cdots. \tag{13.9}$$

The ACGF is then defined as:

$$C(z) = \theta(z)\theta(z^{-1}). \tag{13.10}$$

The coefficient of z^j is the autocovariance at lag j. Thus $\gamma_0 = 1 + \theta_1^2 + \theta_2^2 + \cdots$, $\gamma_1 = -\theta_1 + \theta_1\theta_2 + \cdots$, and so on. The general forms for the seasonal models are cumbersome, and it is convenient to summarize them in somewhat different ways. Thus, for the innovations model, using the canonical reduced form given in Example 11.6, (13.9) becomes

$$\begin{aligned}\theta(z) = 1 &- (1 - \alpha - \beta)z + \beta(z^2 + \cdots + z^{m-1}) - (1 - \beta - \gamma)z^m \\ &+ (1 - \alpha - \gamma)z^{m+1}.\end{aligned}$$

For the MSOE model, it is easier to specify the autocovariances directly. Again using the canonical reduced form, we arrive at the expressions:

$$\begin{aligned}
\gamma_0 &= (m\sigma_\xi^2 + 2\sigma_\eta^2 + 2\sigma_\omega^2 + 6\sigma_\varepsilon^2), \\
\gamma_1 &= (m-1)\sigma_\xi^2 - (\sigma_\omega^2 + 2\sigma_\varepsilon^2), \\
\gamma_j &= (m-j)\sigma_\xi^2, \quad j = 1, 2, \ldots, m-2, \\
\gamma_{m-1} &= \sigma_\xi^2 + \sigma_\varepsilon^2, \\
\gamma_m &= -(\sigma_\eta^2 + 2\sigma_\varepsilon^2), \\
\gamma_{m+1} &= \sigma_\varepsilon^2.
\end{aligned}$$

For these two seasonal models, any attempt to equate autocovariances of the same order leads to more equations than unknowns. No solution exists that matches the autocovariances, other than the degenerate form with $\sigma_\omega^2 = 0$. Thus, the two models are not equivalent. Interestingly, McKenzie (1976) derived an ARIMA representation of this additive Holt-Winters scheme. Careful reading of his paper reveals that he used an innovations form of the state space model to obtain the result. The covariance expressions just derived do not allow a simple mapping from the state space parameters to the ARIMA coefficients. More generally, because the autocovariances are typically quadratic in the moving average parameters, it is only in special cases that explicit solutions are available for the mapping from the MSOE model to its ARIMA reduced form. There can be multiple solutions to such equations, but the requirement of invertibility ensures that there is at most one acceptable solution.

13.4.3 Nonlinear Models

We saw in Chap. 4 that it was possible to specify nonlinear and heteroscedastic schemes using the innovations form, and that the resulting (albeit approximate) Gaussian likelihood was readily obtained, as seen in Chap. 5. Comparable models may be specified in the MSOE framework, but computational difficulties immediately arise. The probability density function now involves terms for each of the unobserved errors and there is no simple way to integrate these out to obtain the likelihood for the unknown parameters. We could make use of Markov Chain Monte Carlo (MCMC) methods, but Gaussian likelihood remains an approximation and adding an *extra* layer of simulations adds to the computational burden.

13.5 Smoothing and Filtering

Harvey and Koopman (2000) showed that the MSOE scheme leads to optimal symmetric two-sided smoothers. These were defined for an infinite series, although applications will clearly involve truncation after a finite number of terms. This smoother corresponds to the Wiener-Kolmogorov (WK) filter. They also noted that when the components are correlated, as in the innovations formulation, the resulting signal extraction filter is asymmetric. Indeed, the perfect correlation among the components of the innovations model led to our observation in Sect. 12.6 that the two-sided filter does not improve the estimates of the state variables asymptotically. However, the WK filter remains available, once we recognize that its role is to smooth the series, not to estimate the state variables as such. Because any innovations model may be expressed in ARIMA form, an appropriate WK filter may be developed within that framework.

The following example illustrates how an appropriate WK smoother can be constructed.

Example 13.2: Local level model

Consider the local level model, written as the reduced ARIMA(0,1,1):

$$(1-L)y_t = [1-(1-\alpha)L]\varepsilon_t.$$

The (doubly infinite) WK filter is given by:

$$\ell_{S,t} = \frac{\alpha^2 y_t}{[1-(1-\alpha)L][1-(1-\alpha)L^{-1}]} = \frac{\alpha}{2-\alpha}\sum_{j=-\infty}^{\infty}(1-\alpha)^{|j|}y_{t-j}.$$

This smoother also corresponds to the two-sided Beveridge–Nelson (BN) filter given by Proietti and Harvey (2000), although it should be noted that the filter is admissible only for $0 < \alpha < 1$. The WK and BN filters often do not have the same form.

As pointed out by Gijbels et al. (1999), when exponential smoothing is interpreted as a kernel estimate, simple exponential smoothing is the natural forecast and the filter given above is the natural smoother.

The approach just described provides a smoothed estimator for the mean of the process, and we now turn to consider the individual components. Key elements in the analysis of economic time series are the creation of the deseasonalized series and the creation of a smoothed trend. Bell (1984) and Burridge and Wallis (1988) extended the WK filter to nonstationary series to enable the extraction of unobserved components.

One way to develop a WK filter for the components of an innovations process is to generate the corresponding ARIMA model and then apply a canonical decomposition, such as that developed by Hillmer and Tiao (1982). However, if we recall from Sect. 13.2.2 that the estimates of the state variables converge to their true values, a much simpler approach is possible. We may construct the seasonally adjusted or detrended series directly, and then smooth the remaining components. This is illustrated in the next example.

Example 13.3: Seasonal levels model

Consider the following innovations model, which should also include the appropriate normalization as described in Chap. 8:

$$\begin{aligned} y_t &= \ell_{t-1} + s_{t-m} + \varepsilon_t, \\ \ell_t &= \ell_{t-1} + \alpha\varepsilon_t, \\ s_t &= s_{t-m} + \gamma\varepsilon_t. \end{aligned}$$

We may generate the approximately detrended series as:

$$z_{1t} = y_t - \ell_{t|n} \approx s_{t-m} + \varepsilon_t.$$

It follows from Example 13.2 that the smoothed seasonal components may be computed as:

$$s_{S,t} \approx \frac{\gamma^2 z_{1t}}{[1-(1-\gamma)L^m][1-(1-\gamma)L^{-m}]} = \frac{\gamma}{2-\gamma}\sum_{j=-\infty}^{\infty}(1-\gamma)^{|j|}z_{1,t-jm}.$$

In turn, the seasonal components lead to the deseasonalized series:

$$z_{2t} = y_t - s_t \approx \ell_t + \varepsilon_t.$$

The smoothed trend is then given by:

$$\ell_{S,t} \approx \frac{\alpha^2 z_{2t}}{[1-(1-\alpha)L][1-(1-\alpha)L^{-1}]} = \frac{\alpha}{2-\alpha}\sum_{j=-\infty}^{\infty}(1-\alpha)^{|j|}z_{1,t-j}.$$

We may iterate between the seasonal and trend components until convergence is obtained, although the differences may be expected to be small provided the series is of reasonable length.

In summary, we observe that while the primary motivation for using the innovations approach is that it is more directly beneficial for forecasting (the focus of this text), smoothing and filtering operations may also be performed within the innovations framework.

13.6 Exercises

Exercise 13.1. Consider a state space model in the general form of (13.3) with

$$\boldsymbol{w}' = (1,0,\ldots,0), \;\; \boldsymbol{g}' = (1,\psi_1,\ldots,\psi_{k-1}) \;\text{ and }\; \boldsymbol{F} = \begin{bmatrix} 0 & 0 & 0 & \ldots & 0 \\ 0 & 1 & 0 & \ldots & 0 \\ \cdots & \cdots & \cdots & \cdots & \cdots \\ 0 & 0 & 0 & \ldots & 1 \\ \phi_k & \phi_{k-1} & \cdots & \cdots & \phi_1 \end{bmatrix}.$$

The state vector is defined as $\boldsymbol{x}_t = (y_t, y_{t+1|t}, \ldots, y_{t+k-1|t})$ and ε_t is deleted from the measurement equation. Show that this model reduces to an ARMA$(k,k-1)$ model.

Exercise 13.2. Consider the innovations model with measurement equation (13.4) used in place of (13.1). Show that the form of the model given by (13.2) still applies, with revised coefficients $\boldsymbol{w}_1 = \boldsymbol{F}'\boldsymbol{w}$ and $\boldsymbol{g}_1 = \frac{\boldsymbol{g}}{1+\boldsymbol{w}'\boldsymbol{g}}$.

Exercise 13.3. Show that the reduced forms of the two MSOE schemes given in Sect. 13.1.1 result in the same ARIMA reduced forms.

Exercise 13.4. Show that the general form of the lag 2 autocovariance for the local trend model (in the notation of Sect. 13.4.1) is

$$v_0^2 + \rho_1 v_0 v_1 + \rho_2 v_0 v_2.$$

Hence show that the first and second order autocorrelations have the same set of conditions for extreme values.

Exercise 13.5. Show that the conditions for forecastability discussed in Sect. 13.4.1 lead to a unique local trend model with a maximal parameter space.

Appendix: Maximizing the Size of the Parameter Space

In a seminal paper on Linear-Fractional Programming (LFP), Charnes and Cooper (1962) showed that the LFP optimization problem

$$\max_u \frac{\sum_j A_j u_j}{\sum_j B_j u_j}, \qquad \text{subject to} \quad 0 \leq u_j \leq c_j \quad \text{for all } j,$$

may be reformulated as a linear program of the form:

$$\max_u \sum_j A_j u_j \qquad \text{subject to} \quad \sum_j B_j u_j = c \quad \text{and} \quad 0 \leq u_j \leq c_j \quad \text{for all } j.$$

In our application, the denominator is always a strictly positive variance term and the $\{u_j\}$ represent either the positive or negative parts of correlation coefficients, so that $c_j = 1$ for all j.

When there are J transition equations and one measurement equation, the joint error distribution involves $K = J(J+1)/2$ correlation coefficients. The LFP optimization is subject to $2K$ constraints and K non-negativity constraints on the correlations plus one equality constraint. By inspection, we can see that $K-1$ of the correlations must each take on one of the three values $(-1, 0, +1)$; the remaining correlation is then determined by the equality constraint. We now proceed to incorporate additional features of the particular problem to arrive at a unique solution:

- A simple reparameterization of the problem (replacing each correlation ρ by $\rho^* = \rho + 1$) serves to demonstrate that the zero values are internal solutions and can be ignored.
- We now have that $K-1$ of the correlations are ± 1; it follows that the remaining correlation must be ± 1.
- We can now return to the state space formulation, because the correlations are generated by the $J+1$ terms $(\varepsilon_t, g_1\varepsilon_t, g_2\varepsilon_t, \ldots, g_J\varepsilon_t)$. The J coefficients g_j give rise to the 2^J possible solutions after setting the coefficient in the measurement equation equal to $+1$, without loss of generality.
- Finally, we may demonstrate that the only solution to satisfy the forecastability conditions is that with all $g_j > 0$. The argument for the local trend model is illustrated in Sect. 13.4.1.

In principle, other formulations may provide parameter spaces of equal size for specific cases, but there is no loss in restricting attention to the innovations models.

14

Time Series with Multiple Seasonal Patterns

Co-authors:[1] **Phillip Gould**[2] **and Farshid Vahid-Araghi**[3]

Time series that exhibit multiple seasonal patterns can arise from many sources. For example, both daily and weekly cycles can be seen in Fig. 14.1 for hourly utility demand data and in Fig. 14.2 for hourly vehicle counts on a freeway. Usually when we discuss seasonal patterns we mean patterns that change with the seasons of the year for weekly, quarterly, or monthly data. In this chapter any periodic pattern of fixed length will be considered to be a seasonal pattern and called a *cycle*. It is easy to think of many examples where multiple seasonal cycles would occur, say for hours of the day within days of the week, such as hospital admissions, demand for public transportation, telephone calls to call centers, requests for cash at ATMs, and accessing computer websites. The seasonal innovations state space models in Tables 2.2 and 2.3 are designed for one seasonal cycle in which the seasonal components are revised only once during every cycle. Thus, the objective of this chapter is to extend some of those models to handle more frequent observations with more than one cycle. Another objective is the ability to revise the seasonal components more often than once every seasonal cycle. For example, in the case of hourly traffic count data we would like to be able to revise the seasonal components for a weekly cycle (a cycle of length 168) more frequently than once a week.

There are several notable features in Fig. 14.1. First, we observe that the daily cycles are not all the same, although it may reasonably be claimed that the cycles for Monday through Thursday are similar, and perhaps Friday also. Those for Saturday and Sunday are quite distinct. In addition, we would expect the patterns for public holidays to be more similar to weekends than to regular weekdays. A second feature of the data is that the underlying levels of the daily cycles may change from one week to

[1] This chapter is based, in part, on material presented in Gould et al. (2008).

[2] Dr. Phillip Gould, Australia and New Zealand Banking Group Limited, Australia.

[3] Professor Farshid Vahid-Araghi, School of Economics, Australian National University, Australia.

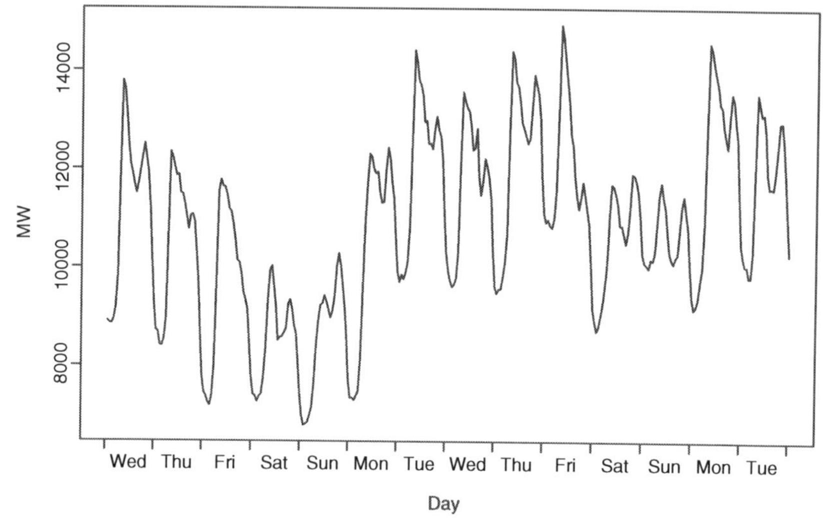

Fig. 14.1. Two-week sub-sample of hourly utility demand data.

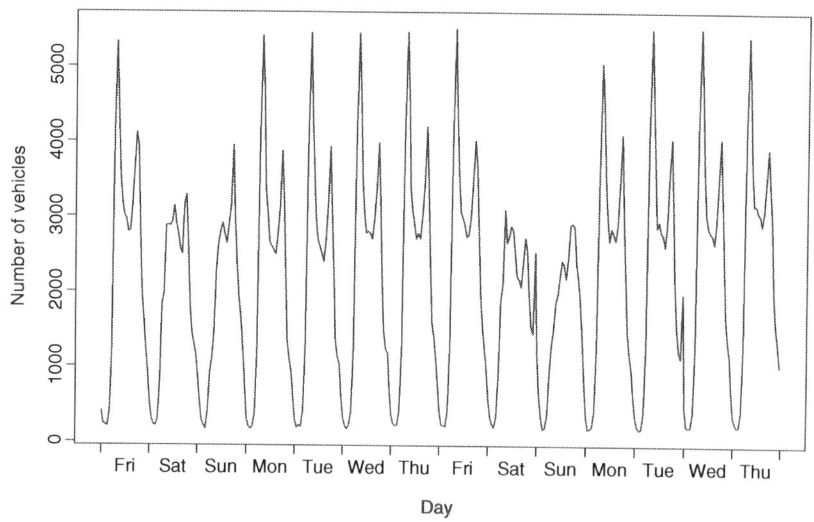

Fig. 14.2. Two-week sub-sample of traffic count data.

the next, yet be highly correlated with the levels for the days immediately preceding. These same characteristics can also be seen to hold for the vehicle count data in Fig. 14.2. An effective time series model must be sufficiently flexible to capture these principal features without imposing too heavy a computational or inferential burden.

The most commonly employed approaches to modeling seasonal patterns within exponential smoothing are the Holt-Winters methods (Winters

1960). These methods correspond to the ETS(A,A,A) and ETS(M,A,M) models in Chap. 2 and might be used for the type of data shown in Figs. 14.1 and 14.2. However, these models suffer from several important weaknesses. In order to capture the weekly cycle, these models would require 168 starting values (24 hours $\times$ 7 days) and would fail to pick up the similarities from day-to-day at a particular time. Also, they do not allow for patterns on different days to adapt at different rates nor for the component for one day to be revised on another day. In a recent paper, Taylor (2003b) has developed a double seasonal exponential smoothing method, which allows the inclusion of one cycle nested within another. His method is described briefly in Sect. 14.1.2. Taylor's method represents a considerable improvement, but assumes the same intra-day cycle for all days of the week. Moreover, updates based upon recent information (the intra-day cycle) are the same for each day of the week.

Two other approaches for modeling seasonal cycles are the Box–Jenkins method for ARIMA models (Box et al. 1994) and the state space method for multiple disturbance models (e.g., Harvey 1989). In ARIMA models, multiple seasonal cycles could be established by including additional seasonal factors. Such an approach again requires the same cyclical behavior for each day of the week. Although the resulting model may provide a reasonable fit to the data, there is a lack of transparency in such a complex model, even in the linear case, compared to the specification provided by Taylor's approach and by the methods we describe later in this chapter. The disadvantages of the multiple disturbance models have previously been described in Chaps. 12 and 13 with respect to the estimation process and the ability to handle nonlinear (multiplicative seasonal) patterns directly.

The additive Holt-Winters (HW) method and Taylor's double seasonal (DS) scheme are outlined in Sect. 14.1. The multiple seasonal (MS) model is introduced and developed in Sect. 14.2; the primary emphasis is on the additive scheme, but the multiplicative version is also briefly described. Applications to hourly data on utility demand and on traffic flows are considered in Sects. 14.3 and 14.4, respectively.

14.1 Exponential Smoothing for Seasonal Data

14.1.1 An Innovations State Space Model for the Holt-Winters (HW) Method

The Holt-Winters (HW) exponential smoothing approach (Winters 1960) includes methods for both additive seasonal patterns (where the size of seasonal variation is not affected by the level of y_t) and multiplicative seasonal patterns (where there is larger seasonal variation at higher values of y_t). Our primary development is in terms of additive seasonality; the corresponding model for the multiplicative case is presented in Sect. 14.2.2. A model for the additive seasonal HW method decomposes the series value y_t into an error ε_t, a level ℓ_t, a trend b_t and a seasonal component s_t. The underlying innovations state space model is the ETS(A,A,A) model:

$$y_t = \ell_{t-1} + b_{t-1} + s_{t-m} + \varepsilon_t, \tag{14.1a}$$
$$\ell_t = \ell_{t-1} + b_{t-1} + \alpha\varepsilon_t, \tag{14.1b}$$
$$b_t = b_{t-1} + \beta\varepsilon_t, \tag{14.1c}$$
$$s_t = s_{t-m} + \gamma\varepsilon_t, \tag{14.1d}$$

where $\varepsilon_t \sim \text{NID}(0, \sigma^2)$, and α, β and γ are smoothing parameters for the level, trend and seasonal terms, respectively. The smoothing parameters reflect how quickly the level, trend and seasonal components adapt to new information. The value of m represents the number of seasons in one seasonal cycle. In this chapter we will need to distinguish an ETS(A,A,A) model with m_1 seasons from one with m_2 seasons. Hence, in this chapter only, we will denote the ETS(A,A,A) model by HW(m) and the seasonal cycle by $c_t = (s_t, s_{t-1}, \ldots, s_{t-m+1})'$. Estimates of $m+2$ different seed values for the unobserved components must be made; one for the level, one for the trend, and m for the seasonal terms (although we constrain the initial seasonal components to sum to 0).

The HW method allows each of the m seasonal terms to be updated only once during the seasonal cycle of m time periods. Thus, for hourly data we might have an HW(24) model that has a cycle of length 24 (a daily cycle). Each of the 24 seasonal terms would be updated once every 24 hours. Or we might have an HW(168) model that has a cycle of length 168 (24 hours × 7 days). Although a daily pattern might occur within this weekly cycle, each of the 168 seasonal terms would be updated only once per week. In addition, the same smoothing constant γ is used for each of the m seasonal terms. In Sect. 14.2 we will show how to relax these restrictions using our model for multiple seasonal (MS) processes.

14.1.2 An Innovations State Space Model for the Double Seasonal (DS) Method

Taylor's double seasonal (DS) exponential smoothing method (Taylor 2003b) was developed to forecast time series with two seasonal cycles: a short one that repeats itself many times within a longer one. It should not be confused with double exponential smoothing (Brown 1959), the primary focus of which is on a local linear trend. Taylor (2003b) developed a method for multiplicative seasonality, which we adapt for additive seasonality.

Like the HW exponential smoothing methods, DS exponential smoothing is a *method*. It was specified without recourse to a stochastic *model*, and hence, it cannot be used in its current form to find estimates of the uncertainty surrounding predictions. In particular, a model is required to find prediction intervals. The problem is resolved by specifying an innovations state space model underpinning the additive DS method. Letting m_1 and m_2 designate the periods of the two cycles, this model is:

$$y_t = \ell_{t-1} + b_{t-1} + s^{(1)}_{t-m_1} + s^{(2)}_{t-m_2} + \varepsilon_t, \tag{14.2a}$$
$$\ell_t = \ell_{t-1} + b_{t-1} + \alpha\varepsilon_t, \tag{14.2b}$$
$$b_t = b_{t-1} + \beta\varepsilon_t, \tag{14.2c}$$
$$s^{(1)}_t = s^{(1)}_{t-m_1} + \gamma_1\varepsilon_t, \tag{14.2d}$$
$$s^{(2)}_t = s^{(2)}_{t-m_2} + \gamma_2\varepsilon_t, \tag{14.2e}$$

where $\varepsilon_t \sim \text{NID}(0, \sigma^2)$, and the smoothing parameters for the two seasonal components are γ_1 and γ_2. We denote this model by DS(m_1, m_2) and the two seasonal cycles by

$$\boldsymbol{c}^{(1)}_t = (s^{(1)}_t, s^{(1)}_{t-1}, \dots, s^{(1)}_{t-m_1+1})' \quad \text{and} \quad \boldsymbol{c}^{(2)}_t = (s^{(2)}_t, s^{(2)}_{t-1}, \dots, s^{(2)}_{t-m_2+1})'.$$

Estimates of $m_1 + m_2 + 2$ seeds must be made for this model.

There are m_2 seasonal terms in the long cycle that are updated once in every m_2 time periods. There are an additional m_1 seasonal terms in the shorter cycle that are updated once in every m_1 time periods. It is not a requirement of the DS(m_1, m_2) model that m_1 be a divisor of m_2. However, if $k = m_2/m_1$, then there are k shorter cycles within the longer cycle. Hence, for hourly data, there would be 168 seasonal terms that are updated once in every weekly cycle of $m_2 = 168$ time periods and another $m_1 = 24$ seasonal terms that are updated once in every daily cycle of 24 time periods. For the longer weekly cycle the same smoothing parameter, γ_2, is used for each of the 168 seasonal terms, and for the shorter daily cycle the same smoothing parameter, γ_1, is used for each of the 24 seasonal terms. In our MS model we will be able to relax these restrictions.

14.1.3 Using Indicator Variables in a Model for the HW Method

We now show how to use indicator variables to express the HW(m_2) model in two other forms when $k = m_2/m_1$. We do this to make it easier to understand the MS model and its special cases in the next section. First we divide the cycle $\boldsymbol{c}_0$ for HW(m_2) into k sub-cycles of length m_1 as follows:

$$\begin{aligned} \boldsymbol{c}_{i,0} &= (s_{i,0}, s_{i,-1}, \dots, s_{i,-m_1+1})' \\ &= (s_{-m_1(k-i)}, s_{-m_1(k-i)-1}, \dots, s_{-m_1(k-i)-m_1+1})', \end{aligned} \tag{14.3}$$

where $i = 1, \dots, k$, and

$$\boldsymbol{c}_0 = (\boldsymbol{c}'_{k,0}, \boldsymbol{c}'_{k-1,0}, \dots, \boldsymbol{c}'_{1,0})'. \tag{14.4}$$

For example, with hourly data, we could divide the weekly cycle of length 168 into $k = 7$ daily sub-cycles of length $m_1 = 24$. At each time period t, $\boldsymbol{c}_{it}$

contains the current values of the m_1 seasonal components for cycle i (i.e., day i) and is defined by

$$\boldsymbol{c}_{it} = (s_{i,t}, s_{i,t-1}, \ldots, s_{i,t-m_1+1})', \qquad i = 1, \ldots, k. \tag{14.5}$$

Next we define a set of indicator variables that indicate which sub-cycle is in effect for time period t. For example, when using hourly data these indicator variables would indicate the daily cycle to which the time period belongs. The indicator variables are defined as follows:

$$x_{it} = \begin{cases} 1 & \text{if time } t \text{ occurs when sub-cycle } i \text{ is in effect} \\ 0 & \text{otherwise.} \end{cases} \tag{14.6}$$

Then the HW(m_2) model may be written as follows:

$$y_t = \ell_{t-1} + b_{t-1} + \sum_{i=1}^{k} x_{it} s_{i,t-m_1} + \varepsilon_t, \tag{14.7a}$$

$$\ell_t = \ell_{t-1} + b_{t-1} + \alpha \varepsilon_t, \tag{14.7b}$$

$$b_t = b_{t-1} + \beta \varepsilon_t, \tag{14.7c}$$

$$s_{it} = s_{i,t-m_1} + \gamma x_{it} \varepsilon_t, \qquad i = 1, \ldots, k. \tag{14.7d}$$

The effect of the x_{it} is to ensure that the $m_2 = k \times m_1$ seasonal terms are each updated exactly once in every m_2 time periods. Equation (14.7d) may also be written in a form that will be a special case of the MS model in the next section as follows:

$$s_{it} = s_{i,t-m_1} + \left(\sum_{j=1}^{k} \gamma_{ij} x_{jt} \right) \varepsilon_t,$$

where $i = 1, \ldots, k$ and

$$\gamma_{ij} = \begin{cases} \gamma & \text{if } i = j \\ 0 & \text{otherwise.} \end{cases}$$

14.2 Multiple Seasonal Processes

14.2.1 An Innovations State Space Model for Multiple Seasonal (MS) Processes

A fundamental goal of our new model for multiple seasonal (MS) processes is to allow for the seasonal terms that represent a seasonal cycle to be updated more than once during the period of the cycle. This goal may be achieved in two ways with our model. We start, as we did for the HW(m_2) model in the previous section, by dividing the cycle of length m_2 into k shorter sub-cycles of length m_1. Then we use a matrix of smoothing parameters that allows the seasonal terms of one sub-cycle to be updated during the time for another sub-cycle. For example, seasonal terms for Monday can be updated on Tuesday. Sometimes this goal can be achieved in a second way by combining sub-cycles with the same seasonal pattern into one common sub-cycle. This latter approach has the advantage of reducing the required number of seed values and, in some cases, the number of parameters. More frequent updates may also provide better forecasts, particularly when the observations m_1 time periods ago are more important than those values m_2 time periods earlier. It is also possible with our model to have different smoothing parameters for different sub-cycles (e.g., for different days of the week).

The existence of common sub-cycles is the key to reducing the number of seed values compared to those required by the HW method and DS exponential smoothing. As described in Sect. 14.1.3, it may be possible for a long cycle to be broken into $k = m_2/m_1$ shorter cycles of length m_1. Of these k possible sub-cycles, $r \leq k$ distinct cycles may be identified. For example, consider the case when $m_1 = 24$ and $m_2 = 168$ for hourly data. By assuming that Monday–Friday have the same seasonal pattern, we can use the same sub-cycle for these 5 days. We can use the same sub-cycle for Saturday and Sunday, if they are similar. Thus, we might be able to reduce the number of daily sub-cycles from $k = 7$ to $r = 2$. The number of seed estimates required for the seasonal terms would be reduced from 168 for the HW method and 192 for the DS method to 48 for the new method.

A set of indicator variables based on the r shorter cycles can be defined by

$$x_{it} = \begin{cases} 1 & \text{if time period } t \text{ occurs when sub-cycle } i \text{ is in effect;} \\ 0 & \text{otherwise.} \end{cases} \tag{14.8}$$

On any given day, only one of the x_{it} values equals 1. Let $\boldsymbol{x}_t = [x_{1t}, x_{2t}, x_{3t}, \ldots, x_{rt}]'$ and $\boldsymbol{s}_t = [s_{1t}, s_{2t}, s_{3t}, \ldots, s_{rt}]'$.

The general MS model for additive seasonality and $r \leq k = m_2/m_1$ is:

$$y_t = \ell_{t-1} + b_{t-1} + \sum_{i=1}^{r} x_{it} s_{i,t-m_1} + \varepsilon_t, \tag{14.9a}$$

$$\ell_t = \ell_{t-1} + b_{t-1} + \alpha\varepsilon_t, \tag{14.9b}$$

$$b_t = b_{t-1} + \beta\varepsilon_t, \tag{14.9c}$$

$$s_{it} = s_{i,t-m_1} + \left(\sum_{j=1}^{r} \gamma_{ij} x_{jt} \right) \varepsilon_t, \tag{14.9d}$$

where $i = 1, \ldots, r$ and $\varepsilon_t \sim \text{NID}(0, \sigma^2)$.

This model is a linear innovations state space model, and the equations can also be written in matrix form:

$$y_t = \ell_{t-1} + b_{t-1} + \boldsymbol{x}_t' \boldsymbol{s}_{t-m_1} + \varepsilon_t, \tag{14.10a}$$
$$\ell_t = \ell_{t-1} + b_{t-1} + \alpha\varepsilon_t, \tag{14.10b}$$
$$b_t = b_{t-1} + \beta\varepsilon_t, \tag{14.10c}$$
$$\boldsymbol{s}_t = \boldsymbol{s}_{t-m_1} + \boldsymbol{\Gamma}\boldsymbol{x}_t\varepsilon_t, \tag{14.10d}$$
$$\hat{y}_{t+1|t} = \ell_{t-1} + b_{t-1} + \boldsymbol{x}_t' \boldsymbol{s}_{t-m_1}. \tag{14.10e}$$

The seasonal smoothing matrix $\boldsymbol{\Gamma}$ contains the smoothing parameters for each of the cycles. The parameter γ_{ii} is used to update seasonal terms during time periods that belong to the same sub-cycle (e.g., days that have the same daily pattern). The parameter $\gamma_{ij}, i \neq j$, is used to update seasonal terms belonging to a sub-cycle during the time periods that occur during another sub-cycle (e.g., seasonal terms for one day can be updated during a day that does not have the same daily pattern). We will denote this model by MS$(r; m_1, m_2)$ and the seasonal cycles by

$$\boldsymbol{c}_{it} = \left(s_{i,t}, s_{i,t-1}, \ldots, s_{i,t-m_1+1}\right)'$$

for $i = 1, \ldots, r$.

This model can also be written in a state space form (see Appendix "First-order form of the model") and estimated using the Kalman filter (Sect. 12.7) or the information filter (Sect. 12.3). Here, as will be discussed in Sect. 14.2.5, we estimate the model using exponential smoothing when maximizing the conditional likelihood in Chap. 5.

14.2.2 A Model for Multiplicative Seasonality

Thus far, we have concentrated upon models for time series that exhibit additive, rather than multiplicative, seasonal patterns. In the additive case the seasonal effects do not depend on the level of the time series, while for the multiplicative case the seasonal effects increase at higher values of the time series. We can adapt the MS$(r; m_1, m_2)$ model to account for a multiplicative seasonal pattern using the approach for nonlinear models in Chap. 4.

The general multiplicative form of the MS model for $r \leq k = m_2/m_1$ is:

$$y_t = (\ell_{t-1} + b_{t-1})\left(\sum_{i=1}^{r} x_{it}s_{i,t-m_1}\right)(1 + \varepsilon_t), \tag{14.11a}$$
$$\ell_t = (\ell_{t-1} + b_{t-1})(1 + \alpha\varepsilon_t), \tag{14.11b}$$
$$b_t = b_{t-1} + \beta(\ell_{t-1} + b_{t-1})\varepsilon_t, \tag{14.11c}$$

$$s_{it} = s_{i,t-m_1}\left[1 + \left(\sum_{j=1}^{r} \gamma_{ij} x_{jt}\right)\varepsilon_t\right] \qquad (i = 1, \ldots, r), \tag{14.11d}$$

where $\varepsilon_t \sim \text{NID}(0, \sigma^2)$. Consequently,

$$\hat{y}_{t+1|t} = (\ell_{t-1} + b_{t-1})\left(\sum_{i=1}^{r} x_{it} s_{i,t-m_1}\right). \tag{14.12}$$

14.2.3 Reduced Form of the MS Model

We now return to the additive form of the MS model. The reduced form of the MS$(r; m_1, m_2)$ model may be derived from (14.9) by applying appropriate transformations to y_t to eliminate the state variables and achieve stationarity. The reduced form of the MS model is

$$\Delta\Delta_{m_2} y_t = \sum_{j=1}^{r}\left(\theta_{jt} L^{jm_1} - \theta_{j,t-1} L^{jm_1+1}\right)\varepsilon_t + \alpha\Delta_{m_2}\varepsilon_{t-1} + \beta\sum_{j=1}^{m_2} L^j \varepsilon_t + \Delta\Delta_{m_2}\varepsilon_t, \tag{14.13}$$

where L is the lag operator and $\Delta_i = (1 - L^i)$ takes the ith difference. In the case where the trend b_t is omitted, the reduced form becomes:

$$\Delta_{m_2} y_t = \left(\sum_{j=1}^{r} \theta_{jt} L^{jm_1}\right)\varepsilon_t + \alpha\sum_{j=1}^{m_2} L^j \varepsilon_t + \Delta_{m_2}\varepsilon_t. \tag{14.14}$$

The θ_{it} value will be a sum of r terms, each of which is a product of a value from $\boldsymbol{x}_t$ and a value from $\boldsymbol{\Gamma}$, but at any time t it will be equal to only one of the values from $\boldsymbol{\Gamma}$.

For example, in the case with no trend, $m_1 = 4$, $m_2 = 12$ and $k = r = 3$ (no repeating sub-cycles), (14.14) can be written as:

$$\Delta_{12} y_t = \left(\sum_{j=1}^{3} \theta_{jt} L^{4j}\right)\varepsilon_t + \alpha\sum_{j=1}^{12} L^j \varepsilon_t + \Delta_{12}\varepsilon_t. \tag{14.15}$$

In this case, $\theta_{1t} = x_{1t}\gamma_{13} + x_{2t}\gamma_{21} + x_{3t}\gamma_{32}$, $\theta_{2t} = x_{1t}\gamma_{12} + x_{2t}\gamma_{23} + x_{3t}\gamma_{31}$ and $\theta_{3t} = x_{1t}\gamma_{11} + x_{2t}\gamma_{22} + x_{3t}\gamma_{33}$. See Appendix "The MS$(r; m_1, m_2)$ model in reduced form" for the derivation of the reduced form.

The reduced form of the model verifies that the MS model has a sensible, though complex, ARIMA structure with time-dependent parameters at the seasonal and near seasonal lags. The advantage of the state space form is that the MS model is more logically derived, more easily estimated, and more interpretable than its reduced form counterpart. In the next section we give the specific restrictions on $\boldsymbol{\Gamma}$ (and hence the θ_{it} values) that may be used to show that the reduced forms of the HW(m_1), HW(m_2) and DS(m_1, m_2) models are special cases of the reduced form of the MS$(r; m_1, m_2)$ model in (14.13).

14.2.4 Model Restrictions

In general, the number of smoothing parameters contained in $\mathbf{\Gamma}$ is equal to the square of the number of separate sub-cycles (r^2) and can be quite large. In addition to combining some of the sub-cycles into a common sub-cycle, restrictions can be imposed on $\mathbf{\Gamma}$ to reduce the number of parameters. We shall see that some of these restrictions produce the HW(m_1), HW(m_2), and DS(m_1, m_2) models as special cases of the MS($r; m_1, m_2$) model in (14.10).

One type of restriction is to force common diagonal and common off-diagonal elements as follows:

$$\gamma_{ij} = \begin{cases} \gamma_1^*, \text{ if } i \neq j & \text{common off-diagonal} \\ \gamma_2^*, \text{ if } i = j & \text{common diagonal.} \end{cases} \tag{14.16}$$

In this case $\theta_{1t} = \theta_{2t} = \cdots = \theta_{r-1,t} = \gamma_1^*$ and $\theta_{rt} = \gamma_2^*$.

Within the type of restriction in (14.16), there are three restrictions of particular interest. We will refer to them as

- *Restriction 1:* $\gamma_1^* = 0$, and $\gamma_2^* \neq 0$
  If $r = k$, this restricted model is equivalent to the HW(m_2) model in (14.1) where $\gamma_2^* = \gamma$. The seed values for the k seasonal cycles in this MS($k; m_1, m_2$) model and the one seasonal cycle in the HW(m_2) model are related as shown in (14.3) and (14.4) of Sect. 14.1.3 (where $t = 0$).
- *Restriction 2:* $\gamma_1^* = \gamma_2^*$
  If the seed values for the r seasonal sub-cycles in the MS($r; m_1, m_2$) model are identical, this restricted model is equivalent to the HW(m_1) model in (14.1) where $\gamma_1^* = \gamma$. Normally in the MS($r; m_1, m_2$) model, the different sub-cycles are allowed to have different seed values. Hence, this restricted model will only be exactly the same as the HW(m_1) model if we also restrict the seed values for the sub-cycles to be equal to each other. However, for sufficiently large n, the forecasts should not be affected by the seed values.
- *Restriction 3:* Equivalent to (14.16)
  If $r = k$, this restricted model is equivalent to the DS(m_1, m_2) model in (14.2) where $\gamma_1^* = \gamma_1$ and $\gamma_2^* = \gamma_1 + \gamma_2$. The seed values for the k seasonal cycles in this MS($k; m_1, m_2$) model and the two seasonal cycles in the DS(m_1, m_2) model are related by

$$\boldsymbol{c}_{i0} = \big(s_0^{(1)} + s_{-m_1(k-i)}^{(2)},\ s_{-1}^{(1)} + s_{-m_1(k-i)-1}^{(2)},\ \ldots,\ s_{-m_1+1}^{(1)} + s_{-m_1(k-i)-m_1+1}^{(2)}\big)'.$$

The MS($r; m_1, m_2$) model allows us to explore a much broader range of assumptions than existing methods, while retaining parsimony. It nests the models underlying the additive HW and DS methods. It contains other restricted forms that stand in their own right. Table 14.1 presents the number of parameters and seed values that require estimates in the

Table 14.1. Number of smoothing parameters and seed values.

Model	Parameters	Seed values
$\text{MS}(r; m_1, m_2)$	$r^2 + 2$	$rm_1 + 2$
$\text{MS}(r; m_1, m_2)$-Rstr. 1	3	$rm_1 + 2$
$\text{HW}(m_2)$	3	$m_2 + 2 = km_1 + 2$
$\text{MS}(r; m_1, m_2)$-Rstr. 2	3	$rm_1 + 2$
$\text{HW}(m_1)$	3	$m_1 + 2$
$\text{MS}(r; m_1, m_2)$-Rstr. 3	4	$rm_1 + 2$
$\text{DS}(m_1, m_2)$	4	$m_2 + 2 = km_1 + 2$[a]

[a] Short cycle seed values may be started at 0.

$\text{MS}(r; m_1, m_2)$ model and some of its restrictions. A procedure for choosing among the possible $\text{MS}(r; m_1, m_2)$ models with and without these restrictions is described in the next section.

14.2.5 Model Estimation, Selection, and Prediction

The estimation, model selection, and prediction described in this section apply to both the additive and multiplicative MS models.

Estimation

Within the exponential smoothing framework, the parameters in an $\text{MS}(r; m_1, m_2)$ model can be estimated by minimizing the one-step-ahead sum of squared errors

$$\text{SSE} = \sum_{i=1}^{n} (y_t - \hat{y}_t)^2,$$

where n is the number of observations in the series, and $\hat{y}_t = \hat{y}_{t|t-1}$. The seed states for the level, trend and seasonal components may be estimated by applying the procedures for $\text{HW}(m_2)$ in Sect. 2.6.1 to the time periods that represent four completions of all the sub-cycles (e.g., the first four weeks for hourly data). The m_1 estimates for each of the k seasonal sub-cycles are then found by using the relationship between the cycles explained in (14.3) and (14.4) of Sect. 14.1.3. If $r < k$, the estimates of the sub-cycles with the same seasonal pattern are averaged. Then the SSE is minimized with respect to the smoothing parameters by using the exponential smoothing equations in (14.9). The smoothing parameters are restricted to values between 0 and 1.

Model Selection

We have seen that various special cases of the $\text{MS}(r; m_1, m_2)$ model may be of interest. We may wish to choose the number of seasonal sub-cycles r to be less than k, restrict the values of the seasonal parameters, or use a combination of the two. We employ a two-step process to make these decisions. First we choose r, and then we determine whether to restrict $\mathbf{\Gamma}$ as follows:

1. Choose the value of r in MS($r; m_1, m_2$)
   (a) From a sample of size n, withhold q time periods, where q is the last 20% of the data rounded to the nearest multiple of m_2 (e.g., whole number of weeks).
   (b) Select a set of values of interest for r (e.g., using common sense and/or graphs), and estimate the parameters for each model using observations 1 to $n-q$.
   (c) For each of the models in 1(b), find one-step-ahead forecasts for time periods $n-q+1$ to n without re-estimating.
   (d) Pick the value of r with the smallest root mean squared forecast error

$$\text{RMSE}(1) = \sqrt{\sum_{t=n-q+1}^{n} (y_t - \hat{y}_t)^2 / q}.$$

2. Choose the restrictions on $\mathbf{\Gamma}$
   (a) Using the value of r selected in part 1 and the same $n-q$ time periods, compute the one-step-ahead forecast errors for Restrictions 1, 2, and 3, no restrictions, and any other restrictions of particular interest over $[n-q+1, n]$.
   (b) Choose the restriction with the smallest RMSE.

Prediction

A point forecast for y_{n+h} at time period n is the conditional expected value:

$$\hat{y}_{n+h|n} = \mathrm{E}(y_{n+h} \mid \boldsymbol{a}_0, y_1, \ldots, y_n),$$

where

$$\begin{aligned}\boldsymbol{a}_0 &= (\ell_0, b_0, s_{1,0}, \ldots, s_{1,-m_1+1}, s_{2,0}, \ldots, s_{2,-m_1+1}, \ldots, s_{r,0}, \ldots, s_{r,-m_1+1})' \\ &= (\ell_0, b_0, \boldsymbol{c}'_{1,0}, \boldsymbol{c}'_{2,0}, \ldots, \boldsymbol{c}'_{r,0})'.\end{aligned}$$

Prediction intervals for h periods in the future from time period n can be found by using the model in (14.9) as follows: simulate an entire distribution for y_{n+h} and pick the percentiles for the desired level of confidence (Ord et al. 1997).

14.3 An Application to Utility Data

In this empirical example, we show that the MS model performs best within the class of exponential smoothing models. Utility demand data was selected to illustrate our MS procedure because it clearly has multiple seasonal cycles. Other approaches to forecasting utility demand may be more appropriate in particular circumstances; see, for example, Ramanathan et al. (1997) and Cottet and Smith (2003).

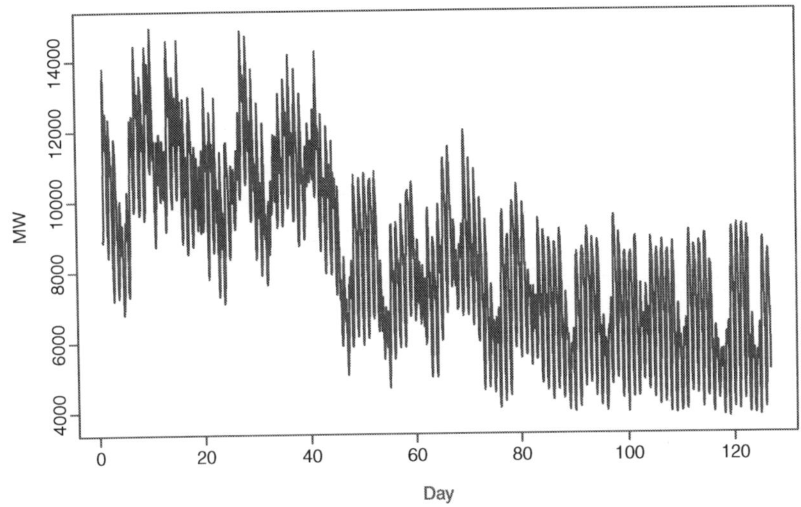

Fig. 14.3. Hourly utility demand.

14.3.1 The Study

The data set consists of 3,024 observations (18 weeks) of hourly utility demand, beginning on January 1, 2003, from a utility company in the Midwestern area of the United States. A graph of the data is shown in Fig. 14.3. This utility data appears to have a changing level rather than a trend so the growth rate b_t is omitted. The data also appear to exhibit an additive seasonal pattern, that is, a seasonal pattern for which the variation does not change with the level of the time series. For this reason the main focus of this application is on additive models, although a multiplicative version of our model is also tested. The data are split into two parts: a fitting sample of ($n = 2{,}520$) observations (i.e., 15 weeks) and post sample data of ($p = 504$) observations (i.e., 3 weeks). There are no weekday public holidays during the period of the post-sample data.

The data have a number of important features that should be reflected in the model structure. There are three levels of seasonality: yearly effects (largely driven by temperatures), weekly effects and daily effects. For this case study, we will only seek to capture the daily and weekly seasonal patterns.

14.3.2 Selecting an MS Model

In this section we follow the procedure for model selection described in Sect. 14.2.5 to select the best MS model. The first step is to choose r in MS(r; 24, 168). To start this step we withhold 504 observations or three weeks of data ($q = 504$) from the fitting sample ($n = 2{,}520$). The value of q is

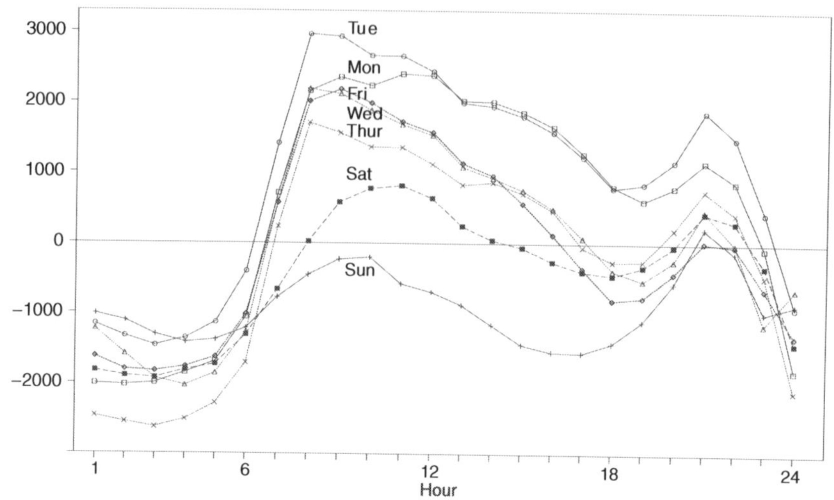

Fig. 14.4. MS(7; 24, 168): Hourly sub-cycles by day, based on the last 168 observations ($t = 2{,}353, \ldots, 2{,}520$).

20% of n and is the same as that of p in this example. Then, we need to re-examine the data to look for common daily patterns for different days of the week. One way to look for potential common patterns is to graph the 24-hour pattern for each day of the week on the same horizontal axis. In Fig. 14.4 we plot the seasonal terms that are estimated for the seven sub-cycles in the MS(7; 24, 168) model during the last week of the sample ($t = n - 167, \ldots, n$). The plot suggests that $r = 7$ may use more daily patterns than is required. The similarity of some weekday sub-cycles indicates that alternative structures could be tested.

Visual inspection of Fig. 14.4 shows that the Monday–Friday sub-cycles are similar, and Saturdays and Sundays are similar. Closer inspection shows that the Monday–Tuesday patterns are similar, Wednesday–Friday patterns are similar, and Saturdays and Sundays display some differences from each other. A third possible approach is to assume that Monday–Thursday have a common pattern and Friday, Saturday and Sunday have their own patterns. This choice is plausible because Fridays should have a different evening pattern to other weekdays as consumers and industry settle into weekend routines. Support for this choice of common sub-cycles can also be seen in Fig. 14.4 where Friday starts to behave more like Saturday in the evening hours. We list these three choices below:

- $r = 4$ Version 1 MS(4; 24, 168):
  Common Monday–Thursday sub-cycle, separate Friday, Saturday and Sunday sub-cycles
- $r = 4$ Version 2 MS(4(2); 24, 168):
  Common Monday–Tuesday, Wednesday–Friday sub-cycles, separate Saturday and Sunday sub-cycles

Table 14.2. Withheld-sample RMSE in MS model selection for utility data.

Model	Restriction	RMSE(1)	Parameters	Seed values
MS(7; 24, 168)	none	234.72	50	168
MS(4; 24, 168)	none	239.67	17	96
MS(4(2); 24, 168)	none	250.34	17	96
MS(2; 24, 168)	none	225.51	5	48
MS(2; 24, 168)	1	246.51	2	48
MS(2; 24, 168)	2	234.49	2	48
MS(2; 24, 168)	3	225.31	3	48

Table 14.3. Comparison of post-sample forecasts for the utility data.

Model	Restriction	RMSE(1)	Parameters	Seed values
HW(24)	na	278.50	2	24
HW(168)	na	278.04	2	168
DS(24, 168)	na	227.09	3	168
MS(7; 24, 168)	none	208.45	50	168
MS(2; 24, 168)	3	206.45	3	48

- $r = 2$ MS(2; 24, 168):
  Common Monday–Friday sub-cycle, common weekend sub-cycle

We finish the first step of the model selection process by comparing the value of RMSE(1) for the MS(7; 24, 168) model to the values for the three submodels listed above. Of these four models, MS(2; 24, 168) has the smallest RMSE(1), as shown in Table 14.2. Thus, we choose this model in the first step. The RMSE(1) values in Table 14.2 are computed for the withheld time periods $n - q + 1$ to n (i.e., 2,017–2,520). In Table 14.2 we say this RMSE(1) compares "withheld-sample" forecasts to distinguish it from the RMSE(1) in Table 14.3, which will be computed for the p post-sample values (i.e., 2,521–3,024) that are not part of the fitting sample.

In the second step of the process from Sect. 14.2.5, we compare Restrictions 1, 2, and 3 from Sect. 14.2.4 for the MS(2; 24, 168) model that was chosen in the first step. The RMSE(1) values for these three additional models are also shown in Table 14.2. The model with the smallest RMSE(1) for the withheld sample test is the MS(2; 24, 168) model with Restriction 3. Hence, this model is our selection for the best MS model for forecasting.

14.3.3 Forecasting with the MS, HW and DS Models

In general, the MS models provide better point forecasts than the HW and DS models. The forecasting accuracy of the models is compared by using the root mean squared forecast error for h periods ahead over p post-sample values. The root mean squared forecast error is defined as:

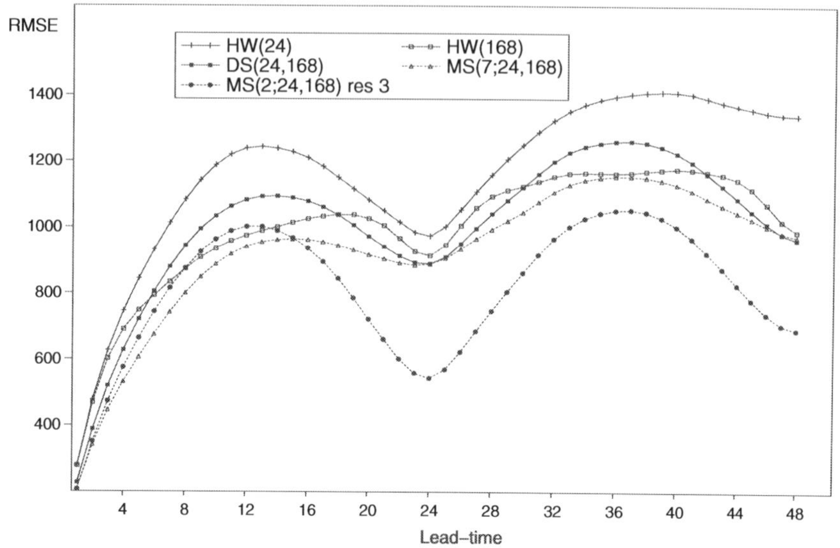

Fig. 14.5. Forecasting accuracy (RMSE) for lead-times from 1 to 48 hours (i.e., 2 days).

$$\text{RMSE}(h) = \sqrt{\frac{1}{p-(h-1)} \sum_{t=n}^{n+p-h} (y_{t+h} - \hat{y}_{t+h|t})^2},$$

where $\hat{y}_{t+h|t}$ is the forecast of y_{t+h} at time t. In this application, the RMSE(h) values are averages based on 3 weeks (i.e., $p = 504$ hours) of post-sample data and lead-times h of 1–48 hours. Table 14.3 contains the post sample RMSE(1) for the two HW models of HW(24) and HW(168), the double seasonal model DS(24, 168), the full unrestricted multiple seasons model MS(7; 24, 168), and the selected multiple seasons model MS(7; 24, 168) with Restriction 3 from Sect. 14.3.2. Figure 14.5 contains the RMSE(h) for these same five models where the forecast horizon, h, ranges from 1 to 48 hours.

The estimation of the parameters and seed values for these five models is done using the fitting sample of size $n = 2{,}520$. The first four weeks of the fitting sample are used to find initial values for the states. For the HW method these values are found by using the approach in Sect. 2.6.1. The 24 additional initial values for the daily seasonal components in the DS method are set equal to 0. The initial values for the MS models are found as described in Sect. 14.2.5. Smoothing parameters for all of the models are estimated by minimizing the SSE for the fitting sample of length $n = 2{,}520$, and all parameters are constrained to lie between 0 and 1.

In examining Table 14.3, we see that MS(2; 24, 168) with Restriction 3 has the smallest RMSE(1), and MS(7; 24, 168) is second best. The MS(2; 24, 168) model also has far fewer parameters (3 versus 50) and seed values (48 versus 168) than the MS(7; 24, 168) model. In Fig. 14.5, the RMSE(h) values are

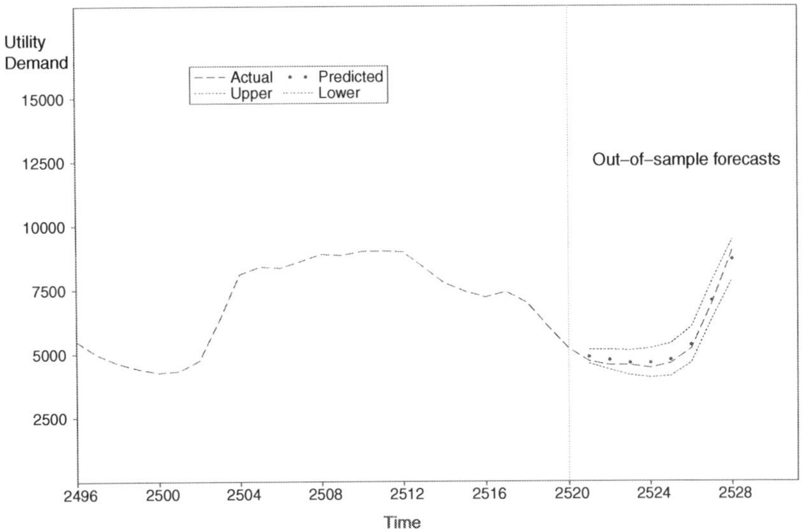

Fig. 14.6. MS(2; 24, 168) Restriction 3: Point forecasts and 80% prediction intervals for the first 8 h in the next week ($t = 2{,}521, \ldots, 2{,}528$) of the utility demand.

consistently lower for the MS models than for the HW and DS alternatives, with the MS model chosen by our selection process being much lower. The more accurate forecasts are the result of the MS models offering a more reliable structure for capturing the changes in seasonality.

Figure 14.6 shows the post-sample forecasting accuracy of the MS(2; 24,168) model with Restriction 3. Forecasts and 80% prediction intervals are provided only for the first 8 h of the post-sample period because the intervals become extremely wide as the time horizon increases. During the 8 h period in Fig. 14.6, the forecasts are very good. The 80% prediction intervals are calculated via simulation. One cause for the wide prediction intervals at longer time horizons is the large estimate of α. Large structural change will require wide prediction intervals. For the utility data the parameter α was estimated to be between 0 and 1, and this constraint was binding in most cases (i.e., $\hat{\alpha} = 1$). In this case, the resulting model corresponds to a purely seasonal model for first differences.

14.3.4 Further Comments

The wide prediction intervals that were found when forecasting the utility data can sometimes be avoided, if one's goal is to forecast more than a few hours ahead. For the longer time horizons, the parameters can be estimated by minimizing the sum of squared h-step-ahead errors instead of the usual one-step-ahead errors. Table 14.4 shows the effect on the estimates of the parameters when the estimation criterion is altered for the utility data in our

Table 14.4. Utility data smoothing parameter estimates.

Estimation done for	$\hat{\alpha}$	γ_1^*	γ_2^*
$h = 1$	1	0.084	0.12
$h = 24$	0	0.83	0.83
$h = 168$	0	0.11	0.13

study. When the sum of squares is minimized for 24-step-ahead or 168-step-ahead errors, the estimate of α is 0. This smaller value of $\hat{\alpha}$ will reduce the width of the prediction intervals at the longer lead-times. Examination of the equations in (14.9), without (14.9c) and when $\alpha = 0$, reveals that the prediction intervals will only increase in width every m_1 periods rather than every period. Figure 14.5 suggests that narrower prediction intervals become possible, especially for $m_1/2 < h \leq m_1$.

An interesting feature of Fig. 14.5 is the way in which the models clearly have lower RMSE(h) values when h is a multiple of 24. This pattern has been seen in other studies of seasonal series (e.g., Makridakis and Hibon 2000), and indicates some degree of model mis-specification. The implication of the numerical results in this case is that the forecasts are more accurate when they are made for a full day ahead for the same time of day (i.e., a purely seasonal model).

In addition to examining the utility data in Fig. 14.3 to decide that additive seasonal models were appropriate, we tested the multiplicative MS(7; 24, 168) model in (14.11) with no trend. We found that the withheld-sample RMSE(1) was 271.48, which is larger than the RMSE(1) of 234.72 for the additive MS(7; 24, 168) model. This provides further support for our choice of additive seasonality. An advantage of the single source of error models is that such nonlinear models can be included in a study.

Because Taylor (2003b) found that adding an AR(1) term improved the forecasting accuracy of the DS model for his utility data, we also examined whether adding an AR(1) would help for our data. We found that forecasts at lead-times longer than one time period are worse when the AR(1) term is included.

14.4 Analysis of Traffic Data

In this section we investigate an application of the MS approach to hourly vehicle counts, and compare the forecasts with those from the HW and DS approaches.

14.4.1 The Study

The fitting sample consists of 1,689 observations (about 10 weeks) of hourly vehicle counts for the Monash Freeway, outside Melbourne in Victoria,

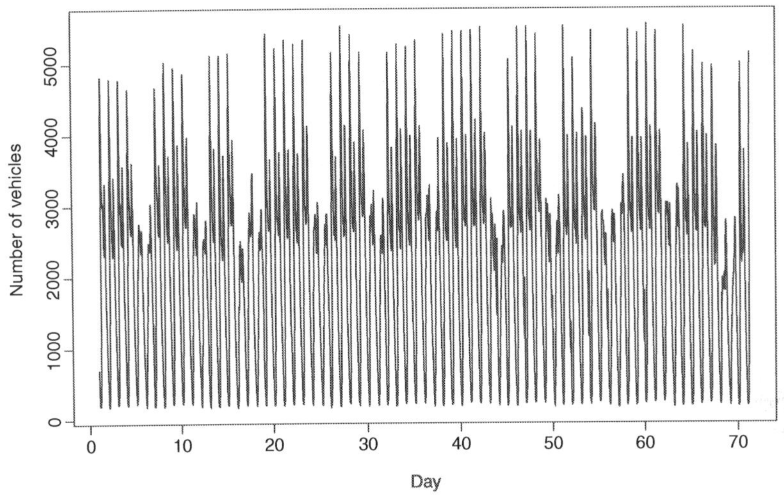

Fig. 14.7. Hourly vehicle counts.

Australia, beginning August 1995. A graph of the data is shown in Fig. 14.7. The observation series has missing values when the data recording equipment was not operational. The gaps in the data are for periods of days (i.e., multiples of 24) and can be handled using unconditional updating of the states. When y_t is not observed, the error cannot be calculated. The error is still unknown and is governed by an $\mathrm{N}(0,\sigma^2)$ distribution. The best predictor of the uncertain error is the mean of its distribution, namely 0. Hence we use 0 as the estimate of ε_t when predicting the next state vector from the old state vector using the transition state equation. Such an approach can be applied to any innovations state space model. In many traffic applications this ability to handle missing values is particularly useful when counting equipment has to be taken off-line for maintenance.

Apart from the missing observations, the traffic data have the same features as the utility data, although the yearly effects are less pronounced. As before, we only seek to capture the daily and weekly seasonal patterns. Because this data appears to have no trend and to exhibit an additive seasonal pattern, we use additive seasonality for the HW, DS, and MS approaches, and omit the equation for the growth rate b_t. Victorian public holidays appear throughout the fitting sample and follow a similar daily pattern to Sundays.

This study of vehicle flows includes the HW(24), HW(168), DS(24,168) and MS models. Models are compared by using the RMSE for h periods ahead over a post-sample period of length $p = 504$, which does not contain any public holidays. We examine lead-times of up to 2 weeks ($h = 1,\ldots,336$), which can be relevant for planning road works. Smoothing parameters and seeds are estimated using the same procedures as in the previous section.

Table 14.5. Comparison of withheld-sample forecasts for the traffic data.

Model	Restriction	RMSE(1)	Parameters	Seed values
MS(7; 24, 168)	none	498.31	50	168
MS(4; 24, 168)	none	428.88	17	96
MS(3; 24, 168)	none	394.42	10	72
MS(2; 24, 168)	none	308.84	5	48
MS(2; 24, 168)	1	310.94	2	48
MS(2; 24, 168)	2	333.85	2	48
MS(2; 24, 168)	3	310.94	3	48
MS(2; 24, 168) public hols.	none	228.68	5	48

An MS model is chosen using the method in Sect. 14.2.5, with $q = 336$ (i.e., 2 weeks of data). Based on a visual inspection of the raw data and plots of the seasonal terms for the MS(7; 24, 168) model, three candidates were tested along with the full MS model.

- $r = 4$ Version 1 MS(4; 24, 168): Common Monday–Thursday sub-cycle, separate Friday, Saturday and Sunday sub-cycles
- $r = 3$ MS(3; 24, 168): Common Monday–Friday sub-cycle, separate Saturday and Sunday sub-cycles
- $r = 2$ MS(2; 24, 168): Common Monday–Friday sub-cycle, common weekend sub-cycle

In Table 14.5 we see that, among the first four models, MS(2; 24, 168) has the smallest RMSE(1), where this RMSE is computed using the withheld values within the original sample. Thus, we choose $r = 2$ in step 1. None of the restrictions are supported. However, if we account for public holidays by using the same indicator as the one for the Saturday/Sunday sub-cycle, the one-step-ahead forecasts for the withheld data are greatly improved. Hence, we choose MS(2; 24, 168) with *public holidays* for our best MS model.

14.4.2 Comparison of the MS Models with the HW and DS Models

In Table 14.6, the post-sample RMSE(1) can be compared for each of the following six models: HW(24), HW(168), DS(24, 168), full MS(7; 24, 168) (with and without public holidays), and MS(2; 24, 168) model with public holidays. We see that the MS(2; 24, 168) model that accounts for public holidays has the smallest RMSE(1), while the RMSE(1) for the MS(7; 24, 168) model is slightly larger than the essentially common value for HW(168) and DS(24, 168). The MS(2; 24, 168) model with public holidays is clearly the best model for forecasting 1 h ahead, offering a reduction of approximately 15% in RMSE over the HW and DS models.

In Fig. 14.8, we can compare the HW(24) model, the HW(168) model, the MS(7; 24, 168) model, and the MS(2; 24, 168) model with public holidays over

Table 14.6. Comparison of post-sample forecasts for the traffic data.

Model	Restriction	RMSE(1)	Parameters	Seed values
HW(24)	na	365.09	2	24
HW(168)	na	228.60	2	168
DS(24, 168)	na	228.59	3	168
MS(7; 24, 168)	none	238.33	50	168
MS(7; 24, 168) public hols.	none	245.25	50	168
MS(2; 24, 168) public hols.	none	203.64	5	48

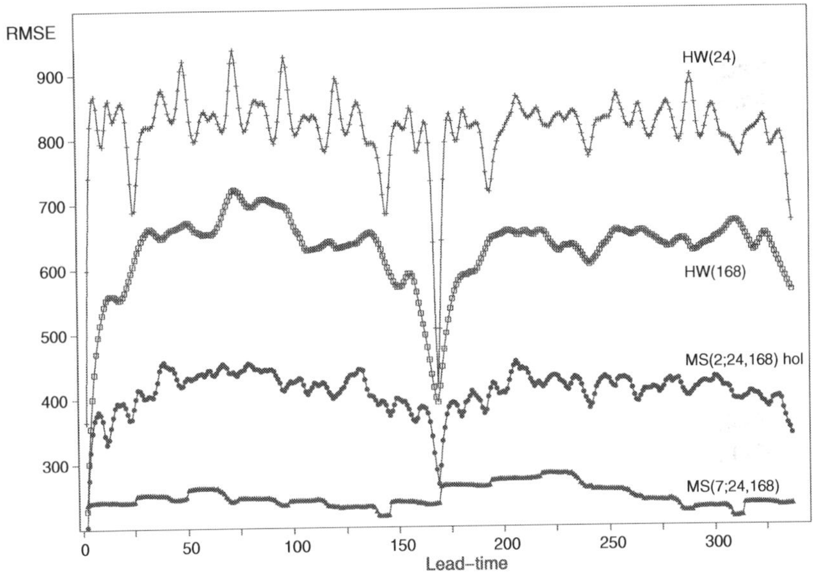

Fig. 14.8. Forecasting accuracy (RMSE) for lead-times from 1 to 336 hours (i.e., 2 weeks).

lead-times of 1–336 hours. The values of RMSE(h) when $h > 1$ in this figure give a different ordering for forecasting accuracy than those in Table 14.6. The estimate of γ_1 when $m_1 = 24$ for DS(24, 168) is effectively zero, making it is equivalent to HW(168). Thus, the DS(24, 168) model is not included, as it is indistinguishable from HW(168). The model selected by our MS selection process, MS(2; 24, 168) with public holidays, is no longer best, but it still provides far more accurate forecasts than the HW and DS models. Clearly, the MS(7; 24, 168) produces the best forecasts (i.e., the smallest RMSE(h)) for forecasting horizons of 2 or more hours ahead.

The unconditional updating of the states during periods of missing data proves to be effective for all models. Generally jumps are observed in the level ℓ_t after periods of missing data. The jumps are more pronounced for

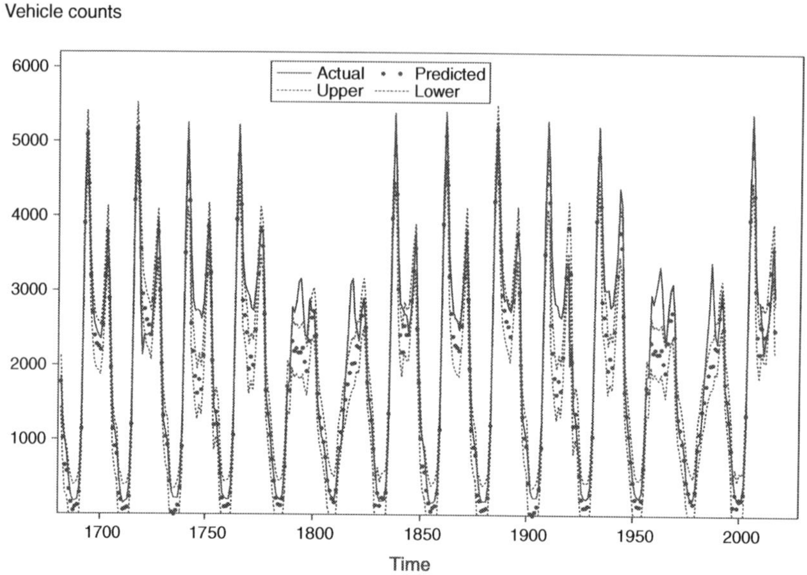

Fig. 14.9. MS(7; 24, 168): Multi-step-ahead point forecasts and 80% prediction intervals for the vehicle counts for each hour in the last 2 weeks of the evaluation sample ($t = 1{,}681, \ldots, 2{,}016$).

the MS(7; 24, 168) model, which has a relatively stable level during periods of no missing data.

Multi-step-ahead forecasts and 80% prediction intervals for the post-sample data using the MS(7; 24, 168) model can be found in Fig. 14.9. The forecasts follow the observed series closely and the prediction intervals are not as wide as those for the utility data. These narrower intervals can be explained by the extremely small estimate of α. For MS(7; 24, 168), $\hat{\alpha} = 0.01$.

14.5 Exercises

Exercise 14.1. Consider the indicator variable models in Sect. 14.1.3 when $m_1 = 4$ and $m_2 = 12$, and show that Restriction 1 applied to Model 14.9 (or the equivalent Model 14.10) yields the Holt-Winters model HW(12).

Exercise 14.2. Repeat Exercise 14.1 to show that Restriction 2 applied to Model 14.9 (or the equivalent Model 14.10) yields the Holt-Winters model HW(4).

Exercise 14.3. Develop an argument for the double smoothing model DS(m_1, m_2) similar to the one in Sect. 14.1.3 to show that Restriction 3 applied to Model 14.9 (or the equivalent Model 14.10) leads to the DS(m_1, m_2) model.

Appendix: Alternative Forms

First-Order Form of the Model

The MS$(r; m_1, m_2)$ model in (14.9) can be written in first-order form where the state variables are lagged by only one period in the state transition equation:

$$\begin{aligned} y_t &= \boldsymbol{w}_t' \boldsymbol{a}_{t-1} + \varepsilon_t, \\ \boldsymbol{a}_t &= \boldsymbol{F} \boldsymbol{a}_{t-1} + \boldsymbol{g}_t \varepsilon_t, \end{aligned}$$

where $\boldsymbol{a}_t$ is the state vector of length $2 + rm_1$ containing level, trend and seasonal terms:

$$\boldsymbol{a}_t = (\ell_t, b_t, s_{1,t}, \ldots, s_{1,t-m_1+1}, s_{2,t}, \ldots, s_{2,t-m_1+1}, \ldots, s_{r,t}, \ldots, s_{r,t-m_1+1})';$$

$\boldsymbol{w}_t$ is a vector of length $2 + rm_1$ containing values of 1 and 0 (depending on which sub-cycle t corresponds to):

$$\boldsymbol{w}_t = (1, 1, 0, \ldots, 0, x_{1t}, 0, \ldots, 0, x_{2t}, 0, \ldots, 0, x_{rt})';$$

$\boldsymbol{F}$ is a block-diagonal $(2 + rm_1) \times (2 + rm_1)$ matrix of the form:

$$\boldsymbol{F} = \begin{bmatrix} \boldsymbol{F}_\ell & \vdots & \boldsymbol{0} \\ \cdots & \vdots & \cdots \\ \boldsymbol{0} & \vdots & \boldsymbol{F}_s \end{bmatrix},$$

where

$$\boldsymbol{F}_\ell = \begin{bmatrix} 1 & 1 \\ 0 & 1 \end{bmatrix}$$

controls the level and trend components, and the seasonal components are controlled by the $rm_1 \times rm_1$ matrix

$$\boldsymbol{F}_s = \boldsymbol{I} \otimes \boldsymbol{F}_1,$$

where $\boldsymbol{I}$ is the $r \times r$ identity matrix and $\boldsymbol{F}_1$ is the $m_1 \times m_1$ matrix of the form

$$\boldsymbol{F}_1 = \begin{bmatrix} 0 & 0 & \ldots & 0 & 1 \\ 1 & 0 & \ldots & 0 & 0 \\ 0 & 1 & \ldots & 0 & 0 \\ \vdots & \vdots & \ddots & \vdots & \vdots \\ 0 & 0 & \ldots & 1 & 0 \end{bmatrix}; \tag{14.17}$$

$\boldsymbol{g}_t$ is a vector of length $2 + rm_1$, the values of which are determined by $\boldsymbol{\Gamma}$, α, β and $\boldsymbol{x}_t$:

$$\boldsymbol{g}_t = \begin{bmatrix} \alpha \\ \beta \\ \sum_{i=1}^{r} \gamma_{1i} x_{it} \\ 0 \\ \vdots \\ \sum_{i=1}^{r} \gamma_{2i} x_{it} \\ 0 \\ \vdots \\ \sum_{i=1}^{r} \gamma_{ri} x_{it} \\ \vdots \\ 0 \end{bmatrix}.$$

The MS($r; m_1, m_2$) Model in Reduced Form

Consider the MS($r; m_1, m_2$) model in (14.9). We restrict our attention to the case where $r = m_2/m_1$. This does not entail any loss of generality, because all cases in which the number of distinct cycles of period m_1 is less than m_2/m_1 can be considered as special cases of the case $r = m_2/m_1$, with some equality constraints on the smoothing parameters and seed values.

The non-seasonal part of the series is a "local linear trend" process, which is an ARIMA(0,2,2); see Sect. 11.2. Hence, the important thing to establish is the reduced form of the seasonal component. Because $s_{i,t-m_1}$ appears in the y_t equation rather than $s_{i,t}$, we start by lagging equation (14.9d) m_1 periods. Then the following is true for $i = 1, \ldots, r$:

$$s_{i,t-m_1} = s_{i,t-2m_1} + \left(\sum_{j=1}^{r} \gamma_{ij} x_{j,t-m_1} \right) \varepsilon_{t-m_1}.$$

Repeated substitution r times leads to:

$$\begin{aligned} s_{i,t-m_1} = s_{i,t-m_2-m_1} &+ \left(\sum_{j=1}^{r} \gamma_{ij} x_{j,t-m_1} \right) \varepsilon_{t-m_1} + \left(\sum_{j=1}^{r} \gamma_{ij} x_{j,t-2m_1} \right) \varepsilon_{t-2m_1} + \cdots \\ &+ \left(\sum_{j=1}^{r} \gamma_{ij} x_{j,t-(r-1)m_1} \right) \varepsilon_{t-(r-1)m_1} + \left(\sum_{j=1}^{r} \gamma_{ij} x_{j,t} \right) \varepsilon_{t-m_2}. \end{aligned}$$

The last term has $x_{j,t}$ rather than $x_{j,t-m_2}$ because $x_{j,t} = x_{j,t-m_2}$. For each value of j, one and only one of the r indicator variables $x_{j,t}, x_{j,t-m_1}, \ldots, x_{j,t-(r-1)m_1}$ is equal to one and the rest are zero, and as j changes, a different one of these indicator variables switches to one. Hence the r terms $(\sum_{j=1}^{r} \gamma_{ij} x_{j,t-m_1})$, $(\sum_{j=1}^{r} \gamma_{ij} x_{j,t-2m_1})$, …, $(\sum_{j=1}^{r} \gamma_{ij} x_{j,t})$ are a circular backward rotation of

$\gamma_{i1}, \gamma_{i2}, \ldots, \gamma_{ir}$, the rth row of the matrix of smoothing parameters. An example of such a backward rotation would be $\gamma_{i2}, \gamma_{i1}, \gamma_{ir}, \gamma_{ir-1}, \ldots, \gamma_{i4}, \gamma_{i3}$. Depending on which sub-cycle t belongs to, the rotation starts from a different point. However, because $s_{i,t-m_1}$ is added to y_t only when $x_{i,t} = 1$, (i.e., when t belongs to sub-cycle i), the relevant rotation starts from $\gamma_{i,i-1}$ (or $\gamma_{i,r}$ if $i = 1$) and circles back and ends with $\gamma_{i,i}$.

Hence, we have

$$(1 - L^{m_2})s_{i,t-m_1} = \sum_{j=1}^{r} \gamma_{ij}^{c}\varepsilon_{t-jm_1} \quad \text{for } t \in \text{sub-cycle } i, \tag{14.18}$$

where $\gamma_{i1}^{c}, \ldots, \gamma_{ir}^{c}$ is the particular backward rotation of $\gamma_{i1}, \gamma_{i2}, \ldots, \gamma_{ir}$ described above. This shows that each of the r seasonal factors is a seasonal ARIMA $(0,1,0)_{m_2} \times (0,0,r-1)_{m_1}$. Using (14.18) and noting that $x_{i,t} = x_{i,t-m_2}$, the seasonal component of y_t can be written as:

$$\begin{aligned}
\sum_{i=1}^{r} x_{i,t}s_{i,t-m_1} &= \sum_{i=1}^{r} x_{i,t}s_{i,t-m_1-m_2} + \sum_{i=1}^{r} x_{i,t} \sum_{j=1}^{r} \gamma_{ij}^{c}\varepsilon_{t-jm_1} \\
&= \sum_{i=1}^{r} x_{i,t-m_2}s_{i,t-m_1-m_2} + \sum_{j=1}^{r} \left(\sum_{i=1}^{r} x_{i,t}\gamma_{ij}^{c} \right) \varepsilon_{t-jm_1} \\
&= \sum_{i=1}^{r} x_{i,t-m_2}s_{i,t-m_1-m_2} + \sum_{j=1}^{r} \theta_{j,t}\varepsilon_{t-jm_1},
\end{aligned}$$

where $\theta_{j,t} \equiv \sum_{i=1}^{r} x_{i,t}\gamma_{ij}^{c}$. This shows that the seasonal component in y_t is a seasonal ARIMA$(0,1,0)_{m_2} \times (0,0,r-1)_{m_1}$ with periodic moving average parameters. Hence y_t is the sum of an ARIMA(0,2,2) and a seasonal ARIMA $(0,1,0)_{m_2} \times (0,0,r-1)_{m_1}$ with moving average parameters that depend on which sub-cycle t belongs to.

To find the reduced form, we first subtract y_{t-m_2} from y_t:

$$y_t - y_{t-m_2} = \ell_{t-1} - \ell_{t-1-m_2} + b_{t-1} - b_{t-1-m_2} + \sum_{j=1}^{r} \theta_{j,t}\varepsilon_{t-jm_1} + \varepsilon_t - \varepsilon_{t-m_2}.$$

Repeated substitution into (14.9b) yields:

$$\ell_{t-1} - \ell_{t-1-m_2} + b_{t-1} - b_{t-1-m_2} = \sum_{j=1}^{m_2} b_{t-j} + \alpha \sum_{j=1}^{m_2} \varepsilon_{t-j},$$

which leads to:

$$\Delta_{m_2} y_t = \sum_{j=1}^{m_2} b_{t-j} + \alpha \sum_{j=1}^{m_2} \varepsilon_{t-j} + \sum_{j=1}^{r} \theta_{j,t}\varepsilon_{t-jm_1} + \Delta_{m_2}\varepsilon_t.$$

If there was no trend in the model, the reduced form would be the above equation without the first term on the right hand side. With the trend, because b_t is integrated, we still have to do an extra round of first differencing to achieve stationarity. Using the facts that $\Delta b_{t-j} = \beta\varepsilon_{t-j}$ and $\sum_{j=1}^{m_2} \varepsilon_{t-j} - \sum_{j=1}^{m_2} \varepsilon_{t-j-1} = \varepsilon_{t-1} - \varepsilon_{t-m_2-1}$, we get

$$\Delta\Delta_{m_2} y_t = \beta \sum_{j=1}^{m_2} \varepsilon_{t-j} + \alpha\Delta_{m_2}\varepsilon_{t-1} + \sum_{j=1}^{r}(\theta_{j,t}\varepsilon_{t-jm_1} - \theta_{j,t-1}\varepsilon_{t-jm_1-1}) + \Delta\Delta_{m_2}\varepsilon_t.$$

This shows that after first and m_2 differencing, y_t is a moving average of order $m_2 + 1$ with non-zero, but periodic, moving average parameters about seasonal lags corresponding to a sub-cycle of period m_1.

15

Nonlinear Models for Positive Data

Co-author:[1] **Muhammad Akram**[2]

In Chap. 4 we considered a class of nonlinear and heteroscedastic innovations state space models and developed their properties. At that time we noted that the Gaussian distribution was not always an appropriate distribution for the error process. Nevertheless, we claimed that the Gaussian likelihood would often provide a reasonable framework for parameter estimation. We also used the Gaussian distribution to construct prediction intervals. Our aim in this chapter is to examine the structure of these nonlinear exponential smoothing state space models in greater detail and to check the conditions under which the use of the Gaussian distribution provides an appropriate approximation.

Why should these issues concern us? First of all, we may note that most of the series that we encounter in business applications are strictly positive, such as sales, prices, etc. Of course, there are many exceptions, most notably the returns on an investment (although the underlying stock or bond price is strictly positive even here). Nevertheless, the linear models of Chap. 3 are widely used for such series, so why should we pay particular attention to the nonlinear models? The first reason is that, under the Gaussian assumption, the forecast variances may be undefined. Second, we find that there are some difficult specification problems associated with models strictly defined on the positive half line; we examine these questions in greater detail in Sect. 15.1. We then explore purely multiplicative models in Sect. 15.2 in order to identify possible solutions to these difficulties. Section 15.3 contains some distributional results for the ETS(M,N,N) model, where the innovations are from a lognormal or gamma distribution. In Sect. 15.4, we examine the extent to which the Gaussian distribution can serve as a reasonable approximation, notwithstanding the theoretical objections noted earlier. We need to consider

[1] This chapter is based, in part, on material presented in Hyndman and Akram (2006) and Akram et al. (2007).

[2] Dr. Muhammad Akram, Department of Econometrics and Business Statistics, Monash University, Australia.

parameter estimation, point forecasting, interval forecasting and simulation. We find that the Gaussian approximation typically works well for the first two issues, has a somewhat mixed record for interval estimation and may lead to problems in long series simulations.

15.1 Problems with the Gaussian Model

Positive time series are very common in business, industry, economics and other fields, and exponential smoothing methods are frequently used for forecasting such series. From a practical viewpoint, this approach often appears to be satisfactory because the process is bounded well away from the origin. However, cases may arise where the prediction intervals developed (following the procedures in Chap. 6) include a sub-interval of negative values. Indeed, as the forecasting horizon is extended, even the point forecasts may become negative.

Because the Gaussian distribution extends over the whole real line, it cannot provide an exact specification for the error process when the series is constrained to be non-negative. When the model is purely multiplicative, a logarithmic transformation seems a reasonable option, and we explore it in Sect. 15.2.2. However, when the model has some additive components, this option is not available. Some authors (e.g., Hyndman et al. 2002) have suggested using a truncated Gaussian distribution for the errors so that the sample space is constrained to take only positive values. Other options include the use of distributions such as the gamma or lognormal which are defined on the positive half-line. In this section we explore the theoretical limitations of the Gaussian model, before exploring other options later in the chapter.

We begin the discussion by considering the various models outlined in Tables 2.2 and 2.3. Fifteen of these 30 ETS models contain additive errors, and the others involve multiplicative errors. It is convenient to divide the 30 models into four classes, although we divide them differently from the classes given in Chap. 6:

Class M: Purely multiplicative models: (M,N,N), (M,N,M), (M,M,N), (M,M,M), (M,M_d,N) and (M,M_d,M)

Class A: Purely additive models: (A,N,N), (A,N,A), (A,A,N), (A,A,A), (A,A_d,N) and (A,A_d,A)

Class X: Models with additive errors and at least one multiplicative component, and models with multiplicative errors and multiplicative trend but additive seasonality: (A,M,$*$), (A,M_d,$*$), (A,$*$,M), (M,M,A), (M,M_d,A), where $*$ denotes any admissible component (11 models)

Class Y: Models with multiplicative errors and additive trend, and the model with multiplicative errors and additive seasonality but no trend: (M,A,$*$), (M,A_d,$*$) or (M,N,A), where $*$ denotes any admissible component (seven models)

Class A corresponds to Class 1 in Chap. 6 and Class X corresponds to Class 5 in Chap. 6. The models in Classes M and Y are divided differently into Classes 2–4 in Chap. 6.

It is evident that only the purely multiplicative models of Class M can guarantee a sample space that is restricted to the positive half-line. Class A contains the purely additive models, widely used in practice for short-term forecasting, but they clearly do not conform to the requirements of non-negative processes unless additional conditions are imposed. The remaining models in Classes X and Y all possess both multiplicative and additive components. If the observational sample space is not restricted to be strictly positive, the Class X models can have an infinite forecast variance for three or more steps ahead (i.e., $h \geq 3$); for some seasonal models the problem occurs for $h \geq m+2$. This problem does not arise, however, for the Class Y models. The following discussion serves to illustrate the general concern.

15.1.1 The Infinite Variance Problem

Consider the ETS(A,M,N) model:

$$\begin{aligned} y_t &= \ell_{t-1}b_{t-1} + \varepsilon_t, \\ \ell_t &= \ell_{t-1}b_{t-1} + \alpha\varepsilon_t, \\ b_t &= b_{t-1} + \beta\varepsilon_t/\ell_{t-1}. \end{aligned}$$

Simulated values from the model ETS(A,M,N) are plotted in the top panel of Fig. 15.1. The Gaussian distribution is used to generate the errors. From this figure, the implications of an infinite forecast variance can be seen quite clearly. As soon as the value of ℓ_{t-1} gets close to zero, the sample path becomes very unstable.

To observe how the behavior of the series changes with the change in the value of ℓ_t (particularly when ℓ_t is close to zero), the first few values of the states have also been plotted in Fig. 15.1. The middle panel shows the level component of the series and the bottom panel shows the slope component of the series. From this figure, it can be seen that the fifth value of the level component is very close to zero. This leads to a rapid decrease in the trend component in the following period. In the next time period the level increases sharply, and it oscillates between successively larger positive and negative values thereafter. As a consequence of these changes in the level and slope components, the value of the sample path becomes unstable from this point onward.

To see that this problem is general in nature, consider the trend equation at time $t = 2$:

$$b_2 = b_1 + \beta\varepsilon_2/\ell_1 = b_0 + \beta\left(\frac{\varepsilon_2}{\ell_1} + \frac{\varepsilon_1}{\ell_0}\right) = b_0 + \beta\left(\frac{\varepsilon_2}{\ell_0 b_0 + \alpha\varepsilon_1} + \frac{\varepsilon_1}{\ell_0}\right).$$

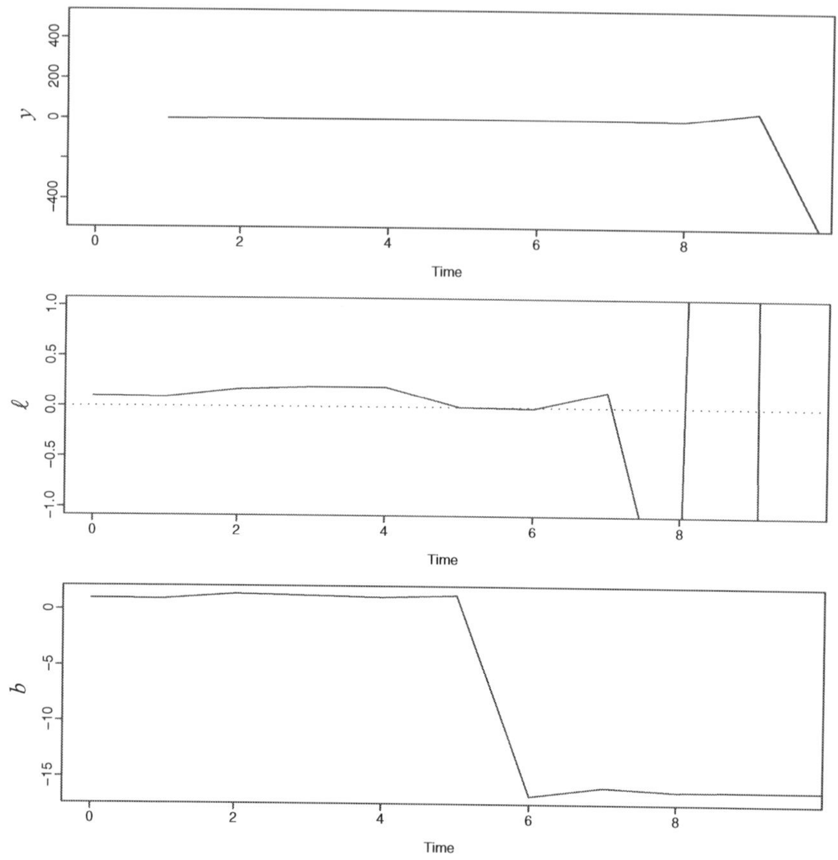

Fig. 15.1. ETS(A,M,N) simulation: $\ell_0 = 0.1, b_0 = 1, \alpha = 0.1, \beta = 0.05$, and $\sigma = 1$.

If ε_t has a Gaussian distribution, the first term in the brackets is a ratio of two Gaussian variables. When $\ell_0 b_0 = 0$ this term has a Cauchy distribution. In general, for all other values of $\ell_0 b_0$, the distribution is not Cauchy but it still has an infinite variance and undefined expectation (see Stuart and Ord 1994, pp. 400, 421). Indeed, these problems arise whenever the level of the series has positive density over an open interval that includes zero. These problems with the trend equation will propagate into the observation equation at time $t = 3$. Similar problems arise with other distributions in Class X; see Hyndman and Akram (2006) for details.

We may try to restrict the range of the errors in such a way as to eliminate the possibility of negative values. A model with multiplicative components (whether E, T or S) is only sensible if observations are defined on the positive half-line, but the Gaussian error structure clearly leads to a non-vanishing probability of a violation sooner or later. Because these models require non-negative observations and state variables, model failure is inevitable in the long-run when the underlying distribution is taken over the whole real line. Essentially, for any model with a Gaussian error process, the first passage time properties will eventually lead to negative values for the series unless there is a strong upward trend.

In order to maintain the strictly positive nature of the model, the error process cannot be specified as Gaussian. A Gaussian approximation may work as the basis for computing point forecasts and short-term prediction intervals, and this method has been widely used over the years. However, such choices cannot lead to exact distributional results.

To find a possible solution, consider the same simple model ETS(A,M,N). In order for the process to remain strictly positive, we require:

$$\ell_{t-1}b_{t-1} + \varepsilon_t > 0.$$

This condition requires the distribution of

$$\varepsilon_t^* = 1 + \frac{\varepsilon_t}{\ell_{t-1}b_{t-1}}$$

to be defined on the positive line; that is, $\varepsilon_t^* \in (0, \infty)$. From a practical perspective, a long series may be needed before the positivity condition is violated; the first passage time depends strongly on the parameters.

15.1.2 The Convergence to Zero Problem

Models with only multiplicative components may appear to be the natural choice for positive data. However, Fig. 15.2 shows three realizations of the ETS(M,N,N) model using the Gaussian distribution, all of which show a tendency to decay towards zero. The reason for this behavior is discussed in Sect. 15.2.1. Again, it is a relatively long-run behavior, and so does not have an immediate impact on short-term forecasting. However, for simulations and long-term forecasting, this behavior needs to be understood.

15.1.3 Non-Constant Innovations Variance

If the error ε_t is to have mean zero and the sample space is to be restricted to the positive real line, then the variance cannot be constant. This is easily seen for the ETS(M,N,N) model by considering the possible values of ε_t when ℓ_t is close to zero. Further, if the process approaches zero, the mean of a truncated distribution becomes more strongly positive, which may cause an uptick in the series; see Exercise 15.1.

Based upon these findings, it would appear that we should consider models with non-negative error structures; we proceed to examine such models in the next section.

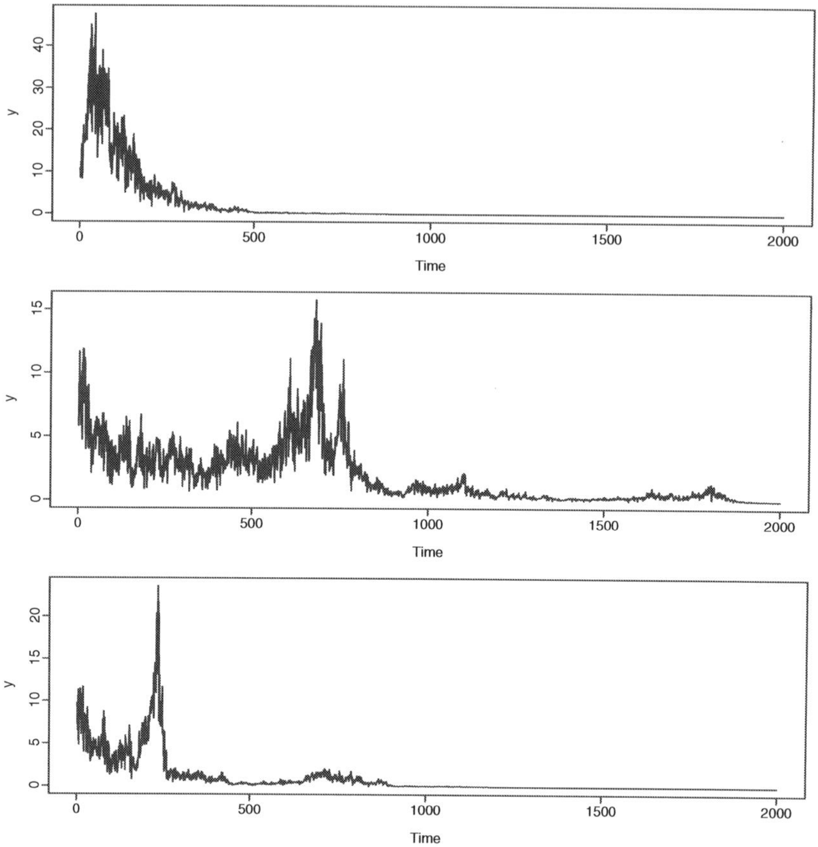

Fig. 15.2. Simulated data from model ETS(M,N,N) with Gaussian errors. The parameter values are $\ell_0 = 10$, $\alpha = 0.3$ and $\sigma = 0.3$.

15.2 Multiplicative Error Models

In the previous section, we concluded that only models with a multiplicative error structure should be considered for strictly positive data. In this section we show that even in these circumstances, the models may fail to perform satisfactorily.

Example 15.1: ETS(M,N,N) model

By way of illustration, we consider the multiplicative simple exponential smoothing model, or ETS(M,N,N), as given below:

$$y_t = \ell_{t-1}(1 + \varepsilon_t), \tag{15.1a}$$
$$\ell_t = \ell_{t-1}(1 + \alpha\varepsilon_t), \tag{15.1b}$$

where ε_t denotes a white noise series with variance σ^2, such that $\varepsilon_t \geq -1$ and $0 < \alpha < 1$ (to ensure the data remain positive). Hyndman et al. (2002) consider the model with $\varepsilon_t \sim \mathrm{N}(0, \sigma^2)$. A truncated Gaussian distribution could be used with positive data to ensure $\varepsilon_t \geq -1$. When σ^2 is very small, the truncation is almost never needed. This assumption is not unreasonable in many applications in business and economics, but we shall not so restrict the discussion here. Other specifications of the error distribution that we consider in this chapter are the lognormal and gamma distributions. We ran simulations on these three alternatives, maintaining the same mean and variance for the errors, using a common random number stream, and selecting plausible sets of parameter values. Somewhat surprisingly, we found that the three specifications produced very similar results, except near zero. The reader is encouraged to explore the behavior of these models under different initial conditions; see Exercise 15.2.

Some models have properties akin to branching processes, in that different realizations may either explode or fade to zero. Even though long series may be necessary for such asymptotic behavior to become manifest, this property is potentially troubling for long forecast horizons or in simulation studies. We will now explore these empirical findings from a theoretical perspective.

15.2.1 Kakutani's Theorem

We are interested in situations where observations are on the positive half-line, but the other features of the innovations model remain unchanged. Therefore, we now assume that the distribution of $\delta_t = 1 + \varepsilon_t$ has mean 1 and variance σ^2, such that the δ_t are defined on the positive half-line and are independent and identically distributed. We continue with the simple case ETS(M,N,N), because the results can be extended directly to deal with more complex models. In this discussion, as a matter of convenience, we assume that $0 < \alpha \leq 1$. (If $1 < \alpha < 2$ we would need to consider $\delta_t = 1 + \alpha\varepsilon_t$ to ensure that the process remains positive. This change would mean that $\delta_t \geq (\alpha - 1)/\alpha$ to guarantee non-negativity, but otherwise the basic argument is unchanged.)

We can write the local level state equation of model (15.1) as

$$\begin{aligned} \ell_t &= \ell_0(1 + \alpha\varepsilon_1)(1 + \alpha\varepsilon_2) \cdots (1 + \alpha\varepsilon_t) \\ &= \ell_0 \prod_{j=1}^{t} (1 + \alpha\varepsilon_j) \\ &= \ell_0 U_t, \end{aligned} \tag{15.2}$$

where $U_t = U_{t-1}(1 + \alpha\varepsilon_t)$ and $U_0 = 1$. Therefore U_t is a non-negative product martingale, because $\mathrm{E}(U_{t+1}|U_t) = U_t$.

Kakutani's theorem for product martingales (see Williams 1991, p. 144) may be stated as follows.

Theorem 15.1. *Let $X_1, X_2, \cdots, X_n$ be positive independent random variables, each with mean 1, and let $a_i = \mathrm{E}\sqrt{X_i}$. Then for $U_n = \prod_{j=1}^{n} X_j$,*

$$U_\infty > 0 \text{ almost surely if } \lim_{n\to\infty} \prod_{i=1}^{n} a_i > 0,$$
$$U_\infty = 0 \text{ almost surely if } \lim_{n\to\infty} \prod_{i=1}^{n} a_i = 0.$$

Note that $a_i \geq 0$ and Jensen's inequality (see Shiryaev 1984, p. 192) gives $a_i \leq 1$. Further, provided the distributions of the X_i are not degenerate, $a_i < 1$. Thus, we may apply Kakutani's theorem to (15.2), and we see that the results in Fig. 15.2 are consistent with the theoretical result. That is, sample paths for ETS(M,N,N) models with the stated properties tend to converge stochastically to zero. This is true regardless of the distribution of $1 + \alpha\varepsilon_t$, provided it has mean one and is non-degenerate.

The results extend to other multiplicative error models under similar conditions; Exercises 15.3 and 15.4 examine the ETS(M,N,M) and ETS(M,$\mathrm{M_d}$,N) models, originally discussed in Hyndman and Akram (2006).

15.2.2 An Alternative Approach

Our results so far indicate that the Gaussian assumption is at best an approximation and that the use of non-Gaussian distributions alone does not resolve the problem when we consider long-term forecasting. Thus, in order to make progress, we must be willing to relax one or more of the underlying assumptions that were made earlier. The result provided by Kakutani's Theorem provides the essential insight. If we are to overcome the tendency to converge to zero, we must allow $\mathrm{E}\sqrt{X_i}$ to take on values equal to or greater than one.

For example, consider a modified ETS(M,N,N) model, which we write as METS(M,N,N;LN) to indicate both the modified form and the dependence on the lognormal distribution:

$$y_t = \ell_{t-1}(1 + \varepsilon_t), \tag{15.3a}$$
$$\ell_t = \ell_{t-1}(1 + \varepsilon_t)^{\alpha}, \tag{15.3b}$$

where $\delta_t = 1 + \varepsilon_t$ is a positive random variable. This form of multiplicative model is chosen primarily for its convenience as it enables us to obtain exact sampling results when we assume that δ_t follows a lognormal distribution. This model also ensures a positive-valued process for all $0 < \alpha < 2$. The model may or may not be an improvement over existing choices, a question we explore in Sect. 15.5.3, but its qualitative behavior is similar and it is more easily explored analytically.

Using a log-transformation, (15.3) can be written as

$$y_t^* = \ell_{t-1}^* + \delta_t^*, \tag{15.4a}$$
$$\ell_t^* = \ell_{t-1}^* + \alpha\delta_t^*, \tag{15.4b}$$

where $y_t^* = \log(y_t)$, $\ell_t^* = \log(\ell_t)$ and $\delta_t^* = \log(\delta_t)$. Thus the log-transformed model in (15.4) is identical to the simple exponential smoothing model ETS(A,N,N).

15.3 Distributional Results

We now proceed to develop some distributional results for each of the models (15.1) and (15.3). If we denote the mean and variance of $\delta_t = 1 + \varepsilon_t$ by M and V respectively, and $\mathrm{E}(\delta_t^k) = M_k$, then the means and variances of the h-step-ahead prediction distributions may be written as:

Model (15.1)

$$\mathrm{E}(y_{n+h|n}) = E_{1A} = \ell_n M(1-\alpha+\alpha M)^{h-1}, \tag{15.5a}$$
$$\mathrm{E}(y_{n+h|n}^2) = E_{2A} = \ell_n^2(M^2+V)\left[(1-\alpha+\alpha M)^2+\alpha^2 V\right]^{h-1}, \tag{15.5b}$$
$$\mathrm{V}(y_{n+h|n}) = E_{2A} - E_{1A}^2. \tag{15.5c}$$

Model (15.3)

$$\mathrm{E}(y_{n+h|n}) = E_{1M} = \ell_n M M_\alpha^{h-1}, \tag{15.6a}$$
$$\mathrm{E}(y_{n+h|n}^2) = E_{2M} = \ell_n^2(M^2+V)M_{2\alpha}^{h-1}, \tag{15.6b}$$
$$\mathrm{V}(y_{n+h|n}) = E_{2M} - E_{1M}^2. \tag{15.6c}$$

Two particular choices of distribution are the lognormal and the gamma and we now consider each in turn.

15.3.1 The Lognormal Distribution

If δ_t^* in (15.4) is Gaussian with mean μ and variance ω, or $\delta_t^* \sim \mathrm{N}(\mu,\omega)$, we may denote the lognormal assumption by $\delta_t \sim \mathrm{logN}(\mu,\omega)$. Standard results for the lognormal distribution (see Stuart and Ord 1994, pp. 241–243) yield:

$$\mathrm{E}(\delta_t^k) = \exp(k\mu + k^2\omega/2), \quad \text{for any } k, \tag{15.7a}$$
$$\mathrm{E}(\delta_t) = \exp(\mu+\omega/2) = E_1, \tag{15.7b}$$
$$\mathrm{V}(\delta_t) = E_1^2[\exp(\omega)-1], \tag{15.7c}$$
$$\text{and} \quad \mathrm{E}(\delta_t^{\alpha/2}) = \exp(\alpha\mu/2 + \alpha^2\omega/8). \tag{15.7d}$$

Table 15.1. Long-term behavior of the prediction distribution for the METS (M,N,N;LN) model, with $0 < \alpha < 1$.

Range	$E(\delta_t^{\alpha})$	$E(\delta_t^{\alpha/2})$	$E(y_h)$	$V(y_h)$
$\mu + \alpha\omega < 0$	<1	<1	Decreasing	Decreasing
$\mu + \alpha\omega = 0$	<1	<1	Decreasing	Finite
$-\alpha\omega < \mu < -\alpha\omega/2$	<1	<1	Decreasing	Increasing
$\mu + \alpha\omega/2 = 0$	=1	<1	Finite	Increasing
$-\alpha\omega/2 < \mu < -\alpha\omega/4$	>1	<1	Increasing	Increasing
$\mu + \alpha\omega/4 = 0$	>1	=1	Increasing	Increasing
$\mu + \alpha\omega/4 > 0$	>1	>1	Increasing	Increasing

The entry "Finite" means that the term approaches a finite limit.

From (15.7d) we can see that the expectation will exceed 1 provided $\mu + \alpha\omega/2 > 0$.

If we now consider forecasting h periods ahead, we may set the forecast origin to $t = 0$ without loss of generality to simplify the notation. Then the prediction distribution for $y_h = \ell_0 z_h$ in model (15.4) is lognormal with $z_h \sim \text{logN}(\mu_h, \omega_h)$, where

$$\begin{aligned} \mu_h &= \mu(1 + (h-1)\alpha), \\ \omega_h &= \omega(1 + (h-1)\alpha^2), \\ E(y_h) &= \ell_0 \exp[\mu_h + \omega_h/2] = E_h, \\ \text{and} \qquad V(y_h) &= E_h^2[\exp(\omega_h) - 1]. \end{aligned}$$

The distributional result is exact, so that we can explore the behavior of the prediction distribution for long lead-times with the help of Kakutani's Theorem. The possible outcomes for different values of the parameters are summarized in Table 15.1. The prediction distributions become increasingly skewed as h increases; when $E(\delta_t^{\alpha/2}) < 1$ and $E(\delta_t^{\alpha}) \leq 1$, $\Pr(y_h > 0) \downarrow 0$.

Individual runs for some parameter combinations are shown in Fig. 15.3. In accordance with Table 15.1, we observe the drift towards zero when $E(\delta_t^{\alpha/2}) < 1$ and $E(\delta_t^{\alpha}) \leq 1$. The reverse is true when $\mu > 0$. Further, the plots show that when the parameter values are close to the boundary conditions, we may need a long series in order to observe the limiting properties. However, we should recall from Fig. 15.2 and the related discussion that different sample realizations may vary considerably.

The sampling distribution for model (15.1) is not exact, but may be approximated by a lognormal distribution with mean and variance given by (15.5) using the expectations given in (15.7).

15.3.2 The Gamma Distribution

Similar results are observed if we use the gamma distribution in place of the lognormal, although exact distributional results are not available. We denote

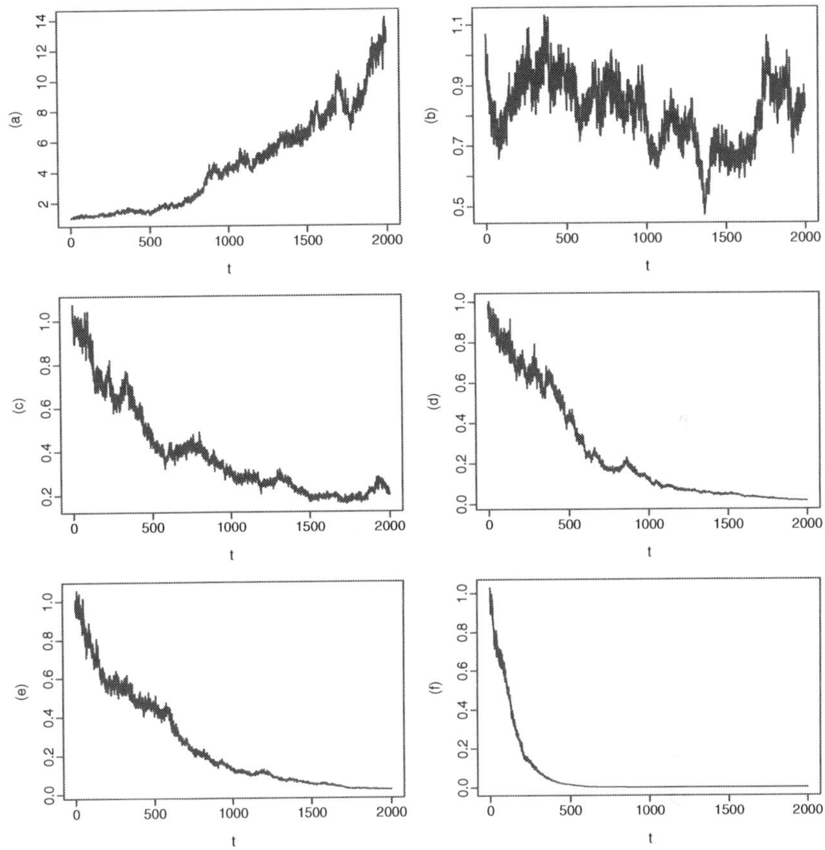

Fig. 15.3. Simulated data from the model METS(M,N,N;LN) with lognormal errors $\delta_t \sim \text{logN}(\mu, \omega)$: **(a)** $\mu = \alpha\omega/4$; **(b)** $\mu = 0$; **(c)** $\mu = -\alpha\omega/4$; **(d)** $\mu = -3\alpha\omega/8$; **(e)** $\mu = -\alpha\omega/2$; and **(f)** $\mu = -3\alpha\omega/4$; where $\ell_0 = 1$, $\omega^{0.5} = \sigma = 0.05$ and $\alpha = 0.3$.

the distribution by $\delta_t \sim \Gamma(r, \lambda)$, where r and λ are the shape and location parameters. The moments of the gamma distribution are (see Stuart and Ord 1994, pp. 81–82):

$$\begin{aligned} \text{E}(\delta_t) &= r\lambda = \mu, \\ \text{E}(\delta_t^k) &= \frac{\lambda^k \Gamma(r+k)}{\Gamma(r)}, \\ \text{and} \qquad \text{V}(\delta_t) &= r\lambda^2 = \mu^2/r. \end{aligned}$$

For large r, the use of Stirling's approximation yields $\text{E}(\delta_t^{0.5}) = \mu^{0.5} \exp(-r/8)$. Thus, for large values of r, μ needs to exceed 1 by only a small amount to avoid convergence to zero.

Table 15.2. Long-term behavior of the prediction distribution for the METS(M,N,N;G) model, where G denotes the gamma distribution with $r = 20$, $\alpha = 0.5$ and $r_0 = [\Gamma(r+\alpha)/\Gamma(r)]^2 = 19.75$.

	$\mathrm{E}(\delta_t^{\alpha})$	$\mathrm{E}(\delta_t^{\alpha/2})$	$\mathrm{E}(y_h)$	$\mathrm{V}(y_h)$
$\lambda < 1/r$	<1	<1	Decreasing	Decreasing
$\lambda = 1/r$	=1	<1	Decreasing	Decreasing
$1/r < \lambda < 1/r_0$	>1	<1	Decreasing	Increasing
$\lambda = 1/r_0$	>1	=1	Decreasing	Increasing
$1/r_0 < \lambda \leq 1/19.66$	>1	>1	Decreasing	Increasing
$\lambda > 1/19.66$	>1	>1	Increasing	Increasing

For comparative purposes, we consider model (15.3); the mean and variance for model (15.1) follow directly from (15.5). The distribution of y_h has mean and variance:

$$\mathrm{E}(y_h) = \ell_0 r \lambda^{(h-1)\alpha+1} \left[\frac{\Gamma(r+\alpha)}{\Gamma(r)}\right]^{h-1},$$

$$\mathrm{V}(y_h) = \ell_0^2 \lambda^{2\alpha(h-1)+2} \left[r(r+1)\left(\frac{\Gamma(r+2\alpha)}{\Gamma(r)}\right)^{h-1} - r^2 \left(\frac{\Gamma(r+\alpha)}{\Gamma(r)}\right)^{2h-2}\right],$$

where $\Gamma(r)$ denotes the gamma function. For large r, $\mathrm{E}(y_h) = \ell_0 \mu^{1+(h-1)\alpha}$ and the variance is approximately

$$\mathrm{V}(y_h) = r^{-1} \ell_0^2 \mu^{2\alpha(h-1)+2}(1 + (h-1)\alpha^2).$$

The last term in parentheses in the approximate variance is precisely that for the local level model given in Table 6.3.

The distribution of y_h can be approximated by the gamma distribution with the same first and second moments. The behavior for the gamma distribution is qualitatively similar to that for the lognormal as summarized in Table 15.2. The results are tabulated only for the case $r = 20$, which is smaller than the values that would typically arise in applications; however, these numbers serve to illustrate the general patterns. Again, long series are often needed before the asymptotic behavior shows in the plots. We note that the ratio r_0/r tends to 1 as $r \to \infty$, so that the region of the parameter space with decreasing mean and increasing variance is vanishingly small. As in the lognormal case, the prediction distributions become increasingly skewed as h increases, and when $\mathrm{E}(\delta_t^{\alpha/2}) < 1$ and $\mathrm{E}(\delta_t^{\alpha}) \leq 1$, $\Pr(y_h > 0) \downarrow 0$.

15.4 Implications for Statistical Inference

We now consider the implications of these results for inference. There are three elements to consider: parameter estimation based upon the likelihood function, prediction distributions for a small to moderate number of steps ahead, and the simulation of (potentially) long series.

15.4.1 The Approximate Likelihood

Once the error distribution is specified, we may examine the form of the distribution to see how close the approximation is to the true version. It is well known that the lognormal density function approaches that of the Gaussian distribution as $\omega \to 0$; see Stuart and Ord (1994, p. 242) for a graphical representation of this limiting relationship. However, our question is somewhat different in that we are concerned with differences in the maximum likelihood estimates, not the density functions. In order to examine this question, we may compare the estimates obtained by:

(a) Applying the Gaussian ML estimators to lognormal data
(b) Evaluating the (correct) estimates using the lognormal likelihood function and then transforming to the mean and variance of the original error process

In analytical terms, it is straightforward to show that the two approaches produce similar results as $\omega \to 0$; the question is: how good is the first form as an approximation to the second? The value of the lognormal parameter μ does not affect the relative bias or variability of the approximate estimates, so we may focus exclusively upon the effect that the value of $\sigma = \omega^{0.5}$ has upon the approximation. We carried out a small simulation study using $M = 100$ replicates for samples of size $n = 25$ with σ set equal to 0.05, 0.10 and 0.20. Values greater than 0.20 are most unlikely in practice in the present context. The results are summarized in the following table, which examines the ratios of the two estimates for each of the mean and standard deviation of the error. The average bias is measured in percentage terms; the bias of the mean of the error is negligible (less than 0.1% in all cases) and so is omitted from the table. The standard deviations of the percentage biases were also computed across the 100 replicates. Again, those for the mean are very small (less than 0.01%) and are omitted. The figures for the variance of the error are reported in the table and it can be seen that they are of a reasonable magnitude, even for $\sigma = 0.2$. The variances of the estimates themselves are almost equal, indicating that the loss in efficiency is very slight in this region of the parameter space.

σ	0.05	0.10	0.20
Percent bias in variance	0.05	0.32	1.54
SD of percent bias in variance	1.98	3.95	7.96

Clearly, much more extensive simulation studies could be run, but the benefits would be marginal. We can be reasonably confident that when the errors follow the lognormal (or gamma) distribution, the Gaussian likelihood function is a reasonable approximation for the region of the parameter space involved. In turn, because the one-step-ahead error distributions are close to the Gaussian form, the approximate one-step-ahead prediction distributions will also be reasonably close to the underlying forms in most cases.

15.4.2 Prediction Distributions and Simulations

We now consider the lognormal model given in (15.4) and examine the prediction distribution. It follows from (15.7) that the h-step-ahead prediction distribution is also lognormal, of the form

$$\text{logN}\left(\log(\ell_0) + \mu[(h-1)\alpha + 1]\,,\ \omega[(h-1)\alpha^2 + 1]\right).$$

As h increases, the divergence between the Gaussian and lognormal models becomes more and more pronounced as the prediction distribution becomes more skewed. Similar results apply for the gamma distribution, but *exact* analytical results are not available. In Table 15.3 we present numerical results for both distributions for typical values of σ and α. Again, we have focussed upon the modified METS(M,N,N;$*$) scheme, but qualitatively similar results will apply more broadly.

The results in the table indicate that there is very little difference between the lognormal and gamma models. We use the standard measures of skewness γ_1 and kurtosis γ_2 based upon the third and fourth moments; $\gamma_1 = \gamma_2 = 0$ for a Gaussian distribution. As expected, the distributions become more skewed and heavy-tailed as the forecasting horizon increases and/or the value of α increases.

For purely multiplicative (Class M) models with lognormal errors, the analytical expressions for point forecasts and prediction intervals for model ETS(A,$*$,$*$) may be used for the log-transformed ETS(M,$*$,$*$) model. That

Table 15.3. Standardized skewness and kurtosis coefficients for predictive distributions for the METS(M,N,N) model with lognormal and gamma errors.

Lognormal	h	$\alpha = 0.5$		$\alpha = 0.8$	
		γ_1	γ_2	γ_1	γ_2
$\sigma = 0.05$	1	0.15	0.04	0.15	0.04
	5	0.21	0.08	0.28	0.14
	10	0.27	0.13	0.39	0.28
$\sigma = 0.10$	1	0.30	0.16	0.30	0.16
	5	0.43	0.33	0.58	0.60
	10	0.55	0.55	0.81	1.19
Gamma	h	$\alpha = 0.5$		$\alpha = 0.8$	
		γ_1	γ_2	γ_1	γ_2
$\sigma = 0.05$	1	0.10	0.02	0.10	0.02
	5	0.20	0.26	0.27	0.23
	10	0.27	0.29	0.39	0.33
$\sigma = 0.10$	1	0.20	0.06	0.20	0.06
	5	0.40	0.49	0.55	0.64
	10	0.54	0.71	0.79	1.21

is, if the forecast on the log scale is F and the lower and upper limits of the prediction interval are L and U, then the forecast on the original scale is e^F with the prediction interval (e^L, e^U). Otherwise, for Class M models, the best approach is to use simulations based upon a careful specification of the underlying distribution, following the methods outlined in Chap. 6.

In order to apply the analytical approach, we must be sure that the underlying model will produce strictly positive values in any realization of the series. The following example illustrates how we may check whether this requirement is met.

Example 15.2: ETS(M,M,M) model

The model equations for the ETS(M,M,M) model are:

$$\begin{aligned}
y_t &= \ell_{t-1} b_{t-1} s_{t-m}(1+\varepsilon_t), \\
\ell_t &= \ell_{t-1} b_{t-1}(1+\alpha\varepsilon_t), \\
b_t &= b_{t-1}(1+\beta\varepsilon_t), \\
s_t &= s_{t-m}(1+\gamma\varepsilon_t).
\end{aligned}$$

For convenience, we will assume that $t = km$ to avoid the notational complexities of partial seasonal cycles. Then repeated substitutions result in the reduced form (taking $t \bmod m = p$):

$$y_t = \ell_0 b_0^t s_{-m+p}(1+\varepsilon_t) \prod_{j=1}^{t-1} \left[(1+\alpha\varepsilon_j)(1+\beta\varepsilon_j)^{t-j}\right] \prod_{i=1}^{k-1}(1+\gamma\varepsilon_i).$$

Inspection of the reduced form shows that the process will remain strictly positive provided all the starting values for the state variables are positive and $\varepsilon_t > \max(-1, -1/\alpha, -1/\beta, -1/\gamma)$ for all t. The most natural way to ensure that this condition is satisfied is to require that $\max(\alpha, \beta, \gamma) < 1$ and that $\varepsilon_t > -1$. Similar conditions apply for the ETS(M,M$_\text{d}$,M) model.

In general, when the model is in Class M, conditions such as those given in Example 15.2 will suffice to maintain a positive path for the process. However, when at least one component is additive (as for the Class A models), an unrestricted sample path may eventually hit negative values. When the series has an overall upward trend, the risk is greatly reduced, but cannot be eliminated as a theoretical possibility.

Because the nonlinear models are applied to series that are non-negative, models with an additive component cannot be formally correct. Nevertheless, they have proved extremely useful, and the implementation problems are minor when considering parameter estimation or predictive statements for relatively short horizons. We only run into difficulties for long horizons or when we are simulating a long series. We may avoid problems either by dropping any realization that becomes negative, or by using the modified series $y_t^* = \max(\Delta, y_t)$ for some small $\Delta > 0$. Neither solution is perfect, and should only be applied in circumstances where violations are infrequent. If negative values occur frequently, this is a sign that the proposed model is inappropriate for the specified set of parameters and starting values.

15.5 Empirical Comparisons

We will now illustrate some of the points discussed in this chapter by examining an annual time series of the number of new freight cars shipped in the USA over the period 1947–1993.[3] The data are plotted in Fig. 15.4. A visual inspection of the series suggests a changing local level, and the AIC comparison of different local models suggests that the ETS(M,N,N) model is the best choice.

15.5.1 Point Forecasts and Estimation

We now compare the performances of the Gaussian-based (M,N,N) and (A,N,N) models to those of the lognormal and gamma-based (M,N,N) models, using fitting samples of 28, 34 and 40 observations and a (non-overlapping) hold-out sample of the next six observations in each case. The

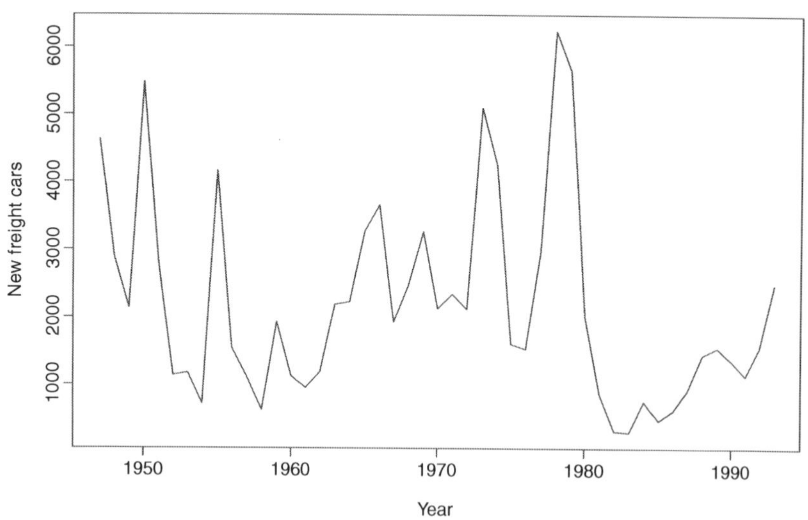

Fig. 15.4. US freight car shipments, 1947–1993.

[3] This series is available as Number N0193 in the M3 Competition data.

Table 15.4. Summary statistics for the US freight cars series.

	(A,N,N)	(M,N,N)	L1	L2	G1	G2
$n = 28$						
α	0.32	0.01	0.43	0.40	0.00	0.29
MAE	1,953	1,668	2,034	2,015	1,810	1,926
MAPE	74	59	79	72	71	72
mean			0.975	1.165	0.615	0.998
$n = 34$						
α	0.21	0.00	0.38	0.29	0.00	0.25
MAE	1,779	2,899	1,271	868	3,211	1,645
MAPE	401	632	286	195	698	371
mean			0.959	1.178	0.578	0.985
$n = 40$						
α	0.42	0.22	1.01	0.73	0.00	0.48
MAE	329	243	294	331	2,266	331
MAPE	24	19	23	25	180	25
mean			1.205	1.202	0.606	1.051

L1 = lognormal model (15.1); L2 = lognormal model (15.3); G1 = gamma model (15.1), r estimated by ML; G2 = gamma model (15.1), $r = 50$.

models were fitted using conditional maximum likelihood, except that we also considered a gamma model with a pre-set value for the index r. There were two reasons for this additional option: first, the likelihood function for the gamma was not well-behaved numerically, suggesting potential problems in more general applications. Second, the results for the fitted gamma model were poor, whereas using a large preset value for r gave better results, while preserving the positive structure of the series.

The results are given in Table 15.4 and show the Mean Absolute Error (MAE) for the one-step-ahead errors for the hold-out sample in each case. In the analysis we report $r = 50$ for the second gamma model, but the results varied only marginally for other "large" values.

Only very limited conclusions may be drawn from a single example, but a few points are worth noting. First, the gamma model with the maximum likelihood estimate of r produces erratic results. Likelihood estimation produces small values for r and consequently gives means that are much less than 1, implying rapid convergence towards zero. Hence, this model is excluded from further discussion. Second, the means of the remaining gamma and lognormal models hover around 1, reflecting the uncertainty about whether or not the series is declining; otherwise the one-step-ahead performance of these three models appears to be similar. However, for longer horizons, the different values of the means imply quite different trajectories. All three models differ somewhat from the ETS(M,N,N) model, but show some similarities with the (A,N,N) results.

Table 15.5. Prediction intervals based upon the gamma and lognormal distributions using models (15.1) and (15.3) with $\ell_0 = 100$ and $V(\delta) = 0.1$.

Distribution h:	Means			Lower PI			Upper PI		
	1	5	10	1	5	10	1	5	10
$\alpha = 0.3$									
Lognormal (15.1)	100	100	100	54	48	42	186	208	236
Lognormal (15.3)	100	96.1	91.4	52	44	37	175	182	189
Gamma (15.1)	100	100	100	48	41	33	171	186	203
Gamma (15.3)	100	95.9	90.9	48	39	31	171	178	186
ETS(A,N,N)	100	100	100	38	28	17	162	172	183
ETS(M,N,N)	100	100	100	38	27	14	162	172	186
$\alpha = 0.8$									
Lognormal (15.1)	100	100	100	54	29	15	186	351	657
Lognormal (15.3)	100	97.0	93.4	52	26	14	175	256	326
Gamma (15.1)	100	100	100	48	16	03	171	259	356
Gamma (15.3)	100	96.9	93.1	48	16	03	171	251	332
ETS(A,N,N)	100	100	100	38	−17	−61	162	217	261
ETS(M,N,N)	100	100	100	38	−25	−88	162	225	288

These results raise more questions than they resolve, but support the general contention that estimation properties and short-term point forecasts are not seriously affected by the long-run behavior discussed earlier in the chapter.

15.5.2 Prediction Intervals

One of the principal reasons for the introduction of the gamma and lognormal models is the concern about prediction intervals. To illustrate how the positivity constraint affects these intervals, we provide some numerical examples in Table 15.5. As expected, the prediction intervals based upon the Gaussian distribution for (A,N,N) and (M,N,N) grow progressively more misleading as α becomes larger or the forecast horizon is extended. The results for models (15.1) and (15.3) are fairly similar, although the slightly longer upper tail of the lognormal distribution becomes evident for model (15.1) at $h = 10$. Note that point forecasts for model (15.1) are constant because we set $E(\delta_t) = 1$; this result would not hold otherwise.

15.5.3 Forecasting Jewelry Sales

In order to explore further the relative merits of formulations (15.1) and (15.3), we fitted these models to 314 series that describe weekly sales of costume jewelry items over the period week 5, 1998 to week 24, 2000. The data were provided by a leading company in that field. Products that were either launched or discontinued during that period were removed from the study. Most products had very high sales over the Christmas period, so we partitioned the data as follows:

Estimation sample:	weeks 5–45, 1998 and weeks 2–20, 1999	$(n = 60)$;
Test sample:	weeks 21–45, 1999	$(n^* = 25)$.

The gap in the estimation sample did not cause any problems because the differences in levels before and after the Christmas period were minor; the random fluctuations were generally much larger than any level changes.

Three ETS(M,N,N) models were fitted to each series by maximum likelihood:

- Model 1: (15.1) assuming a Gaussian error distribution with mean 0
- Model 2: (15.1) assuming a lognormal error distribution with median 1
- Model 3: (15.3) assuming a lognormal error distribution with median 1

We calculated the one-step-ahead forecasting errors for each series over the test samples and created summaries using the MSE, MAPE and MASE measures introduced in Sect. 2.7. Although the results sometimes differ for individual series, the overall picture is consistent across the three measures, and only the MASE results are reported here. A plot of MASE values by individual series is shown in Fig. 15.5, which indicates that the three models often behave similarly, although Model 1 appears to have a greater number of large MASE values overall. Plots of pairwise comparisons of MASE values for the

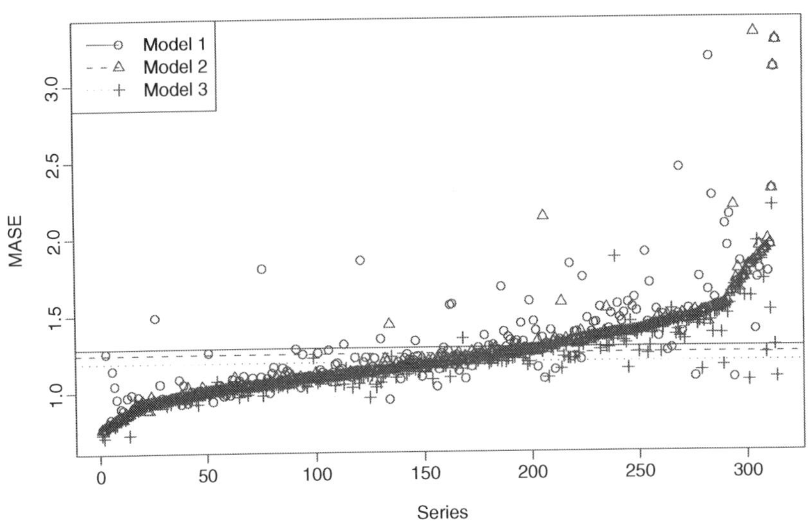

Fig. 15.5. MASE of the three ETS(M,N,N) models fitted to the jewelry data. The series are ordered by the median of the three MASE values to show the differences more clearly. The average MASE values for each of the models are shown as *horizontal lines*.

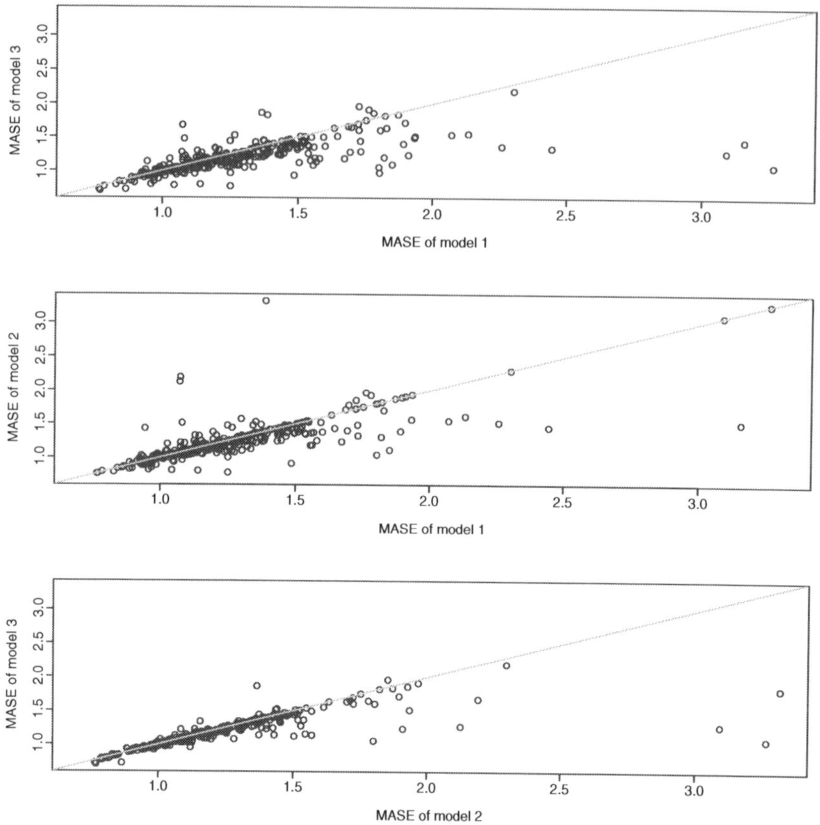

Fig. 15.6. MASE comparison of the three ETS(M,N,N) models. On the diagonal line the two models have the same MASE.

different models are given in Fig. 15.6. Further study is clearly necessary, but the limited results suggest that model 1 is inferior to the other two. Of the two lognormal models, (15.3) appears to be marginally preferable.

15.6 An Appraisal

In this chapter we have undertaken an exploration of models defined on the positive half-line. One of the attractions of the innovations approach is that it enables an exact specification of such models that can lead to explicit results for the prediction distribution. Nevertheless, we have uncovered certain properties that make the use of such models more intricate than conventional practice might suggest. We now summarize our findings to date, while recognizing that this is an area where further research is needed.

Parameter estimation using the Gaussian likelihood appears to be a viable option for the ranges of the parameters that we typically encounter. Further, the point forecasts generated from such fitted models appear to be satisfactory. However, when we turn to prediction intervals, the Gaussian approximation becomes progressively less reasonable as h increases. The lognormal and gamma assumptions appear to provide very similar results.

For simulation purposes there is no substitute for an appropriate non-Gaussian model. At this stage, we are inclined to recommend the lognormal over the gamma on the grounds of operational simplicity. Because only the purely multiplicative models have a sample space restricted to the positive half-line, model simulations with other schemes may need to provide a floor below which the series cannot go. Clearly, this is an area where the investigator must proceed with caution.

15.7 Exercises

Exercise 15.1. Assume that the distribution of ε_t follows a Gaussian distribution with parameters μ and σ, and with left-truncation at -1. Show that the mean of ε_t is

$$\mu + \sigma\frac{\phi(c)}{\Phi(c)},$$

where ϕ and Φ represent the density function and distribution function of the standard Gaussian distribution, and $c = (1+\mu)/\sigma$. Numerically or otherwise, show that the adjustment to the mean is negligible for reasonable values of σ; for example, $\sigma < 0.2$.

Exercise 15.2. Simulate the model ETS(M,N,N) with lognormal errors, where $\sigma = 0.05$, $\alpha = 0.3$ and $\ell_0 = 10$. Generate a series with $n = 5{,}000$ to obtain a clear picture of the behavior of the series:

a. Run several simulations with different sets of random numbers and observe the different types of realization that can occur.
b. Keeping the same set of random numbers, change the values of σ and α and see how the results change.
c. Use Gaussian errors in place of the lognormal errors and compare the two formulations when generating realizations with the same set of random numbers.

Exercise 15.3. The ETS(M,N,M) model has the form:

$$\begin{aligned}
y_t &= \ell_{t-1}s_{t-m}(1+\varepsilon_t),\\
\ell_t &= \ell_{t-1}(1+\alpha\varepsilon_t),\\
s_t &= s_{t-m}(1+\gamma\varepsilon_t),
\end{aligned}$$

where ε_t denotes a white noise series such that $\varepsilon_t \geq -1$, and m is the number of seasons in a year (e.g., $m = 4$ for quarterly data, $m = 12$ for monthly data, etc.).

Using the approach of Sect. 15.2.1, write down the $m + 1$ product martingales, one for the level equation and m for the seasonal components.

Show that almost all of the sample paths for the model converge to zero.

Exercise 15.4. Consider the ETS(M,M$_\mathrm{d}$,N) model:

$$\begin{aligned} y_t &= \ell_{t-1} b_{t-1}^{\phi}(1 + \varepsilon_t), \\ \ell_t &= \ell_{t-1} b_{t-1}^{\phi}(1 + \alpha\varepsilon_t), \\ b_t &= b_{t-1}^{\phi}(1 + \beta\varepsilon_t), \end{aligned}$$

where ε_t denotes a white noise series such that $\varepsilon_t \geq -1$.

Following the same approach as in the previous exercise, show that almost all of the sample paths for the model converge to zero. [Note that the results for the ETS(M,M,N) model follow on setting $\phi = 1$.]

16

Models for Count Data

Time series are often formed from counts. The number of accidents per month at an intersection, the number of cardiac cases per day presenting at an emergency center, the number of power failures each month in a geographical region, and the weekly demand for a slow moving inventory are all examples of time series of counts. Such data are non-negative and integer-valued.

The models in earlier chapters can be used with count data when counts are large because a Gaussian distribution typically provides a good fit to an empirical distribution of large counts. The latter is typically symmetric, and although a Gaussian distribution spills over into the negative part of the real line, the probability of a negative value implied by a fitted Gaussian distribution is usually very small.

However, the earlier models are not appropriate when counts are small. Counts cannot be negative, yet the probability of a negative value implied by a fitted Gaussian distribution is usually not negligible in this circumstance. Moreover, empirical distributions of low count data are typically positively skewed rather than symmetric. A common practice that retains a role for the Gaussian distribution in the presence of low counts is to model the data after a log transformation. However, this approach fails on count time series which contain zeros, and does not take account of the discrete nature of the sample space.

A Poisson distribution is often recommended for count data in place of the Gaussian distribution (Brown 1959). A random variable Y has a Poisson distribution if it takes the values $y = 0, 1, 2, \dots$ with probabilities given by $\Pr(Y = y) = \lambda^y e^{-\lambda}/y!$. Its mean and variance are both equal to λ. Data governed by a Poisson distribution are said to be equi-dispersed because the variance is equal to the mean. In practice, count time series are often over-dispersed; that is, they have a variance which is greater than their mean.

Consequently, the negative binomial distribution is sometimes used because its variance is always greater than its mean (Stuart and Ord 1994). Another common option is to retain the Poisson distribution but introduce further randomness by allowing λ to be a random variable. Moreover, because count time series are often autocorrelated as well as over-dispersed, λ is assumed to change over time according to an autoregressive process. One possibility is to assume that $y_t \mid \lambda_{t-1} \sim \text{Poisson}(\lambda_{t-1})$, where $\lambda_t = a + b\lambda_{t-1} + cy_t$. The parameters a, b and c are constrained to be non-negative and to satisfy the stationarity condition $b + c < 1$ (Heinen 2003; Jung et al. 2006). Another possibility is to introduce an additional source of randomness and assume that the Poisson parameter is governed by a recurrence relationship $\log(\lambda_t) = r + \kappa \log(\lambda_{t-1}) + \eta_t$, where $\eta_t \sim \text{NID}(0, \sigma_\eta^2)$. Both d and κ are parameters, the latter satisfying the stationarity condition $|\kappa| < 1$ (Chan and Ledolter 1995; Jung et al. 2006). These two autoregressive approaches have been compared in Snyder et al. (2008), where it has been shown that the latter dual source of randomness approach applies to a wider range of samples than does the former single source of error approach, but that the approaches are mutually complementary in the sense that the samples to which they may be applied do not overlap.

Time series are rarely stationary in practice, so the focus of this chapter is on models that are related to exponential smoothing and which are suitable for application to nonstationary count data. Thus we consider the Poisson analogue of the Gaussian innovations local level model and a similar approach where the Poisson distribution is replaced by a negative binomial distribution (Harvey and Fernandes 1989). These models are introduced in Sect. 16.1.

For time series which contain a large number of zeros, Croston (1972) proposed an approach whereby the non-zero values were forecast separately from the time between them. His proposal was specifically designed to forecast intermittent demand data, and it has become a popular method in business applications. We discuss Croston's method in Sect. 16.2, with particular attention to the possible models underlying the method.

In Sect. 16.3, we compare the forecast efficiency of these methods via an empirical study.

16.1 Models for Nonstationary Count Time Series

16.1.1 Local Poisson Model

The Poisson analogue of the Gaussian innovations local level model is

$$y_t \mid \lambda_{t-1} \sim \text{Poisson}(\lambda_{t-1}), \tag{16.1}$$

where $\lambda_t = (1 - \alpha)\lambda_{t-1} + \alpha y_t$ and $0 < \alpha < 1$. The only source of randomness is the Poisson distribution itself. Nevertheless, the associated time series is over-dispersed because additional randomness feeds into the process through random changes in the local level.

Simple exponential smoothing is used to calculate the levels λ_t. The seed level λ_0 and the smoothing parameter α may be selected to maximize a Poisson likelihood. Point predictions are obtained by extrapolating the final level λ_n. The one-step-ahead prediction distribution for $y_{t+1} \mid \lambda_t$ is Poisson(λ_t). The prediction distributions for multiple steps ahead are not available in closed form. It may be shown, however, that $\mathrm{E}(y_{t+h} \mid \lambda_t) = \lambda_t$ and $\mathrm{V}(y_{t+h}) = [1 + \alpha^2(h-1)]\lambda_t$. A negative binomial distribution, fitted using the method of moments, is likely to form a good approximation to such a prediction distribution. When more precise results are required, it is necessary to resort to simulation methods.

16.1.2 Local Negative Binomial Model

Another approach (Harvey and Fernandes 1989) relies on a negative binomial distribution in place of a Poisson distribution. Instead of selecting the seed level to maximize the likelihood function, an adaptation of simple exponential smoothing for finite samples is used. Gilchrist (1967) demonstrated, for any real time series, that the underlying level that minimizes the *discounted* sum of squared errors may be computed from two quantities c_t and b_t with $\lambda_t = c_t / b_t$. The required quantities for this calculation are obtained recursively by

$$c_t = (1-\alpha)c_{t-1} + y_t \qquad \text{and} \qquad b_t = (1-\alpha)b_{t-1} + 1,$$

where $0 < \alpha < 1$ and $1 - \alpha$ is the discount factor. These recurrence relationships are seeded with $c_0 = b_0 = 0$, and so their solutions are

$$c_t = \sum_{j=0}^{t}(1-\alpha)^j y_{t-1} \qquad \text{and} \qquad b_t = \sum_{j=0}^{t}(1-\alpha)^j.$$

Thus, λ_t is a weighted average of the observations, where the weights decline exponentially with increases in the age index j. As t increases, λ_t converges to the exponentially weighted average associated with the traditional simple exponential smoothing formula $\lambda_t = (1-\alpha)\lambda_{t-1} + \alpha y_t$.

This use of the discounted average allows us to bypass the determination of a seed level and to focus on the single parameter α which may be estimated by maximizing a likelihood function. The likelihood is based on the product of the one-step-ahead prediction distributions

$$p(y_t \mid y_1, \ldots, y_{t-1}, \alpha) = \frac{\Gamma(\delta c_{t-1} + y_t)}{y_t!\,\Gamma(\delta c_{t-1})} \left(\frac{\delta b_{t-1}}{1 + \delta b_{t-1}}\right)^{\delta c_{t-1}} \left(\frac{1}{1 + \delta b_{t-1}}\right)^{y_t},$$

where $\delta = 1 - \alpha$.

Harvey and Fernandes (1989) use Bayesian-like arguments to obtain their method without direct recourse to a state space model. Later they infer that the model underlying their method must have a measurement distribution

$$y_t \mid \lambda_t \sim \text{Poisson}(\lambda_t).$$

It must also have the transition equation $\lambda_t = (1-\alpha)^{-1}\lambda_{t-1}\eta_t$, where η_t has a beta distribution with parameters $(1-\alpha)c_{t-1}$ and αc_{t-1}. Because this model relies on two probability distributions, it may be viewed as the multiple source of randomness analogue of the local Poisson model in the previous section.

16.1.3 Convergence Problem

The local level models described in this section have a fixed point of $\lambda_t = 0$. Moreover, this fixed point is an attractor: there is a finite probability that λ_t will drop to zero. The problem with this particular fixed point is that all subsequent values of the time series are forced to be zero. That is, all sample paths will converge to zero. Figure 16.1 illustrates this phenomena with some data generated by a local Poisson model where $\lambda_0 = 2$ and $\alpha = 0.5$. One must be aware of this problem when using the model to simulate prediction distributions.

Grunwald et al. (1997) provide a detailed explanation of this phenomenon and show that this is a much more general problem. They prove that if the sample space of a model defined as an exponentially weighted moving average is bounded to any subset of $[0,\infty)$, (e.g., taking only positive values or non-negative integers), then the original process will converge almost surely to a constant.

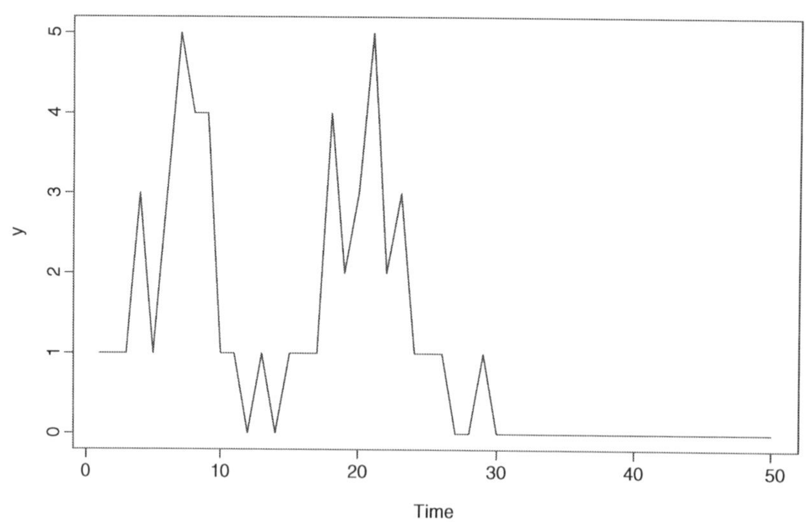

Fig. 16.1. Time series generated from a Poisson local level model: $\lambda_0 = 2$; $\alpha = 0.5$.

This result means that the count models defined in this section are only suitable for short-term forecasting, and should not be used for long-term forecasting or simulation.

16.2 Croston's Method

Time series of demand are often recorded as counts and can sometimes contain many zeros; such data are usually called "intermittent demand" and are associated with "slow-moving" items. A popular method for forecasting the demand of such data was developed by Croston (1972). It involves separating the original demand time series into two derived series: the non-zero demands, and the time gaps between the periods with non-zero demands. The method then involves applying simple exponential smoothing separately to each derived time series, with the same smoothing parameter being used in each case. The point forecast of demand is the ratio of the forecasts of the non-zero demand and the time gap.

To be specific, let y_t be the demand occurring during the time period t and x_t be the indicator variable for non-zero demand periods; i.e., $x_t = 1$ when demand occurs at time period t and $x_t = 0$ when no demand occurs. Furthermore, let j_t be the number of periods with non-zero demand during interval $[0, t]$; that is, $j_t = \sum_{i=1}^{t} x_i$ is the index of the non-zero demand. For ease of notation, we will usually drop the subscript t on j. Then we let q_j represent the quantity of the jth non-zero demand and τ_j the number of periods between the occurrence of q_{j-1} and q_j. Using this notation, we can write $y_t = x_t q_j$.

Croston's method forecasts $\{q_j\}$ and $\{\tau_j\}$ separately using simple exponential smoothing, with forecasts being updated only after demand occurrences. Let $q_{j+1|j}$ and $\tau_{j+1|j}$ be the forecasts of the $(j+1)$th non-zero demand and the $(j+1)$th time gap respectively, based on data up to and including non-zero demand j. Then Croston's method gives

$$\hat{q}_{j+1|j} = (1-\alpha)\hat{q}_{j|j-1} + \alpha q_j, \quad j = 1, 2, \ldots, j_n, \tag{16.2a}$$

$$\hat{\tau}_{j+1|j} = (1-\alpha)\hat{\tau}_{j|j-1} + \alpha \tau_j, \quad j = 2, 3, \ldots, j_n. \tag{16.2b}$$

The smoothing parameter α usually takes values between 0 and 1, and is assumed to be the same for both q_j and τ_j. Then the mean demand rate, which is used as the h-step-ahead forecast for the demand at time $n+h$, is estimated by the ratio

$$\hat{y}_{n+h|n} = \hat{q}_{j_n+1|j_n} \Big/ \hat{\tau}_{j_n+1|j_n}, \tag{16.3}$$

where j_n is the last period with a non-zero demand. Several variations on this procedure have been proposed including Johnston and Boylan (1996) and Syntetos and Boylan (2001, 2005).

16.2.1 An Underlying Model?

Croston (1972, Appendix B) states that the assumptions of this method are:

(1) The non-zero demand sizes q_j are governed by an ARIMA(0,1,1) model
(2) The distribution of times gaps τ_j is iid geometric
(3) Demand sizes q_j and time gaps τ_j are mutually independent

However, as indicated in Snyder (2002), assumption (2) is incorrect because it would result in the use of a simple average rather than an exponentially weighted average for the forecasts of the time gaps. Moreover, this mistake has been compounded in much of the published empirical analyses of Croston's method (e.g., Willemain et al. 1994; Syntetos and Boylan 2001, 2005) where an independence assumption is also imposed on the non-zero demands in place of Assumption 1.

Note that (16.2a) can be rewritten as an exponentially weighted average of past values:

$$\hat{q}_{j+1|j} = \sum_{k=0}^{j-1} \alpha(1-\alpha)^k q_{j-k} + (1-\alpha)^j \lambda_0. \tag{16.4}$$

A similar equation can be obtained for τ_j. This immediately means that the underlying models must be nonstationary (e.g., Abraham and Ledolter 1983, Sect. 3.3).

Shenstone and Hyndman (2005) used this result, along with the convergence problem noted above, to show that there is no possible model that would lead to (16.3) as the optimal forecast equation unless we allow a sample space that includes both negative values and non-integer values.

However, if we are prepared to ignore the convergence problem, we can derive a model that gives *one*-step forecasts that are equivalent to Croston's method (16.3). We assume two separate local level models underlying the processes $\{q_j\}$ and $\{\tau_j\}$. For the process of non-zero demands, $\{q_t\}$, we will assume that they are governed locally by a Poisson distribution where the domain of the distribution is shifted up by one. Thus,

$$q_j \mid \lambda_{j-1} \sim \text{Poisson}(\lambda_{j-1} - 1) + 1, \quad \text{where } \lambda_j = (1-\alpha)\lambda_{j-1} + \alpha q_j.$$

We also assume that a *local* Bernoulli distribution governs whether there is a positive or zero demand. It is assumed that the Bernoulli distribution in period t is conditionally independent of its predecessors, and that the Bernoulli distribution remains unchanged between periods with non-zero demands. Thus, the gaps between non-zero demands are governed locally by a geometric distribution:

$$\tau_j \mid \ell_{j-1} \sim \text{Geometric}(\ell_{j-1}), \quad \text{where } \ell_j = (1-\alpha)\ell_{j-1} + \alpha\tau_j.$$

Equation (16.2) gives the one-step forecasts of these processes, but not the multi-step-ahead forecasts. In fact, the convergence problem noted in Sect. 16.1.3 applies to both models, and sample paths in both cases will converge to 1.

The probability of a non-zero demand in period t is given by $\pi_t = 1/\ell_{j-1}$, and so the probability of demand in period t is given by

$$\Pr(y_t = k \mid \lambda_{j-1}, \ell_{j-1}) = \begin{cases} 1 - \pi_t & \text{if} \quad k = 0 \\ \pi_t p_{k,j} & \text{if} \quad k > 0, \end{cases} \tag{16.5}$$

where $k = 1, 2, \ldots,$ and

$$p_{k,j} = \Pr(q_j = k \mid \lambda_{j-1}) = (\lambda_{j-1} - 1)^{k-1} e^{1-\lambda_{j-1}}/(k-1)!\,.$$

16.2.2 Estimation

Croston indicated, on the basis of experience, that the smoothing parameter α should take a value between 0.1 and 0.2. He had little to say about the choice of seed values for the non-zero demands and gap recurrence relationships.

Given the model derived in the previous section, we can estimate the parameters using a likelihood approach. The likelihood function is the product of the mass functions (16.5) for periods $1, 2, \ldots, n$. This function may be maximized with respect to λ_0, ℓ_1 and α. For numerical stability, this solution is typically found by maximizing the associated log-likelihood.

16.3 Empirical Study: Car Parts

The predictive capabilities of the Poisson model, the negative binomial model and Croston's method were compared on 2,674 time series supplied by a US car company. The time series, representing the monthly sales for slow moving parts, cover a period of 51 months. The 2,509 time series without missing values have an average gap between positive demands of 2.9 months and an average positive demand of 2. Eighty-nine percent of the time series were over-dispersed, and the dispersion ratio (i.e., the ratio of the variance to the mean), averaged across all time series, was 2.3.

The time profile of aggregate demand for all of the car parts is shown in Fig. 16.2. It indicates that there is a tendency for demands to decline as the age of a part increases. Demands appear to be nonstationary.

Although a downward trend is discernable in the aggregate data, it is far from clear that such trends always operate at the individual parts level. It is important to allow for other possible trajectory shapes which may only be observed at the individual part level. One possibility, ignoring the zero demands, is a gradual increase to some peak and then a slow decline. Because such patterns are not known in advance, there is a need for an

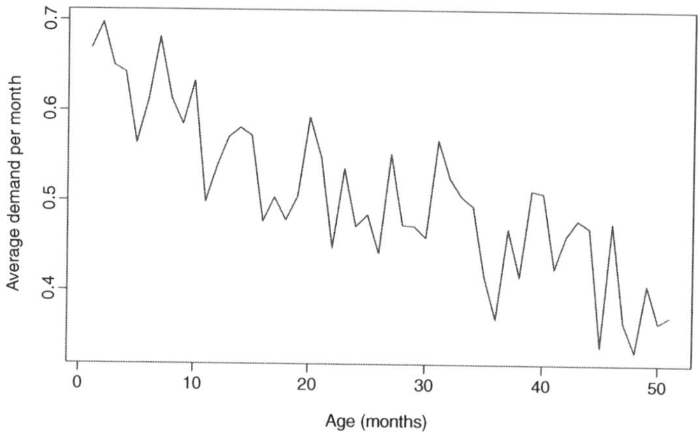

Fig. 16.2. Time profile of demands averaged across 2,674 time series denoting demand for car parts.

approach that adapts to whatever pattern emerges as a part ages. Given the uncertainty over the trajectory, it is best to treat the underlying level as a random variable and assume that its evolution over time is governed by a stochastic process.

To minimize the computational problems that arise with time series containing a small number of positive values, the database was further culled to eliminate those time series which:

(a) Possessed fewer than ten positive monthly demands
(b) Had no positive demand in the first 15 and final 15 months

There were 1,046 time series left after this additional cull.

Six approaches to forecasting were compared in the study. The simplest, designated by Z, was to set all the predictions to the value zero, on the grounds that the empirical distributions of many time series have a mode of zero. The second was based on a *global* Poisson (GP) distribution with iid observations where it is optimal to use a simple average of observed demands. The others were the methods described in the previous section that allow for random changes in the underlying level: the *local* Poisson (LP) model; the *local* negative binomial (LNB) model; and Croston's method. Maximum likelihood estimation was used in each case. However, because the folklore of exponential smoothing (Brown 1959; Croston 1972) suggests that it is best to use fixed values of α across an entire range of products, particularly with small samples, this possibility was also considered. In those cases where a fixed parameter was used, the seed levels continued to be estimated by maximizing the likelihood function. In other words, a *partial* maximum likelihood approach was employed.

The models were estimated using the first 45 observations of each time series. Their forecasting performances were compared over periods 46–51

Table 16.1. Forecast performance for each method applied to the 1,046 car parts time series. Summary statistics are for MASE values.

Model/method	α	Mean	Median	Stdev
Z		0.42	0.30	0.47
LP	0.3	0.63	0.55	0.42
LP	0.2	0.64	0.56	0.38
LNB	0.8	0.64	0.56	0.38
LNB	ML	0.65	0.59	0.40
Croston	0.3	0.65	0.60	0.40
Croston	0.2	0.65	0.61	0.39
LP	ML	0.68	0.64	0.36
Croston	0.1	0.68	0.65	0.38
Croston	ML	0.68	0.67	0.40
Croston	0.0	0.70	0.70	0.39
GP		0.82	0.75	0.31

Summary statistics are for MASE values.

using the MASE statistic (see Sect. 2.7.3). The means, medians and standard deviations of the MASEs calculated across the 1,046 time series are given in Table 16.1. The approaches are ordered by the median.

The zero method (Z) had the most remarkable performance, its median MASE being substantially lower than the other methods. However, the standard deviation of its MASE was much higher, suggesting a strong lack of consistency of performance over all the time series. Intriguingly, the traditional *global* Poisson distribution had the worst performance. The associated simple average, which places an equal weight on all the observations (including the zeros), has little value for the type of count data represented by car parts demands.

Of the approaches labeled "maximum likelihood" (ML), the local negative binomial distribution was best, followed by the local Poisson distribution, with the Croston method surprisingly coming last. The folklore about the use of fixed parameter values was also confirmed. Fixed value approaches did better than their maximum likelihood counterparts. The performances of the local Poisson and local negative binomial distributions were reversed, but Croston's method continued to have the poorest performance.

For the case of Gaussian observations considered in earlier chapters, each multiple source of randomness model has an equivalent single source of randomness model. This no longer appears to be true for count data. A deeper analysis of the results indicates that the local Poisson model may or may not work better than the local negative binomial model at the level of individual time series. The LNB 0.8 was better than LP 0.2 for 56% of the time series and they tied for 28% of the time series. However, the maximum difference in the MASE was only 0.15%, so the two approaches are really quite similar with regard to point forecasts.

In a further attempt to separate the LP and LNB models, 90% prediction intervals were simulated for periods 46–51. In both cases, about 95% of the withheld values were found to lie within these prediction intervals. This result was obtained whether or not the smoothing parameter (discount factor) was optimized. In these circumstances, the dual source of randomness approach (LNB) appears to have little advantage over the single source of randomness approach (LP) on this particular data set. Curiously, the prediction intervals generated from these models are a little too wide.

In general it was observed that reductions in the median MASE were accompanied by increases in its standard deviation. This suggests that a multi-model approach might work better than any single model approach by tailoring the model choice to the individual structures of the time series. However, it is not clear how this can be done. Sample sizes, in practice, are often too small to withhold data for a prediction validation approach. And the models are based on different probability distributions, something that precludes the use of an information criterion approach. This is an issue that warrants further investigation.

16.4 Exercises

Exercise 16.1. A time series is governed by a Poisson local level model (Sect. 16.1.1) with $\lambda_0 = 2$. Simulate series of length 100 for $\alpha = 0, 0.2, 0.5$ and 1. What conclusions can you make about the effect of α? Do such series converge to the fixed point of zero?

Exercise 16.2. Demonstrate that the value of λ_t which minimizes the discounted sum of squared errors $\sum_{j=0}^{t-1} \delta^j (y_{t-j} - \lambda_t)^2$ may be calculated with the ratio a_t/b_t where $a_t = \delta a_{t-1} + y_t$ and $b_t = \delta b_{t-1} + 1$.

Exercise 16.3. Consider the Poisson model with a time-dependent mean, which may be written as $y_t|\lambda_{t-1} \sim \text{Poisson}(\lambda_{t-1})$ where $\lambda_t = a + b\lambda_{t-1} + cy_t$. The parameters a, b and c must be non-negative to avoid negative values for the conditional mean. Given the starting value λ_0, show that

$$\text{E}(y_t \mid \lambda_0) = a + a(b+c) + \cdots + a(b+c)^{t-1} + (b+c)^t \lambda_0.$$

Note that if $b + c < 1$ the conditional expected value of $y_t \mid \lambda_0$ converges to the limiting value $a/(1-b-c)$; otherwise it increases without limit.

Exercise 16.4. The data set `partx` contains a history of monthly sales of an automobile part. Compare the following forecasting methods applied to these data: (1) a progressive simple average (where the average is changed each period to reflect the effect of each additional observation); (2) a local Poisson model; (3) a local negative binomial model; and (4) Croston's method. For models (2)–(4), parameters should be estimated by either maximizing the likelihood or minimizing the sum of squared errors.

17

Vector Exponential Smoothing

Co-author:[1] **Ashton de Silva**[2]

In earlier chapters we have considered only univariate models; we now proceed to examine multi-series extensions and to compare the multi-series innovations models with other multi-series schemes. We shall refer to our approach as the vector exponential smoothing (VES) framework. The innovations framework is similar to the structural time series models advocated by Harvey (1989) in that both rely upon unobserved components, but there is a fundamental difference: in keeping with the earlier developments in this book, each time series has only one source of error.

The VES models are introduced in Sect. 17.1; special cases of the general model are then discussed in Sect. 17.2. An inferential framework is then developed in Sect. 17.3 for the VES models, building upon our earlier results for the univariate schemes.

The most commonly used multivariate time series models are those defined within the ARIMA framework. Interestingly, this approach also has only one source of randomness for each time series. Thus, the vector versions of the ARIMA framework (VARIMA), and special cases such as vector autoregression (VAR) and vector moving average (VMA), may be classified as innovations approaches to time series analysis (Lütkepohl 2005). We compare the VES framework with existing approaches in Sect. 17.4. As in Chap. 11, when we consider equivalences between vector innovations models and the VARIMA forms, we will make the infinite start-up assumption.

Finally we compare the performance of VES models to VAR and other existing state space alternatives, first in an empirical study of exchange rates (Sect. 17.5), and then across a range of different time series taken from a large macroeconomic database, in Sect. 17.6.

---

1 This chapter is based on de Silva et al. (2007), which should be consulted for further details.

2 Dr. Ashton de Silva, School of Economics, Finance & Marketing, RMIT University, Melbourne, Australia.

17.1 The Vector Exponential Smoothing Framework

The general vector exponential smoothing model (VES) is introduced in this section. Conceptually, the model builds directly upon the univariate framework; see, for example, Anderson and Moore (1979). We stack the univariate observations into an N-vector and then assume that the vector of observations $\boldsymbol{y}_t$ is a linear function of a k-vector of unobserved components $\boldsymbol{x}_{t-1}$ plus error. That is, we have the *measurement equation*

$$\boldsymbol{y}_t = \boldsymbol{W}\boldsymbol{x}_{t-1} + \boldsymbol{\varepsilon}_t, \tag{17.1a}$$

where $\boldsymbol{W}$ is an $N \times k$ matrix of coefficients that are often known, as in the univariate case, and $\boldsymbol{\varepsilon}_t$ is an N-vector. The innovations $\{\boldsymbol{\varepsilon}_t\}$ follow a common multivariate Gaussian distribution with zero means and variance matrix $\boldsymbol{\Sigma}$, but we assume that $\{\boldsymbol{\varepsilon}_t\}$ and $\{\boldsymbol{\varepsilon}_{t+i}\}$ are independent for all $i \neq 0$.

The evolution of the unobserved components is governed by the first-order Markovian relationship

$$\boldsymbol{x}_t = \boldsymbol{F}\boldsymbol{x}_{t-1} + \boldsymbol{G}\boldsymbol{\varepsilon}_t. \tag{17.1b}$$

This is called the *transition equation*. The fixed $k \times k$ matrix $\boldsymbol{F}$ is referred to as the *transition matrix*. The $k \times N$ matrix $\boldsymbol{G}$ typically has elements that are unknown; they determine the effects of the innovations on the process beyond the period in which they occur. When $\boldsymbol{G} = \boldsymbol{0}$, the components are deterministic. When $\boldsymbol{G}$ is block-diagonal with some non-zero elements within each block, each innovation has an effect only on its own series. When $\boldsymbol{G}$ has non-zero elements outside these blocks, an innovation will have an effect on other series as well as its own.

The general model given in (17.1) is rather opaque and will often be "parameter-heavy." A common formulation separates out the state variables for each series, so that the elements of (17.1) may be written as:

$$\boldsymbol{x}_t = \begin{bmatrix} \boldsymbol{x}_{1t} \\ \vdots \\ \boldsymbol{x}_{Nt} \end{bmatrix}, \qquad \boldsymbol{W} = \begin{bmatrix} \boldsymbol{w}_1' & 0 & 0 \\ 0 & \ddots & 0 \\ 0 & 0 & \boldsymbol{w}_N' \end{bmatrix},$$

$$\boldsymbol{F} = \begin{bmatrix} \boldsymbol{F}_1 & 0 & 0 \\ 0 & \ddots & 0 \\ 0 & 0 & \boldsymbol{F}_N \end{bmatrix} \quad \text{and} \quad \boldsymbol{g} = \begin{bmatrix} \boldsymbol{g}_{11} & \cdots & \boldsymbol{g}_{1N} \\ \boldsymbol{g}_{21} & \cdots & \boldsymbol{g}_{2N} \\ & \cdots & \\ \boldsymbol{g}_{N1} & \cdots & \boldsymbol{g}_{NN} \end{bmatrix}.$$

That is, the state variables are updated as functions of the random errors of all series but there are no common states. We now examine several special cases.

17.1.1 Seemingly Unrelated Models

When $g_{ij} = 0$ for all $i \neq j$, the series are related only through the error terms, an example of a seemingly unrelated regression (SUR) model, as defined by Zellner (1962). In the time series context, Harvey (1989) refers to such models as SUTSE (seemingly unrelated time series equations) models. No special properties arise in this case, save that joint estimation will produce more efficient estimators than treating each series separately. An iterative scheme that involves estimating the parameters in the transition equations and then the variance matrix will often work well.

17.1.2 Homogeneous Coefficient Models

As a further specialization of the SUTSE models, we now allow the coefficients to be the same across all series. That is, we set $\boldsymbol{w}_i = \boldsymbol{w}_0$, $\boldsymbol{F}_i = \boldsymbol{F}_0$ and $\boldsymbol{g}_i = \boldsymbol{g}_0$. We refer to these models as *homogeneous* because each series has an equal number of states with the same updating mechanism. These models were introduced by Harvey (1986). The models have two important features: first, the ability to pool estimates across series leads to more efficient estimation. Second, and perhaps more importantly, a homogeneous system allows the aggregation of individual series to a total such that the single series and aggregate forecasts are in complete agreement. Such a feature is desirable, for example, when forecasts for individual products must match up with the total for that product class.

Because the series may be not be measured in the same units, we introduce coefficients a_i and define: $y_{0,t} = \sum_i a_i y_{i,t}$, $\boldsymbol{x}_{0,t} = \sum_i a_i \boldsymbol{x}_{i,t}$ and $\varepsilon_{0,t} = \sum_i a_i \varepsilon_{i,t}$. The zero subscripts are included to emphasize that the aggregation process has unequal coefficients. These summations lead to the aggregate model

$$\begin{aligned} y_{0,t} &= \boldsymbol{w}_0' \boldsymbol{x}_{0,t-1} + \varepsilon_{0,t}, \\ \boldsymbol{x}_{0,t} &= \boldsymbol{F}_0 \boldsymbol{x}_{0,t-1} + \boldsymbol{g}_0 \varepsilon_{0,t}. \end{aligned}$$

17.1.3 Models with Group Seasonality

As an illustration of a model that contains both specific and common state variables, we consider a model with group seasonality, developed in detail by Ouwehand et al. (2007). The model is designed to cover situations where the set of series has a common seasonal pattern. Such circumstances often arise when forecasting a group of related products; the sales patterns may be evolving differently, but the overall product group may be subject to similar seasonal variations. Because individual series are often rather short, this model enables the forecaster to improve the estimates of the seasonal factors by pooling across series.

For example, we may extend the additive Holt-Winters' seasonal or ETS(A,A,A) model given in Sect. 3.4.3 in the following way:

$$\begin{aligned} y_{i,t} &= \ell_{i,t-1} + b_{i,t-1} + s_{t-m} + \varepsilon_{i,t}, \\ \ell_{i,t} &= \ell_{i,t-1} + b_{i,t-1} + \alpha_i \varepsilon_{i,t}, \\ b_{i,t} &= b_{i,t-1} + \beta_i \varepsilon_{i,t}, \\ s_t &= s_{t-m} + \gamma \sum_i c_i \varepsilon_{i,t}, \end{aligned}$$

where $i = 1, 2, \ldots, N$, $\ell_{i,t-1}$ and $b_{i,t-1}$ denote the level and trend terms for the ith series, and s_t denotes the common seasonal term. The c_i are non-negative constants that define the weights on the different series. Thus, the model contains $2N + m$ state variables in all.

17.2 Local Trend Models

We now consider a class of local trend vector models, which are developed by analogy with the univariate linear models formulated in Chap. 3. In practice, the growth rates may be more appropriately modeled as a stationary process rather than a random walk (Gardner and McKenzie 1985), so we consider the *vector damped local trend* model in the form:

$$y_t = \ell_{t-1} + \Phi b_{t-1} + \varepsilon_t, \tag{17.3a}$$

$$\ell_t = \ell_{t-1} + \Phi b_{t-1} + A\varepsilon_t, \tag{17.3b}$$

$$b_t = \Phi b_{t-1} + B\varepsilon_t, \tag{17.3c}$$

where Φ denotes a matrix of damping factors. When Φ is diagonal, we apply separate damping factors to each series, as in the univariate case. In the present context, the formulation is more general, because the slopes are (potentially) functions of all the other series' slopes, damped at different rates. This construction may sound rather artificial, but it turns out that in the reduced form of the model, Φ is the matrix of first-order autoregressive coefficients; see Exercise 17.1. The system is forecastable provided the eigenvalues λ in the determinantal equation $|\Phi(I - A) - (I + \Phi - A - \Phi B)\lambda + \lambda^2 I| = 0$ all have modulus less than 1. This expression clearly reduces to N similar univariate statements when A and B are diagonal.

Various special cases follow from the vector damped local trend model:

- Vector local trend model, when $\Phi = I$
- Vector local level model, when $\Phi = 0$

17.3 Estimation

The matrices W, F and G in the vector exponential smoothing model potentially depend on a vector of unknown parameters designated by θ. We also need to estimate Σ. We develop a maximum likelihood procedure following the same general argument as in Chap. 5 for the univariate case. As in Sect. 5.1, we condition upon the starting values $\boldsymbol{x}_0$. This conditional likelihood, viewed as a function of $\boldsymbol{\theta}, \boldsymbol{\Sigma}$ and $\boldsymbol{x}_0$, can be represented by $\mathcal{L}(\boldsymbol{\theta}, \boldsymbol{\Sigma}, \boldsymbol{x}_0 \mid \boldsymbol{y}_1, \boldsymbol{y}_2, \ldots, \boldsymbol{y}_n) = p(\boldsymbol{y}_1, \boldsymbol{y}_2, \ldots, \boldsymbol{y}_n \mid \boldsymbol{\theta}, \boldsymbol{\Sigma}, \boldsymbol{x}_0)$. In turn, the likelihood may be written as the product of the one-step-ahead prediction density functions:

$$\mathcal{L}(\boldsymbol{\theta}, \boldsymbol{\Sigma}, \boldsymbol{x}_0 \mid \boldsymbol{y}_1, \boldsymbol{y}_2, \ldots, \boldsymbol{y}_n) = \prod_{t=1}^{n} p(\boldsymbol{y}_t \mid \boldsymbol{y}_1, \boldsymbol{y}_2, \ldots, \boldsymbol{y}_{t-1}, \boldsymbol{\theta}, \boldsymbol{\Sigma}, \boldsymbol{x}_0).$$

The moments of the prediction distributions are

$$\mathrm{E}(\boldsymbol{y}_t \mid \boldsymbol{y}_1, \boldsymbol{y}_2, \ldots, \boldsymbol{y}_{t-1}, \boldsymbol{\theta}, \boldsymbol{\Sigma}, \boldsymbol{x}_0) = \boldsymbol{W}\boldsymbol{x}_{t-1}$$

and

$$\mathrm{V}(\boldsymbol{y}_t \mid \boldsymbol{y}_1, \boldsymbol{y}_2, \ldots, \boldsymbol{y}_{t-1}, \boldsymbol{\theta}, \boldsymbol{\Sigma}, \boldsymbol{x}_0) = \boldsymbol{\Sigma}.$$

The state vectors are calculated using the general linear recursions (where $\boldsymbol{e}_t$ denotes the estimated errors):

$$\begin{aligned} \hat{\boldsymbol{y}}_t &= \boldsymbol{W}\boldsymbol{x}_{t-1}, \\ \boldsymbol{e}_t &= \boldsymbol{y}_t - \hat{\boldsymbol{y}}_t, \\ \boldsymbol{x}_t &= \boldsymbol{F}\boldsymbol{x}_{t-1} + \boldsymbol{G}\boldsymbol{e}_t. \end{aligned}$$

The log-likelihood function is

$$\log \mathcal{L}(\boldsymbol{\theta}, \boldsymbol{\Sigma}, \boldsymbol{x}_0 \mid \boldsymbol{y}_1, \ldots, \boldsymbol{y}_n) = -\frac{n}{2}\left(N \log(2\pi) + \log|\boldsymbol{\Sigma}|\right) - \frac{1}{2}\sum_{t=1}^{n} \boldsymbol{e}_t' \boldsymbol{\Sigma}^{-1} \boldsymbol{e}_t.$$

The variance matrix may be concentrated out of the likelihood because its ML estimator is:

$$\hat{\boldsymbol{\Sigma}} = n^{-1} \sum_{t=1}^{n} \boldsymbol{e}_t \boldsymbol{e}_t'.$$

The concentrated log-likelihood function then becomes

$$\log \mathcal{L}(\boldsymbol{\theta}, \hat{\boldsymbol{\Sigma}}, \boldsymbol{x}_0 \mid \boldsymbol{y}_1, \ldots, \boldsymbol{y}_n) = -\frac{n}{2}(N \log(2\pi) + N \log|\hat{\boldsymbol{\Sigma}}|) - \frac{1}{2}\sum_{t=1}^{n} \boldsymbol{e}_t' \hat{\boldsymbol{\Sigma}}^{-1} \boldsymbol{e}_t.$$

The vector $\boldsymbol{\theta}$ is restricted to satisfy the various forecastability conditions that are specific to the particular model under consideration. Because this framework assumes a finite start-up, it is not necessary to impose stationarity conditions, although the investigator may prefer to do so.

One approach to starting the numerical search procedure is to fit the univariate models separately and to use these estimates as starting values. A potential limitation of this approach is that it does not provide starting values for inter-series parameters. However, setting the initial values of these parameters to zero seems to work well in practice. A more general difficulty is that the likelihood function is not concave so that individual runs may lead to local optima. To avoid these difficulties, de Silva et al. (2007) used multiple runs with randomly assigned starting values centered upon the values given below, subject to the choices being within the feasible region.

Values for $\boldsymbol{x}_0$*:*

Vector Local Level Model: Start up values for the initial levels $\boldsymbol{\ell}_0$ equal the average of the first ten observations for each series.

Vector Local Trend Model: The first ten observations of each series are regressed against time. The intercept and slope estimates provide approximations of the values of $\boldsymbol{\ell}_0$ and $\boldsymbol{b}_0$ respectively.

Values for the elements of the parameter matrices:

$\boldsymbol{A}$	Diagonal elements set to 0.33,	off-diagonal elements set to 0
$\boldsymbol{B}$	Diagonal elements set to 0.50,	off-diagonal elements set to 0
$\boldsymbol{\Phi}$	Diagonal elements set to 0.90,	off-diagonal elements set to 0

17.3.1 Prediction

Following the univariate discussion in Chap. 6, we may develop prediction distributions under the assumption that the errors follow a multivariate Gaussian distribution. Let $\boldsymbol{\mu}_{n+j|n}$ denote the mean of the jth-step-ahead prediction distribution with the forecast origin being at the end of period n; and let $\boldsymbol{V}_{n+j|n}$ be the variance matrix. Also, let $\boldsymbol{m}_{n+j|n}$ and $\boldsymbol{U}_{n+j|n}$ be the conditional mean vector and variance matrix of the state vector in period $n+j$. Then the moments of the prediction distributions can be computed recursively using the formulae for $j = 1, 2, \ldots, h$:

$$\begin{aligned}
\boldsymbol{\mu}_{n+j|n} &= \boldsymbol{W}\boldsymbol{m}_{n+j-1|n} \\
\boldsymbol{V}_{n+j|n} &= \boldsymbol{W}\boldsymbol{U}_{n+j-1|n}\boldsymbol{W}' + \boldsymbol{\Sigma} \\
\boldsymbol{m}_{n+j|n} &= \boldsymbol{F}\boldsymbol{m}_{n+j-1|n} \\
\boldsymbol{U}_{n+j|n} &= \boldsymbol{F}\boldsymbol{U}_{n+j-1|n}\boldsymbol{F}' + \boldsymbol{G}\boldsymbol{\Sigma}\boldsymbol{G}'.
\end{aligned}$$

Note that $\boldsymbol{m}_{n|n} = \boldsymbol{x}_n$ and $\boldsymbol{U}_{n|n} = \boldsymbol{0}$. As in our earlier discussion, no attempt has been made to incorporate the effects of sampling error into these distributions.

17.3.2 Model Selection

Following the discussion in Chap. 7, we recommend the Akaike information criterion (AIC) as the best of the common information criteria for model selection. Letting q designate the number of unknown parameters, the multivariate AIC is specified as:

$$\text{AIC} = -2\log \mathcal{L}(\hat{\boldsymbol{\theta}}, \hat{\boldsymbol{x}}_0 \mid \boldsymbol{y}_1, \ldots, \boldsymbol{y}_n) + 2q,$$

where $\hat{\theta}$ and $\hat{x}_0$ denote the maximum likelihood estimates. As before, we note that the AIC may be used to choose among different models, provided the estimates are derived from the conditional likelihood, described above.

17.4 Other Multivariate Models

The most widely used framework for the study of multivariate time series is the vector autoregressive integrated moving average (VARIMA) class of models. The VARIMA models have the general form

$$\boldsymbol{\Phi}(L)\boldsymbol{z}_t = \boldsymbol{\Theta}(L)\boldsymbol{\varepsilon}_t, \tag{17.4}$$

where L is the lag operator, and $\boldsymbol{z}_t = (1 - L)^d \boldsymbol{y}_t$ denotes the dth order difference. The operators $\boldsymbol{\Phi}(L)$ and $\boldsymbol{\Theta}(L)$ are matrix polynomial functions of the lag operator satisfying the usual stationarity and invertibility conditions. These models represent the direct vector extension of the univariate models presented in Sect. 11.1. As in Sect. 11.1.4, extensions can be made to include seasonal factors, but we prefer to keep the notation general and as simple as possible. It is worth noting that $\boldsymbol{\varepsilon}_t$ is an innovation vector that corresponds to the innovation vector used in the VES model. The frameworks ostensibly differ in that (17.4) contains no unobserved components. However, as we shall see in Sect. 17.4.1, the two approaches have close links.

The full VARIMA model (17.4) has not been widely used in applications, largely because order selection and estimation is difficult, although recent work by Athanasopoulos and Vahid (2008a) goes some way towards overcoming these problems.

A popular special case of the VARIMA model is the vector autoregressive (VAR) scheme:

$$\boldsymbol{\Phi}(L)\boldsymbol{z}_t = \boldsymbol{\varepsilon}_t,$$

introduced by Sims (1980). VAR models have proved very popular in econometrics and related areas, partly because they can be fitted by ordinary least squares provided the errors are taken to be independent. Another attractive feature is their ability to capture the dynamic relationships among the variables, an ability which is shared by all the models considered in this chapter. However, the restriction to VAR rather than VARMA may harm forecast accuracy (Athanasopoulos and Vahid 2008b). A potential drawback of VAR models is that when p lags are used, N^2p parameters must be estimated. State space models are usually more parsimonious in this respect.

Because the processes of interest are often nonstationary, one common approach in VAR model-building is to difference each nonstationary series before fitting the model. However, this amount of differencing may prove excessive because some series may move in tandem, such as the spot price and futures price series examined in Chap. 9. In these circumstances, we may consider cointegrated models (Engle and Granger 1987; Johansen 1988). For the purpose of exposition, we assume that all the series become stationary after a single differencing; the usual terminology is that the y_t are integrated of order 1 or $I(1)$. The differenced series, $z_t = y_t - y_{t-1}$, are then $I(0)$. The VAR model may then be rewritten in the form:

$$z_t = \boldsymbol{\Phi}_1^* z_{t-1} + \boldsymbol{\Psi} y_{t-1} + \varepsilon_t. \tag{17.5}$$

The matrix $\boldsymbol{\Psi}$ cannot be of full rank given the initial assumption that the z_t are $I(0)$. In general, if the rank of $\boldsymbol{\Psi}$ is $R < N$ we may write $\boldsymbol{\Psi} = \boldsymbol{\Gamma}\boldsymbol{\Lambda}'$, where $\boldsymbol{\Gamma}$ and $\boldsymbol{\Lambda}$ are $N \times R$ matrices of rank R. The vector $\boldsymbol{\Lambda}' y_t$ defines the R *cointegrating relationships*. The modified model

$$z_t = \boldsymbol{\Phi}_1^* z_{t-1} + \boldsymbol{\Gamma}\boldsymbol{\Lambda}' y_{t-1} + \varepsilon_t \tag{17.6}$$

is then known as the vector error-correction model (VECM). For further details of the theoretical developments see, for example, Hendry (1995, Chap. 8).

The performance of VAR and VECM models in econometric forecasting has been somewhat mixed; for a recent review see Allen and Morzuch (2006). Because the VECM models do not add new perspectives to vector innovations models, we will not examine them further, but concentrate upon the VAR models in our empirical comparisons.

17.4.1 Reduced Forms

The reduced forms of the VES models are obtained by eliminating the state variables between the measurement and transition equations, as in the univariate case. For example, for the local level model we eliminate the levels and arrive at the VARIMA(0,1,1) model $(1 - L)y_t = \varepsilon_t - \boldsymbol{\Theta}\varepsilon_{t-1}$, where $\boldsymbol{\Theta} = \boldsymbol{I} - \boldsymbol{A}$ and $\boldsymbol{I}$ is an identity matrix. A unique value of $\boldsymbol{\Theta}$ is associated with a given matrix $\boldsymbol{A}$, and vice versa. Thus, the vector local level model is equivalent to the VARIMA(0,1,1) model.

Likewise, the reduced form for the local trend model is found by double differencing the measurement equation, and then using the transition equations to eliminate the levels and growth rates, to give the VARIMA(0,2,2) model

$$(1 - L)^2 y_t = \varepsilon_t - \boldsymbol{\Theta}_1\varepsilon_{t-1} - \boldsymbol{\Theta}_2\varepsilon_{t-2},$$

where $\boldsymbol{\Theta}_1 = 2\boldsymbol{I} - \boldsymbol{A} - \boldsymbol{B}$ and $\boldsymbol{\Theta}_2 = \boldsymbol{A} - \boldsymbol{I}$. Again, both $\boldsymbol{\Theta}_1$ and $\boldsymbol{\Theta}_2$ are uniquely determined in terms of $\boldsymbol{A}$ and $\boldsymbol{B}$, and vice versa. Thus, the VES local trend and VARIMA(0,2,2) models are equivalent. Finally, we note that the vector damped local trend model and the VARIMA(1,1,2) model are equivalent. The proof is left as Exercise 17.1. The similarities with the univariate cases are evident.

17.4.2 Structural Models

The VES models may be contrasted with the multi-series structural time series model developed by Harvey (1989) that has multiple sources of randomness for each series. It takes the general form

$$\begin{aligned} \boldsymbol{y}_t &= \boldsymbol{\Omega} \boldsymbol{x}_t + \boldsymbol{u}_t, \\ \boldsymbol{x}_t &= \boldsymbol{F} \boldsymbol{x}_{t-1} + \boldsymbol{v}_t, \end{aligned}$$

where the N-vector $\boldsymbol{u}_t$ and the k-vector $\boldsymbol{v}_t$ are disturbances that act as $N + k$ primary sources of randomness. Unlike the innovations form, the unobserved components vector is not lagged in the measurement equation. Typically, the structure matrices of the two models are related by $\boldsymbol{W} = \boldsymbol{\Omega} \boldsymbol{F}$. The disturbance vectors are contemporaneously and inter-temporally uncorrelated so that the variance matrices of the disturbance vectors are diagonal.

Thus, the local level model becomes:

$$\begin{aligned} \boldsymbol{y}_t &= \boldsymbol{\ell}_t + \boldsymbol{u}_t, \\ \boldsymbol{\ell}_t &= \boldsymbol{\ell}_{t-1} + \boldsymbol{v}_t. \end{aligned}$$

Although there are some close parallels with the innovations form, the links are not as direct. The levels can be eliminated to give the reduced form $(1 - L)\boldsymbol{y}_t = \boldsymbol{u}_t - \boldsymbol{u}_{t-1} + \boldsymbol{v}_t$. The right hand side of this reduced form is the sum of two moving average processes. According to the Granger–Newbold (1986) theorem, the sum of moving average processes is itself a moving average process. Thus, this reduced form is also a VARIMA(0,1,1) process.

Despite these processes having the same reduced form, the multiple source of randomness specification is more restrictive than the innovations approach. In particular, a comparison of the auto-covariance structure shows that the multi-disturbance specification has a smaller parameter space and is of lower dimension. Furthermore, these restrictions become even tighter as more components are added (Harvey 1989, p. 432).

From this observation and our earlier comments, the following conclusions may be drawn:

1. The multi-disturbance vector local level model is equivalent to a *restricted* VARIMA(0,1,1) process
2. The VES local level model is equivalent to a VARIMA(0,1,1) process without restrictions apart from the usual invertibility conditions
3. The VES local level model is more general than the multi-disturbance vector local level model
4. The multi-disturbance vector local level model always has an equivalent innovations local level model

The multi-disturbance vector local trend model is

$$\begin{aligned} \boldsymbol{y}_t &= \boldsymbol{\ell}_t + \boldsymbol{u}_t, \\ \boldsymbol{\ell}_t &= \boldsymbol{\ell}_{t-1} + \boldsymbol{b}_{t-1} + \boldsymbol{v}_t, \\ \boldsymbol{b}_t &= \boldsymbol{b}_{t-1} + \boldsymbol{w}_t. \end{aligned}$$

This model can also be reduced to the VARIMA(0,2,2) form; see Exercise 17.2. As for the local level model, the parameter space is restricted. Similar comments apply to the damped local trend model.

17.5 Application: Exchange Rates

To gauge the forecasting capacity of the VES framework and to compare it with commonly used alternatives, we summarize the results of an empirical study reported in de Silva et al. (2007). These authors developed a VES model for the monthly exchange rate time series of the UK pound (UKP) and US dollar (USD) against the Australia dollar (AUD); see Fig. 17.1. The expectation was that changes in economic conditions in Australia could affect both exchange rates simultaneously, and so create interdependencies between them. However, we believe that these effects should manifest themselves through the state variables after a lag, rather than contemporaneously, so we employ unrestricted coefficient matrices but a diagonal variance matrix.

The data comprised 77 monthly observations spanning the period January 2000 to May 2006. The natural logarithm of the series was taken before the models were fitted. Models were fitted to the first 60 observations.

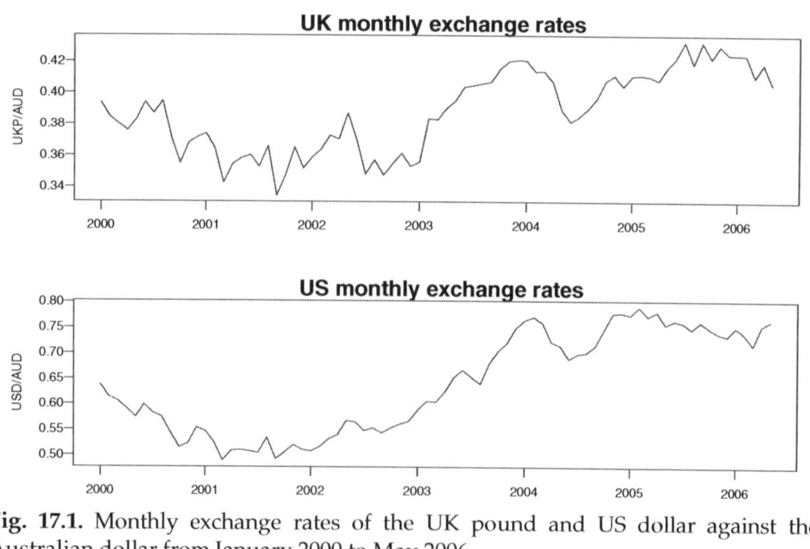

Fig. 17.1. Monthly exchange rates of the UK pound and US dollar against the Australian dollar from January 2000 to May 2006.

The forecasting performances of each model were evaluated on the 17 withheld observations using the mean absolute scaled error (MASE) defined in Sect. 2.7.3 (p. 26). We define the absolute scaled error (ASE) for the kth replicate of the ith series as

$$\text{ASE}(i,k) = \frac{|e_{i,n+k}|}{\frac{1}{n-1}\sum_{t=2}^{n}|y_{i,t} - y_{i,t-1}|},$$

where $e_{i,n+k} = y_{i,n+k} - \hat{y}_{i,n+k}$. The overall MASE for the system of equations, averaged across all forecasting horizons, is then given by

$$\text{MASE} = \frac{1}{Nh}\sum_{i=1}^{N}\sum_{k=1}^{h}\text{ASE}(i,k),$$

with $N = 2$ and $h = 17$.

The study examined seven distinct models: random walk, local level, local trend, damped local trend, and VAR models of orders one, two and three. The series were differenced before fitting the VAR models. For the local level and trend models, both univariate and multivariate models were considered. In the multivariate case, both the traditional and innovations models were also explored. The results are summarized in Table 17.1.

The first conclusion that can be drawn is that there is indeed some structure in the series, so that improvements can be achieved over the random walk model. Second, the VES models out-perform their traditional counterparts due to the ability of VES models to encompass a larger parameter space. Third, the VES damped trend model forecasts better than the VAR approach for this particular set of data. As shown in Table 17.2, the AIC measure would lead to the selection of this model from the three innovations schemes.

The poor performance of the local trend models probably reflects the fact that these series, like most macroeconomic series, are I(1) rather than I(2). Further, the two series generally moved in the same direction over the fitting sample, but in opposite directions in the test sample; see Fig. 17.2.

Table 17.1. Comparison of approaches: overall MASE by model.

	Multi-series traditional	Uni-series innovation	Multi-series innovation
Random walk		0.166	
Local level	0.174	0.127	0.150
Local trend	0.589	0.119	0.329
Damped local trend	0.265	0.130	**0.114**
VAR(1)			0.157
VAR(2)			0.143
VAR(3)			0.172

The bolded figure denotes the minimum value.

Table 17.2. Akaike Information Criterion for each of the fitted VES models.

Model	AIC
Vector local level	−13.608
Vector local trend	−13.628
Vector damped local trend	−13.763

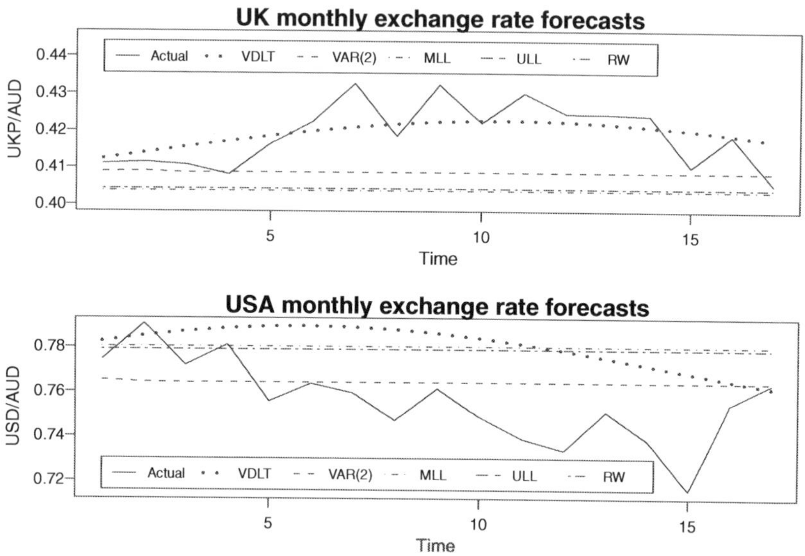

Fig. 17.2. Predicted monthly exchange rates.

The parameter estimates of the vector damped local trend model are:

$$\boldsymbol{\Sigma}(\times 1{,}000) = \begin{bmatrix} 0.787 & 0 \\ 0 & 0.898 \end{bmatrix}, \quad \boldsymbol{A} = \begin{bmatrix} 0.476 & 0.445 \\ -0.086 & 1.246 \end{bmatrix},$$

$$\boldsymbol{\Phi} = \begin{bmatrix} 0.827 & 0.203 \\ -0.272 & 1.136 \end{bmatrix}, \quad \boldsymbol{B} = \begin{bmatrix} 0.102 & -0.238 \\ 0.041 & -0.135 \end{bmatrix}.$$

In general, the parameters within $\boldsymbol{\Phi}$, $\boldsymbol{A}$ and $\boldsymbol{B}$ denote various elasticities. It is difficult to draw any specific conclusions using the parameter estimates because the states and series interact. Importantly, the modulus of the largest eigenvalue is less than one (albeit marginally), and therefore the dampening characteristic of the trend is captured. de Silva et al. (2007) use impulse response functions (see Lütkepohl 2005) to examine the various inter-relationships in the model in greater detail.

17.6 Forecasting Experiment

The conclusions in the previous section were based upon just one pair of series. In order to gain a deeper understanding of the properties of VES models, de Silva et al. (2007) conducted a detailed study using 1,000 different data sets. The variables and their starting dates were randomly chosen from the Watson (2003) macroeconomic database. The Watson (2003) database comprises eight groups which can be loosely considered to represent different economic sectors. The number of variables in each group ranges from 13 to 27. All variables are real and non-seasonal, with observations from January 1959 to December 1998. Every data set in the de Silva et al. (2007) study consisted of two variables, chosen randomly from different economic sectors. The starting date was randomly chosen, where the only restriction was that there must be enough observations to fit and evaluate out-of-sample forecasts up to twelve periods ahead.

Two sizes of estimation sample were considered: 30 and 100. These sample sizes were chosen because they resemble small and large sample sizes that occur in practice. All variables were standardized by dividing by the standard deviation of their first differences.

The overall conclusions from the de Silva et al. (2007) study may be summarized as follows:

1. There are definite advantages to using a multivariate rather than a univariate framework.
2. The VES models perform slightly better than their traditional multi-disturbance time series counterparts (a function of the relevant parameter spaces).
3. The predictive ability of the VES models is comparable to that of the VAR models.

17.7 Exercises

Exercise 17.1. Show that the VES damped local trend model given in (17.3) reduces to the VARIMA(1,1,2) scheme with moving average matrices $\boldsymbol{\Theta}_1 = \boldsymbol{I} + \boldsymbol{\Phi} - \boldsymbol{A} - \boldsymbol{B}$ and $\boldsymbol{\Theta}_2 = \boldsymbol{\Phi}(\boldsymbol{A} - \boldsymbol{I})$.

Exercise 17.2. The multi-disturbance vector local trend model is

$$\begin{aligned} \boldsymbol{y}_t &= \boldsymbol{\ell}_t + \boldsymbol{u}_t, \\ \boldsymbol{\ell}_t &= \boldsymbol{\ell}_{t-1} + \boldsymbol{b}_{t-1} + \boldsymbol{v}_t, \\ \boldsymbol{b}_t &= \boldsymbol{b}_{t-1} + \boldsymbol{w}_t. \end{aligned}$$

Show that this model can be reduced to an equivalent VARIMA(0,2,2) model, $(1 - L)\boldsymbol{y}_t = \boldsymbol{w}_t + (\boldsymbol{v}_t - \boldsymbol{v}_{t-1}) + (\boldsymbol{u}_t - 2\boldsymbol{u}_{t-1} + \boldsymbol{u}_{t-1})$. Use the Granger–Newbold addition theorem to establish that the first-order autocovariance is always non-positive, and the second-order autocovariance is always positive.

Exercise 17.3. The auto-covariance structure for the VARIMA(0,1,1) model is:

$$\begin{aligned}
\boldsymbol{\Gamma}(0) &: \mathrm{E}(\Delta\boldsymbol{y}_t, \Delta\boldsymbol{y}_t) = \boldsymbol{\Sigma} + \boldsymbol{\Theta}\boldsymbol{\Sigma}\boldsymbol{\Theta}', \\
\boldsymbol{\Gamma}(1) &: \mathrm{E}(\Delta\boldsymbol{y}_t, \Delta\boldsymbol{y}_{t-1}) = -\boldsymbol{\Sigma}\boldsymbol{\Theta}', \\
\boldsymbol{\Gamma}(k) &: \mathrm{E}(\Delta\boldsymbol{y}_t, \Delta\boldsymbol{y}_{t-k}) = \boldsymbol{O}, \quad k > 1,
\end{aligned}$$

where $\Delta\boldsymbol{y}_t = -\Theta\boldsymbol{e}_{t-1} + \boldsymbol{e}_t$.

a. Derive the auto-covariance structure for the multi-disturbance structural time series model:
$$(1 - L)\boldsymbol{y}_t = \boldsymbol{u}_t - \boldsymbol{u}_{t-1} + \boldsymbol{v}_t.$$
b. Compare the parameter space of the first auto-covariance derived in (a) to the unrestricted VARIMA(0,1,1) model.
c. Calculate the number of parameters to be estimated for the VARIMA (0,1,1) and the multiple disturbance structural time series model (set $N = 3$).

Part IV

Applications

18

Inventory Control Applications

Since the pioneering work of Brown (1959), it has been a common practice to use exponential smoothing methods to forecast demand in computerized inventory control systems. It transpired that exponential smoothing often produced good point forecasts. However, the methods proposed to measure the risk associated with the predictions typically ignored the effect of random changes to the states, and so seriously underestimated the level of risk as a consequence (Johnston and Harrison 1986; Snyder et al. 1999; Graves 1999). The innovations state space model provides the statistical underpinnings of exponential smoothing and may be used to derive measures of prediction risk that are properly consistent with the use of these forecasting methods, and which, as a consequence, allow for random changes in the states.

There is often, however, a twist to the problem of predicting demand in inventory systems. Predictions are typically made from sales data that are recorded for accounting purposes. Sales, however, may be lost during shortages, in which case sales are a corrupted reflection of demand. Without proper precautions, the use of sales data can lead to forecasts of demand with a downward bias. This problem is considered in Sect. 18.1.

Once obtained, predictions of demand and the uncertainty surrounding them are used as inputs to replenishment decisions. The details of how this is done depends in part on the decision rules employed to determine the timing and size of replenishment orders. There is an extensive literature on inventory control; see, for example, Silver et al. (1998) for a comprehensive coverage. Section 18.2 provides some insights into the problem of properly integrating the demand models underlying the exponential smoothing methods of forecasting with common inventory models.

18.1 Forecasting Demand Using Sales Data

The methods of estimation described in Chap. 5 may not always be suitable for application in an inventory control context. They were implicitly based on the assumption that the series, in this context the demand series, is observed without error. This assumption may not always be true.

Businesses record transactions with their customers for accounting purposes, but, because there are no transactions during shortages, lost sales often go unrecorded. Thus, sales data are typically an incomplete record of demand. During those periods when there are shortages, sales understate demand. In the parlance of statistics, this is the problem of forecasting with "censored data."

We conducted a small simulation study to gauge the effects of truncation when the estimation methods of Chap. 5 are applied. In this study, demand series of lengths $n = 36$ and $n = 72$ were generated from a local level model with $\ell_0 = 100$ and an innovations standard deviation of $\sigma = 10$. Experiments were completed for a grid of values of the persistence parameter α from 0.1 to 1.0 in increments of 0.1, and a variety of truncation levels. Truncation occurred at $z = 0, 0.67, 1.65$ and 100 standard deviations above the underlying levels (local means). The last case $z = 100$ was included as a benchmark corresponding to the case where there is (effectively) no truncation.

The primary aim of the study was to measure the effect of truncation on the moments of the prediction distributions of *aggregate* or lead-time demand $Y_n(h) = y_{n+1} + \cdots + y_{n+h}$ over various possible lead-times h. For a local level model, the aggregate demand has a mean (Sect. 6.6.1)

$$\mathrm{E}(Y_n(h) \mid \ell_n) = h\ell_n$$

and variance [(6.16)]

$$\mathrm{V}(Y_n(h) \mid \ell_n) = \sigma^2 h\Big[1 + \alpha(h-1) + \tfrac{1}{6}\alpha^2(h-1)(2h-1)\Big].$$

The cases $h = 5$ and $h = 9$ periods were examined in the study. Moreover, the special case $h = 1$ was also included to cover the moments of the conventional one-step ahead prediction distribution.

The prediction distributions from the estimated models were benchmarked against the prediction distributions from the "true" models of demand. More specifically, each moment from an estimated model was divided by the corresponding moment from the true model to give an error index. Each experiment was replicated 1,000 times, and the median index, a measure of bias, was calculated. These median indexes, averaged over all the αs, are reported in Table 18.1. A value of 1 means that there is no bias; a value less than 1 indicates that there is a downward bias.

Table 18.1 indicates, somewhat surprisingly, that mean lead-time demands are effectively unbiased. All standard deviations, on the other hand,

Table 18.1. The effect of using sales data instead of demand data on predictions of lead-time demand using conventional simple exponential smoothing, as reflected by bias ratios.

	Lead-time	z	$n = 36$	$n = 72$
Mean	1	0	0.98	0.98
		0.67	1.00	1.05
		1.65	1.00	1.01
		100	1.00	0.99
	5	0	0.98	0.96
		0.67	1.00	0.99
		1.65	1.00	1.00
		100	1.00	1.00
	9	0	0.98	0.99
		0.67	1.00	1.00
		1.65	1.00	1.00
		100	1.00	1.00
Stdev	1	0	0.77	0.79
		0.67	0.91	0.93
		1.65	0.96	0.98
		100	0.96	0.98
	5	0	0.81	0.85
		0.67	0.87	0.92
		1.65	0.91	0.95
		100	0.91	0.95
	9	0	0.83	0.88
		0.67	0.87	0.92
		1.65	0.88	0.94
		100	0.88	0.94

have a downward bias, but this bias contracts as the degree of truncation decreases or the sample size n increases.

Problems involving censored data have been considered by Robinson (1980) and Park et al. (2007) for ARMA processes. In an inventory control context, Nahmias (1994) and Agrawal and Smith (1996) have proposed methods of estimation in the lost sales case. None of these approaches relate directly to the case where demand processes are represented by linear innovations state space models, so a new estimation procedure is needed.

Because demand can be understated, we develop an approach which augments sales by an adjustment factor in those periods where there is a shortage. The logic of the adjustment factor is as follows. At the *start* of period t, demand y_t is uncertain and has a distribution represented by $f_{t|t-1}(\cdot)$, with a mean corresponding to the one-step-ahead prediction $\hat{y}_{t|t-1}$ and a variance σ^2. At the *end* of period t, the sales value ζ_t is observed. If there is no shortage, demand becomes the fixed quantity $y_t = \zeta_t$. If there is a shortage, demand remains uncertain but its density becomes

$$\frac{f_{t|t-1}(y)}{\int_{\zeta_t}^{\infty} f_{t|t-1}(y)\, dy}$$

over the domain $y_t \geq \zeta_t$. The expected understatement of demand in such a period is given by

$$c_t = \mathrm{E}(y_t - \zeta_t) = \frac{\int_{\zeta_t}^{\infty} (y - \zeta_t) f_{t|t-1}(y) dy}{\int_{\zeta_t}^{\infty} f_{t|t-1}(y) dy}. \tag{18.1}$$

In the case where the innovations have a Gaussian distribution, c_t becomes (Exercise 18.1)

$$c_t = \frac{\sigma \left[\phi(z_t) - z_t \int_{z_t}^{\infty} \phi(u) du \right]}{\int_{z_t}^{\infty} \phi(u) du}, \tag{18.2}$$

where $z_t = (\zeta_t - \hat{y}_{t|t-1})/\sigma$ and $\phi(u)$ is a standard Gaussian density function.

Algorithm 1 *The parameters of the demand model are estimated using a minimum sum of squared errors criterion employing the associated version of exponential smoothing to generate the required errors. They are initially estimated directly from the sales data to give an initial estimate of the standard deviation σ and initial values for the one-step ahead predictions $y_{t|t-1}$. This process is then repeated on a* demand *series (not the sales series) constructed as follows:*

1. *Demand is set equal to sales for those periods without a shortage.*
2. *Demand is set equal to sales* plus *the non-negative correction factor c_t calculated with equation* (18.2) *for those periods with a shortage, the current estimate of the standard deviation and the current one-step ahead predictions being used for the calculations.*

Both steps are repeated as each new set of estimates emerge. They are continued until the estimates of the standard deviation converge to a fixed value.

The simulation study, based on a local level model, was extended to evaluate this augmented sales approach. The algorithm was terminated after the change in the estimates of the standard deviation dropped below 5%. The results are presented in Table 18.2. Again, the mean seems to be effectively unbiased. The standard deviation still has a downward bias, but a comparison with the original results in Table 18.1 suggests that this has been reduced.

The study shows that the bias depends on the persistence parameter α. Table 18.3 summarizes the situation for both conventional and augmented exponential smoothing for a selection of values of α. First, even in the case where exponential smoothing is applied to demand rather than sales data,

Table 18.2. Biases for augmented simple exponential smoothing.

	Lead-time	z	$n = 36$	$n = 72$
Mean	1	0	1.02	1.03
		0.67	1.00	1.05
		1.65	1.00	1.01
		100	1.00	0.99
	5	0	1.06	1.07
		0.67	1.00	0.99
		1.65	1.00	1.00
		100	1.00	1.00
	9	0	1.03	1.02
		0.67	1.00	1.00
		1.65	1.00	1.00
		100	1.00	1.00
Stdev	1	0	1.03	1.06
		0.67	0.97	0.99
		1.65	0.96	0.98
		100	0.96	0.98
	5	0	0.92	0.97
		0.67	0.91	0.96
		1.65	0.91	0.96
		100	0.91	0.95
	9	0	0.89	0.95
		0.67	0.89	0.94
		1.65	0.89	0.94
		100	0.88	0.94

there is a tendency to under-estimate the standard deviation of lead-time demand. This problem is exacerbated when sales data are used, becoming quite acute when there is severe truncation. The augmented method succeeds in reducing the bias, particularly for lower levels of α. Nevertheless, the bias is not completely eliminated.

The augmentation strategy is a step forward in the quest for more reliable forecasting practices when only sales data, rather than demand data, are available. Nevertheless, the simulation results indicate that there is still room for improvement. The most concerning thing revealed by the simulation study is that the problems are most acute for small values of α. Brown (1959) and others have suggested that small values of α are typical in inventory applications. We have here a serious practical problem that is yet to be properly resolved.

A useful step towards a resolution of the problem, from a practical point of view, is to measure lost sales during shortages. In systems where customer demands are revealed through a computerized ordering process, it is possible to detect shortages in real time and record demands even when they are not satisfied. In this way it is possible to obtain a precise measurement of

Table 18.3. The relationship between bias in standard deviation (averaged across lead-times and sample sizes) and the persistence parameter α.

z	α	Method	
		Conventional	Bias-corrected
0	0.1	0.61	0.79
	0.3	0.75	0.90
	0.7	0.88	1.05
	1	0.95	0.95
0.67	0.1	0.77	0.88
	0.3	0.87	0.94
	0.7	0.96	0.96
	1	0.95	0.95
1.65	0.1	0.88	0.90
	0.3	0.91	0.92
	0.7	0.96	0.96
	1	0.95	0.95
100	0.1	0.89	0.89
	0.3	0.92	0.92
	0.7	0.96	0.96
	1	0.95	0.95

demand, and so avoid the use of sales data altogether. Moreover, with modern computer technologies, the additional cost of operating a scheme like this is typically negligible. Nevertheless, the study reveals that even when using demand data, the downward bias problem does not completely disappear.

18.2 Inventory Systems

18.2.1 Basic Features of Inventory Control Systems

Decision rules used to determine the timing and size of replenishment orders for inventory systems must account for a number of features. First, demands are uncertain unless customers place forward orders and are prepared to wait, in which case demands are known with complete certainty. In the common case where forward orders are not placed, it makes little sense to match supply exactly to predicted demand. The predicted demand corresponds to the mean of a prediction distribution, and if this distribution is symmetric, as with the Gaussian distribution, demand will exceed supply 50% of the time. A service level of this order is unacceptable, and so an inventory decision rule should be designed to ensure that the stock lies above the predicted level of demand. The additional stock is referred to as safety stock. The problem is to determine its size.

Although supply is normally above the predicted demand, there are circumstances where a supply below the predicted demand makes sense (Snyder 1980).

Second, it should be recognized that an inventory system operates over time and that replenishment decisions may be made either periodically or after each customer transaction. The act of checking the state of a system is called a *review*. If reviews occur daily, weekly, monthly or at some other fixed interval, the system is said to use a *periodic review* control policy. If they occur after each transaction, it is said to use a *perpetual (continuous) review* policy. Because most of the methods of forecasting considered in this book have been predicated on measurements over fixed intervals, the focus of this chapter is restricted to periodic review systems. It is assumed that periods over which demand is collated correspond to the review periods.

It should also be recognized that replenishment orders need not be placed at a review if the state of the inventory system is considered to be satisfactory. There may be no restrictions on the size of the replenishment order, in which case it normally makes sense to order at each review. In some circumstances, however, inventories may be delivered in standard quantities governed by pack sizes and the capacities of available distribution technologies. Reordering and delivery costs also imply the need for larger order quantities. Moreover, costs associated with resetting machines in batch manufacturing may lead to long production runs and large batch sizes. In addition, suppliers may offer discounts for purchasing in larger quantities. In each of these situations, it may be necessary to defer the placement of replenishment orders until a standard replenishment quantity Q is needed.

The ideal initial stock is referred to as the *order-up-to level* S. When there are standard replenishment quantities, replenishment orders are deferred until the stock drops below a critical level called the *reorder level* R. In such situations, the gap between the order-up-to level and the reorder level equals the standard replenishment quantity Q. It is thus possible to distinguish two basic kinds of periodic inventory control systems: those that have no restrictions on the replenishment quantity and that rely on an order-up-to level alone; and those that rely on a reorder level in addition to the order-up-to level. The two systems are referred to as the *order-up-to level inventory control system* and the *reorder level inventory control system* respectively. An order-up-to level inventory system is a special case of a reorder level system where $Q = 1$.

If required, a replenishment order is placed the instant that a review is completed. Typically, however, such an order is not delivered immediately. The delay until delivery, designated by L, is called the *delivery lead-time*, and it may extend over many review periods. It is convenient to assume that it is an exact multiple of the review period and that a delivery, possibly from a much earlier order, can occur in the instant following the placement of an order.

The implication of a delivery lead-time is that the decision maker must think ahead. An order placed at time t is not delivered until time $t + L$ and so cannot have any effect on stock levels until the period beginning at time $t + L$. The decision maker has no control in the intervening period covered by the delivery lead-time.

Another detail concerns what customers do during shortages. One possibility, considered in the previous section, is the *lost sales* case. Another possibility is that customers are prepared to wait until their demands can be satisfied following a future replenishment of the system. The quantity of demand awaiting satisfaction is called the *backlog*. Many businesses are confronted with a mix of these customer behaviors. For modeling purposes, however, it is simplest to assume that there is no mix. The primary focus of this chapter will continue to be on the lost sales case.

18.2.2 System Dynamics

Stock changes over time. It is decreased by sales and increased by deliveries. A delivery d_t immediately following a review at time t causes a jump in the stock, so that pre-review stock $\bar{s}_t$ and post-review stock s_t are related by

$$s_t = \bar{s}_t + d_t. \tag{18.3}$$

During the subsequent review period $t + 1$, sales ζ_{t+1} deplete the stock, so that by the end of the review period, at time $t + 1$, the stock is given by

$$\bar{s}_{t+1} = s_t - \zeta_{t+1}. \tag{18.4}$$

Sales cannot exceed the initial stock s_t or the demand y_{t+1}, so that

$$\zeta_{t+1} = \min(s_t, y_{t+1}). \tag{18.5}$$

The quantity

$$P_t = \bar{s}_t + \sum_{j=0}^{L} q_{t-j} \tag{18.6}$$

is called the *provision*. It consists of the pre-review stock $\bar{s}_t$ at the current time t, the outstanding orders $q_{t-1}, \ldots, q_{t-L}$ and the new order q_t. The provision immediately before the review, designated by $\bar{P}_t$, is given by

$$\bar{P}_t = \bar{s}_t + \sum_{j=1}^{L} q_{t-j}. \tag{18.7}$$

The pre-review provision is known to the decision maker at time t. It can be increased by an order q_t placed by the decision maker at time t to

$$P_t = \bar{P}_t + q_t. \tag{18.8}$$

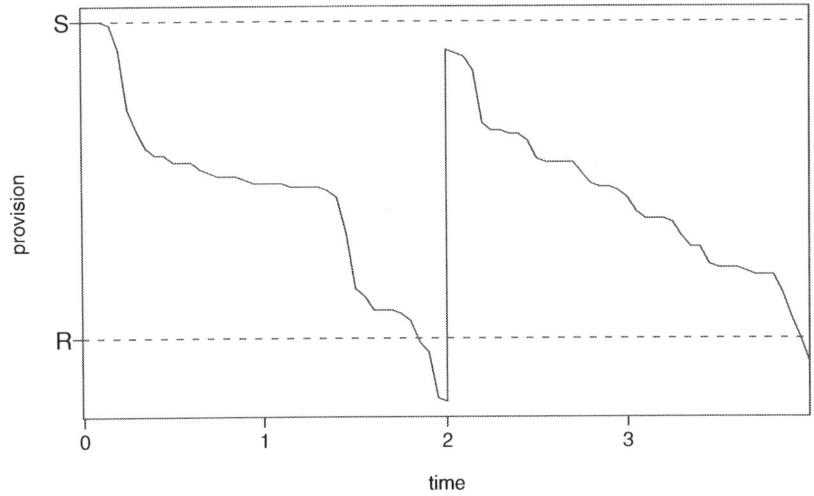

Fig. 18.1. Stationary reorder level system.

The provision is therefore not only known, but it is also *controllable*. The provision represents a system's capacity to meet future demands. By increasing the provision at time t through the placement of a new order, the chance of a shortage in the period immediately after its delivery is reduced. It therefore makes sense to use the provision to signal the state of the inventory system.

The operation of a typical reorder level system is depicted in Fig. 18.1. In reorder level systems, orders are placed at those reviews where the provision is found to be below the reorder level R. Enough is then ordered to raise the provision above the reorder level but not above the order-up-to level. Orders are a multiple of Q. At those reviews where no order is placed, the period begins with a provision determined by what happened in earlier periods.

18.2.3 Nonstationary Demands

Constant ordering parameters such as R and S make sense when demands are stationary and independent. However, demands are typically nonstationary and autocorrelated, and so the ordering parameters need to change over time in response to changing underlying conditions. When demands are seasonal, for example, it makes sense to adapt R and S to the demand levels that prevail in each season. Why carry high stocks during those seasons when demands are low? When demands are autocorrelated there is then a tendency for low demands to be followed by low demands. It makes sense to use lower reorder levels during such periods.

Graves (1999) considered the operation of an order-up-to level system when demands are governed by a local level model under the assumption of backlogging of demands during shortages. The order-up-to level was made time-dependent in such a way that the order q_t at the beginning of period $t+1$ is determined by the formula $q_t = y_t + h\alpha\varepsilon_t$, where $h = L+1$. As in the stationary case, the order replaces the quantity demanded in period t. However, the term $h\alpha\varepsilon_t$ is now added to account for the effect of the permanent change in the underlying demand level on the order-up-to level.

Here we also propose an adaption of traditional stationary inventory theory. We consider a reorder level system, of which the order-up-to level system is a special case, where the reorder level R is made to depend on time. In addition, we consider the lost-sales rather than the backlog situation. In doing this, another theory of inventory control emerges that is compatible with the traditional exponential smoothing methods of forecasting.

The reorder level is determined by the formula

$$R_t = \hat{Y}_t(L+1) + \Delta, \tag{18.9}$$

where $\hat{Y}_t(L+1)$ is the predicted demand over the period $(t, t+L+1)$ and Δ is the safety stock. The prediction is a quantity that varies over time and induces the required change in the reorder level. The safety stock, when positive, is the extra provision needed to meet demand above the predicted level. A consequence of increasing the size of Δ is to reduce the likelihood and size of lost sales in the period following a delivery.

The order-up-to level is always Q units above the reorder level, so that

$$S_t = R_t + Q. \tag{18.10}$$

The order quantity is determined with the rule

$$q_t = \left\lceil \frac{(R_t - \bar{P}_t)^+}{Q} \right\rceil Q, \tag{18.11}$$

where $\lceil\rceil$ is the ceiling operator[1] and the superscript $+$ is the positive part operator.[2]

It is a common practice to use the fill rate as a measure of the performance of an inventory system. The fill rate is the proportion formed by the ratio of sales to demand. Thus, a fill rate of 90% means that sales represent 90% of demand, in which case lost sales have amounted to 10% of demand. A common practice in inventory theory based on stationary demands, is to focus on

[1] The ceiling operator rounds a number to its nearest integer value in an upwards direction. Thus, $\lceil 3.2 \rceil = 4$.

[2] $x^+ = x$ if $x \geq 0$ and 0 otherwise.

the fill rate of a representative review period when the system has reached a steady state. However, inventory systems with nonstationary demands never reach a steady state, and the expected fill rate can change from one period to the next as the expected review period demand changes. In these circumstances we recommend using a fill rate measured over the period of a year to indicate the performance of a system. In the case of demands with seasonal effects, it allows the system to be evaluated over the full span of a seasonal cycle. Even for non-seasonal demands there is an advantage. It typically allows a number of inventory cycles to occur so that an *average* performance is gauged.

Therefore, in the case of the reorder level system, the aim is to find a value of the safety stock Δ that achieves a specified target for the annual fill rate. When there are τ periods per year, the annual fill rate r_n is governed by the formula

$$r_n = \sum_{t=n+L+1}^{n+L+\tau} \zeta_t \Big/ \sum_{t=n+L+1}^{n+L+\tau} y_t,$$

where n is the size of the sales sample. The lead-time L is included because any order placed at the beginning of period $n+1$ cannot influence the system performance until it is delivered at time $n+L$. Once the safety stock is determined, it can be used in conjunction with the predictions of demand over the planning horizon, to find the ordering parameters R_t and S_t with (18.9).

The situation is too complex to permit the development of an analytical method for determining the safety stock. Lost sales mean that no simple formula can be derived that summarizes the way the provision affects lost sales. Moreover, in nonstationary cases, the provision can become temporarily stranded above the order-up-to level, implying that a more complex multi-period approach is necessary in place of the traditional representative period approach. It appears that simulation is the only viable option.

We now outline a procedure for safety stock determination that may be applied using either the order-up-to or reorder level rules. It is predicated on the assumption that sales have been recorded for periods $1, 2, \ldots, n$, and it consists of the following basic steps.

Algorithm 2 (Safety Stock Determination)

Step 1: *Fit the relevant additive innovations models from Chap. 2 to the sales $\zeta_1, \zeta_2, \ldots, \zeta_n$ using the estimation method (Algorithm 1 from Sect. 18.1), and, where appropriate, employ a model selection procedure from Chap. 7 to select the best model.*

Step 2: *Simulate M series of demands over the future periods $n+1, n+2, \ldots, n+L+\tau$ with the selected model from Step 1. Denote the ith replication of the series value in future period t by y_{ti}.*

Step 3: *For a trial value of the safety factor* Δ, *and for each series* $i = 1, 2, \ldots, M$ *from Step 2, calculate the the annual fill rate*

$$r_{ni} = \sum_{t=n+L+1}^{n+L+\tau} \zeta_{ti} \Big/ \sum_{t=n+L+1}^{n+L+\tau} y_{ti},$$

where ζ_{ti} *is the sales in period t for replicated series i.*

Step 4: *Estimate the mean annual fill rate by averaging the fill rates obtained at Step 3.*

Step 5: *Repeat Steps 3 and 4 for a succession of trial values of the safety factor without changing the future demands from Step 2, until a value is found that achieves the specified sample fill-rate.*

Step 3 of this procedure relies on an inventory model consisting of (18.3), (18.4), (18.5), (18.7), (18.8), (18.9), (18.10) and (18.11). The inventory equations are initialized with the actual stock and outstanding orders at the end of period n. This step must also account for the potential future role of the decision maker. Part of this role involves the prediction of demand using exponential smoothing as an input to the determination of the reorder level. It is necessary to recognize in the simulation that the decision-maker can observe only sales, as determined by (18.5), rather than demand. It is therefore assumed that the decision-maker adjusts sales in those future periods with a shortage using the correction factor 18.2 before revising the prediction with exponential smoothing.

In Step 5, the fill rate is a function of the safety stock Δ. The choice of Δ to meet a target fill rate involves the solution of a nonlinear equation. A bisection search procedure (Press et al. 2002) typically works well. As the fill-rate function is continuous, this method always finds a solution. However, because the function may not be monotonic, it is conceivable that it has multiple solutions. In this situation, one would ideally want to select the smallest solution, but this is not guaranteed by the search procedure. Experience suggests that any dips in the function that destroy its monotonicity tend to be relatively small, and so multiple solutions (if they exist) are likely to be fairly close together; any over-stocking caused by this problem is likely to be relatively small in practice.

Algorithm 2 involves considerable computational loads when repeated many times on thousands of different inventories. A business might contemplate the prospect of applying it annually to re-estimate model parameters and to revise the level of the ideal safety stock for each inventory. In between, stocks might be reviewed on a weekly basis. At the beginning of each week the reorder level would be recalculated with the rule (18.9). The states of the demand model would be updated to reflect any new information provided by the previous week's sales using the augmented version of exponential smoothing (Algorithm 1). The prediction of lead-time demand needed for reorder level determination would be calculated. The decision rule (18.11) would then be used to determine the order quantity.

18.3 Exercises

Exercise 18.1. Derive (18.2) using the relationship $\int_z^\infty u\phi(u)du = \phi(z)$.

Exercise 18.2. The monthly sales for a product and shortage indicators are provided in the data set `msales`:

a. Fit a local linear model directly to the sales data to obtain initial estimates of the innovations standard deviation and the conditional means (one-step ahead predictions).
b. Using the results from (a), together with the adjustment formula (18.2), construct a demand series.
c. Fit a local linear model directly to the demand data from (b) to obtain revised estimates. Has the estimate of the standard deviation increased?

Exercise 18.3. Weekly demand for a product is *stationary*, having a Gaussian distribution with a mean of 2,000 and a standard deviation of 100. Replenishment orders are delivered immediately so there is no delivery lead-time. An order-up-to level inventory system is to be used with the a goal of a 90% fill-rate in a representative future week. The purpose of this exercise is to demonstrate that the appropriate value of the order-up-to level is about 1,799. Undertake the following steps to achieve this end:

a. Simulate a sample of 100 possible demands for a representative week from a Gaussian distribution with a mean of 2,000 and standard deviation of 100.
b. Set a trial value for the order-up-to level (supply) of 1,500 and derive the corresponding shortages; demonstrate that the average fill-rate is about 75%.
c. Now use a solver to find that value for the order-up-to level which achieves the 90% fill-rate goal.

Exercise 18.4. When the process generating monthly demands is non-stationary, it is no longer possible to focus on a representative week. Nor does it make sense to use a fixed order-up-to level: the focus changes to the determination of a fixed level of safety stock. Furthermore, the goal is changed to achieving a specified average fill-rate over a year. Suppose that monthly demand is governed by a local level model with $\ell_n = 1{,}000$, $\sigma = 100$ and $\alpha = 0.2$:

a. Use the local level model to simulate ten possible demand series over the next 12 months.
b. Set a trial value of 100 for the safety stock and show that the fill rate averaged across the 10 future demand scenarios is about 96% when the delivery lead-time is zero (stocks above an order-up-to level can be returned to the supplier without incurring additional costs).
c. Demonstrate that a safety stock of about -30 achieves a 90% fill-rate. (Yes, safety stock can be negative!).

19

Conditional Heteroscedasticity and Applications in Finance

In 1900, Louis Bachelier published the findings of his doctoral research on stock prices; his empirical results indicated that stock prices behaved like a random walk. However, this study was overlooked for the next 50 years. Then, in 1953, Maurice Kendall published his analysis of stock market prices in which he suggested that price changes were essentially random. Such a claim ran counter to the perceived wisdom of the times, but the empirical studies that followed confirmed Kendall's claim and ultimately led to the path-breaking work of Black and Scholes (1973) and Merton (1973) on the *Efficient Market Hypothesis*. In essence, the Black–Scholes theory states that prices will move randomly in an efficient market. Intuitively, we may argue that if prices were predictable, trading would quickly take place to erode the implied advantage. Of course, the theory does not apply to insider knowledge exploited by the few!

Why did it take so long for these ideas to take hold? The lack of empirical research is clearly part of the story, but there is an interesting statistical effect that also obscured the picture. Until the 1950s, many stock prices were published as daily averages, and it was not until the paper of Working (1960) that it was realized that such averaging induces spurious autocorrelations, and hence apparent predictability among successive observations. This finding led to the now standard practice of publishing prices at particular times, such as closing prices, rather than averages. Working's result is stated in Exercise 19.1.

We discuss the Black–Scholes model briefly in Sect. 19.1 and relate it to our analysis of discrete time processes. This development leads naturally to conditionally heteroscedastic processes, which is the subject of Sect. 19.2. Then, in Sect. 19.3 we examine time series that evolve over time in both their conditional mean and conditional variance structures. We conclude with a re-analysis of the US gasoline price data considered earlier in Chap. 9, which illustrates the value of conditionally heteroscedastic models in the construction of prediction intervals.

19.1 The Black–Scholes Model

Merton, Black and Scholes (hereafter MBS) were concerned with the valuation of options, typically the option to buy (sell) an asset at some pre-specified future date, known as a call (put) option. The value of such options depends upon the expected level of price *volatility*, typically measured by the variance of future returns. To take a trivial example, when this variance is zero, the future price is perfectly known, so we would have no interest in hedging against possible fluctuations. As the variance increases, holding the stock becomes more risky and our interest in risk-mitigating options increases.

As in earlier chapters, we denote the time series by y_t, $t = 1, 2, \ldots$. We assume that y_t is a martingale process so that $\mathrm{E}(y_t \mid y_1, \ldots, y_{t-1}) = y_{t-1}$. A random walk has the same conditional mean property, but the martingale assumption is more general and enables us to consider a non-constant variance later.

The basic MBS formulation rests upon a stochastic partial differential equation of the form:

$$\mathrm{d} \ln y_t = \mu \, \mathrm{d}t + \sigma \, \mathrm{d}W_t, \tag{19.1}$$

where μ denotes the rate of change in the mean, or the expected rate of return, also known as the *drift* in a general time series setting. The parameter σ^2, also denoted by v on occasion, is the rate of change in the variance. Thus W_t denotes a standard Wiener process, so that y_t follows a geometric Brownian motion. When we condition on the value of the asset at time zero, y_0, (19.1) yields a lognormal distribution for $y_t \mid y_0$, where the conditional mean of $\ln y_t$ is $\ln y_0 + t\mu$ and its conditional variance is $t\sigma^2$. When the parameters (μ, σ) are replaced by time-dependent terms (μ_t, σ_t) in (19.1), a lognormal distribution still results but the conditional moments become functions that must be integrated over time.

Once we allow the parameters to be time-dependent, we must describe how they evolve over time, and a stochastic description will usually be appropriate. Hull and White (1987) develop a time-dependent version of MBS:

$$\mathrm{d} \ln y_t = \mu \, \mathrm{d}t + \sigma_t \, \mathrm{d}W_{t,1}, \tag{19.2a}$$

$$\mathrm{d} \ln v_t = \eta \, \mathrm{d}t + \xi \, \mathrm{d}W_{t,2}, \tag{19.2b}$$

where $v_t = \sigma_t^2$ and $(\mathrm{d}W_{t,1}, \mathrm{d}W_{t,2})$ are, to use Hull and White's terminology, "possibly correlated" Wiener processes.

Our purpose here is not to describe the theory of options pricing, but rather to recognize the implications of this work for building forecasting models. Clearly, we can make the usual transition from continuous to discrete time by using differences in place of the differential elements. Also, we can extend model (19.2) by incorporating state equation(s) for the mean, as in earlier chapters. Moreover, we may go further and allow the Wiener processes to be (perfectly) correlated. Better yet, because we are interested in forecasting rather than an explicit theoretical solution, we may consider some (nonlinear) functional dependence between the processes. When we incorporate all of these considerations, we arrive at the model

$$\ln y_t = \ell_{t-1} + \varepsilon_t, \tag{19.3a}$$

$$\ell_t = \mu + \ell_{t-1} + \alpha\varepsilon_t, \tag{19.3b}$$

$$\ln v_{t+1} = u_0 + u_1 \ln v_t + u(\varepsilon_t), \tag{19.3c}$$

where $v_t = \sigma_t^2$ and $\varepsilon_t \sim \mathrm{N}(0, v_t)$.

We have selected the logarithmic form for the variance in (19.3c) for two reasons. First of all, the logarithmic form guarantees a positive value for the variance without placing constraints upon the parameters (u_0, u_1) or the function u. Second, this form tends to dampen the impact of extreme values of the error process, making it somewhat more robust. Although $\ln y_t$ is used in model (19.3) because of the MBS framework, the model is generally applicable to random variables measured in their original units, or any suitable transformation.

The function u is open to choice and the selection may well be application-specific. However, a reasonable choice is the logarithm of the absolute value of the error term. Because we are dealing with logarithms, this produces results equivalent to using the logarithm of the squared error. To avoid complications when the error is (close to) zero, we could add a small positive constant, say u_3, so that we have a function such as $u(\varepsilon_t) = u_2 \ln(|\varepsilon_t| + u_3)$.

In addition to the variance specification, there are many variations on the basic form of (19.3). We may include slope and seasonal state equations in the usual way or allow the variance to depend upon the state variables used to describe the mean.

19.2 Autoregressive Conditional Heteroscedastic Models

Model (19.3) shows the need to integrate the specification of the conditional variance into the overall framework. The first formulation of this type was the ARCH (Autoregressive Conditional Heteroscedastic) model proposed by Engle (1982). The ARCH models represent the conditional variance in a purely autoregressive way, and the GARCH model, proposed by Bollerslev (1986) and now generally preferred, may be thought of as an ARMA formulation, although the details are somewhat more involved. Both of these approaches model v_t directly, so that conditions need to be placed on the parameter space to ensure that the process remains positive. As with models of the mean, the question of stationarity is important in the ARMA world. Stationarity in the variance may be imposed upon a GARCH model, although nonstationary versions, known as integrated or IGARCH models, are also useful. This scheme is typically used only for first-order or local level variance models when it reduces to a local level model with drift.

Nelson (1991) proposed the exponential or EGARCH model based on $\ln v_t$; Nelson also incorporated a term to allow for asymmetric movements in asset returns, but we do not explore that extension here.

Since these early developments, a number of variations on the basic GARCH model have appeared in the literature, as researchers sought to incorporate the particular features associated with different types of asset. Tsay (2005, Chap. 3) provides an excellent guide to these developments. The main points for our discussion are that conditional heteroscedasticity is important and that such processes are readily incorporated into the innovations framework.

19.2.1 Estimation

Much of the work on asset prices assumes that the returns follow a martingale process so that, from the forecasting perspective, we need to worry only about the state equation for the variance. This reduction makes specification of the likelihood function very straightforward (cf. Tsay 2005, pp. 106–108). In the formulation of the Hull–White model (19.2), we noted that the Wiener processes were "possibly correlated"; thus, as a limiting case, we may assume perfect correlation and specify an innovations model. This approach enables us to use functionally related processes in the state equations like (19.3c). In turn, this step enables us to employ the estimation procedures developed in Chap. 5. The extension to heavy-tailed distributions such as Student's t (cf. Tsay 2005, p. 108) adds to the computational effort but presents no conceptual problems.

19.2.2 GARCH Model for the Dow Jones Index

We now illustrate the general method using the series of monthly closing prices for the Dow Jones Index (DJI) over the period January 1990 to March 2007. The series is plotted in Fig. 19.1.

We first fitted four variants of the local level model to the series, ignoring any changes in variance; these versions were: random walk (RW), local level (LL), random walk with drift (RWD) and local level with drift (LLD). We then incorporated the following GARCH-type model based upon (19.3):

$$\ln y_t = \ell_{t-1} + \varepsilon_t, \tag{19.4a}$$

$$\ell_t = \mu + \ell_{t-1} + \alpha\varepsilon_t, \tag{19.4b}$$

$$\ln v_{t+1} = u_0 + u_1 \ln v_t + u_2 \ln|\varepsilon_t|. \tag{19.4c}$$

The results are given in Table 19.1. As in earlier chapters, the AIC is used to select among models, with a penalty equal to twice the number of parameters. The AIC values given in Table 19.1 are adjusted by subtracting the

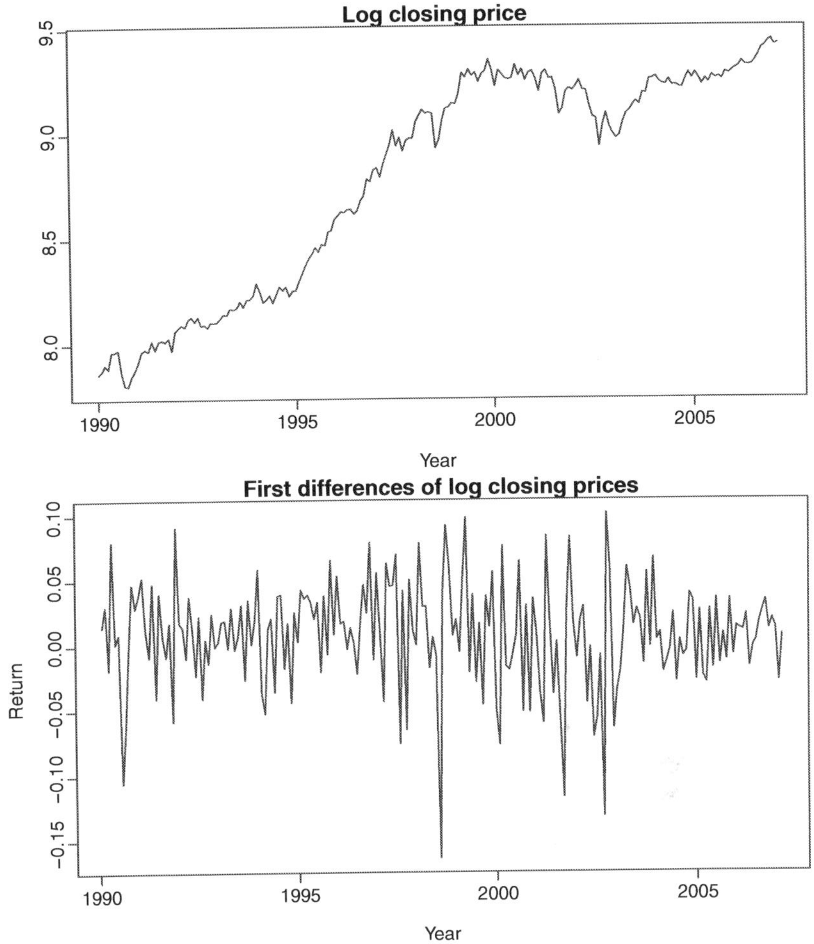

Fig. 19.1. Dow Jones Index: (**a**) logarithms of monthly closing prices (Jan 1990–Mar 2007); (**b**) returns for the same period.

value for the RW scheme so that comparisons are easier to make. The results are much as we would expect. The positive drift term reflects the average monthly return that an investor would receive if holding a portfolio that matched the DJI (with appropriate rebalancing by buying and selling stocks to ensure that the portfolio matched the DJI). Models with drift clearly outperform those without that term. In the LL models, the parameter α was allowed to range over $(0, 2)$. The resulting estimates of α were slightly less than 1, but there is no real evidence against the random walk hypothesis. Finally, we note that the GARCH-type model leads to an improvement in the

Table 19.1. Comparison of conditional heteroscedastic models for the Dow Jones Index, January 1990–March 2007.

Model	α	Drift	AIC
Constant variance			
Random walk (RW)	1.000	–	0.00
Local level (LL)	0.966	–	1.76
RW + drift	1.000	0.0076	−5.13
LL + drift	0.927	0.0076	−4.21
Heteroscedastic			
RW + GARCH	1.000	–	−2.60
LL + GARCH	0.965	–	−5.56
RW + drift + GARCH	1.000	0.0091	−17.73
LL + drift + GARCH	0.960	0.0074	−21.22

overall fit. In Exercise 19.3, the reader is encouraged to compare this model with a similar model for the DJI, without the logarithmic transformation.

The final form of the GARCH-type equation, corresponding to the LL + drift + GARCH model, is

$$\begin{aligned} \ln y_t &= \ell_{t-1} + \varepsilon_t, \\ \ell_t &= 0.0074 + \ell_{t-1} + 0.960\varepsilon_t, \\ \ln v_{t+1} &= 0.043 + 0.932 \ln v_t + 0.125 \ln |\varepsilon_t|. \end{aligned}$$

Replacing $|\varepsilon_t|$ by $(|\varepsilon_t| + u_3)$ for small values of u_3 produced no changes worthy of note.

19.3 Forecasting

We now revert to the more general notation used in earlier chapters, and use the measurement equation to define the one-step-ahead forecast as $y_{t+1|t} = \boldsymbol{w}'\boldsymbol{x}_t$. To this model, we add the equation for the one-step-ahead conditional variance

$$\ln v_{t+1} \equiv \ln v_{t+1|t} = u_0 + u_1 \ln v_t + u_2 \ln |\varepsilon_t|. \tag{19.6}$$

The forecast and conditional variance for multiple steps ahead may then be written as $y_{t+j|t} = \boldsymbol{w}'\boldsymbol{F}_{j-1}\boldsymbol{x}_t$ and

$$\ln v_{t+j+1|t} = u_0 + u_1 \ln v_{t+j|t} + u_2 \mathrm{E}(\ln |\varepsilon_{t+j}|), \; j = 1, 2, \ldots. \tag{19.7}$$

Because $\varepsilon_t \sim N(0, v_t)$, it may be shown that the conditional expectation of the logarithm of the absolute error term is $\mathrm{E}(\ln |\varepsilon_t|) = 0.5 \ln(v_t) - 0.635$ (see Abramowitz and Stegun 1964, p. 943). If u_3 is included in the model, the exact expectation is not available, but the bias will be modest when the constant is small, as is usually the case. The prediction of the error variance at time $t+j+1$ is then obtained iteratively from (19.6) and (19.7). An explicit solution appears in Exercise 19.2.

19.3.1 Gas Price Data Revisited

We return to the data on gasoline prices examined in Sect. 9.2.2. In addition to the constant variance models considered earlier, we also fitted GARCH-type models with and without the constant u_3. The results are given in Table 19.2. The full model, with all elements included, appears to be preferable based upon the AIC values.

The GARCH-type equation for the (full) fitted model is:

$$\ln v_t = 0.009 + 0.987 \ln v_{t-1} + 0.019 \ln |\varepsilon_t + 0.006|.$$

We observe that each of the models produces an estimate of α greater than 1, suggesting some persistence in the direction of price movements, but also implying that the innovations model provides a better fit than the multiple source of error version. We then generated the forecasts for the months January 2002–November 2006. In general, the point forecasts from the model with local level, regression and seasonal terms (LLRS) and those from the complete model were very close; 54 of 59 forecasts were within 3 cents of each other and the largest discrepancy was 5.5 cents. However, the picture was very different for the prediction intervals. We considered 90% intervals for the 59 observations. As reported in Table 19.3, the constant variance model had almost one-third of the observations outside the intervals, whereas the heteroscedastic model was right on target. Because options pricing and other decisions are based upon price volatility, the benefit from using conditional heteroscedastic models is clearly seen.

Table 19.2. Comparison of fitted models for the US gasoline price data, January 1991–December 2001.

Model	α	Spot price (1)	AIC
Local level + regression (LLR)	1.61	0.112	0.0
LLR + seasonal (LLRS)	1.49	0.144	−13.1
LLRS + GARCH	1.48	0.100	−29.9
LLRS + GARCH + u_3	1.49	0.145	−34.0

Table 19.3. Coverage of one-step-ahead prediction intervals for US gasoline price data, January 2002–November 2006.

Model	$A < L_{90}$	$A > U_{90}$	Total
LLR + seasonal (LLRS)	7	12	19
LLRS + GARCH + u_3	4	2	6

A = Actual, L_{90} = lower 90% limit, U_{90} = upper 90% limit

19.3.2 Stochastic Volatility

Although GARCH type models have had some measure of success, the search for better descriptions of inherent volatility has continued. In particular, Harvey et al. (1994) and others have introduced an additional stochastic element into the variance equation to create what is known as a *stochastic volatility* model. Thus, (19.3c) may be extended to:

$$\ln v_t = u_0 + u_1 \ln v_{t-1} + \eta_t,$$

where η_t is another Gaussian variable and is independent of ε_t. Parameter estimation procedures for such models lead to a considerable increase in computational complexity and we refer the interested reader to Tsay (2005, pp. 134–140). For some other recent developments in modeling heteroscedasticity, see Andersen et al. (2004).

19.4 Exercises

Exercise 19.1. Assume that the price of a stock follows a random walk, represented as:

$$y_t = \ell_0 + \varepsilon_1 + \varepsilon_2 + \cdots + \varepsilon_t.$$

Each day, the price is recorded at m equally spaced points in time and the m values for that day are averaged. That is, the average for successive days may be written as:

$$A_1 = (y_1 + y_2 + \cdots + y_m)/m,$$
$$A_2 = (y_{m+1} + y_{m+2} + \cdots + y_{2m})/m,$$

and so on. We now define the "change" in price by the differences $D_t = A_t - A_{t-1}$. If $\mathrm{V}(\varepsilon_t) = \omega$, show that $\mathrm{V}(D_t) = \omega(2m^2 + 1)/3m$ and $\mathrm{Cov}(D_t, D_{t-1}) = \omega(m^2 - 1)/6m$. Hence show that for large m, the first order autocorrelation between the differences approaches 0.25.

(Working 1960)

Exercise 19.2. Show that (19.6) yields the solution

$$\ln v_{t+j+1|t} = (u_0 - 0.635u_2)(1 + u_1 1 + \cdots + u_{11}^{j-1}) + u_{11}^{j} \ln v_{t+1|t},$$

where $u_{11} = u_1 + 0.5u_2$ and $j \geq 1$.

Exercise 19.3. Using the Dow Jones series `dji` described in Sect. 19.2.2, fit a model similar to (19.4) but without the logarithmic transformation on y_t. Compare your results with those in Table 19.1.

Exercise 19.4. Using the `gasprice` data set discussed in Sect. 19.3.1, transformed via logarithms, develop models similar to those in Table 19.2 and compare the results.

20

Economic Applications: The Beveridge–Nelson Decomposition

Co-authors: Chin Nam Low[1] and Heather M. Anderson[2]

Two features that characterize most macroeconomic time series are sustained long run growth and fluctuations around the growth path. These features are often called "trend" and "cycle" respectively, and the macroeconomic, econometric and statistical literatures contain a variety of techniques for decomposing economic time series into components that are roughly aligned with these notions. Most popular in the macroeconomic literature is the use of the Hodrick–Prescott (1980) filter for trend-cycle decomposition, followed by the Beveridge–Nelson (1981) decomposition, the decomposition implied by Harvey's (1985) unobserved component model, and a myriad of other contenders. Some, but not all, of these decomposition methods are based on statistical models, and there is vigorous debate about which of these methods leads to series that best capture the concepts of economic growth and business cycles. Canova (1998) provides an excellent survey of various methods that are used to decompose economic data, and he also outlines the motivational and quantitative differences between them. He is careful to point out that the "cycles" which result from statistical filters need not have a close correspondence with the classical ideas that underlie business cycle dating exercises undertaken by think-tanks such as the National Bureau of Economic Research in the USA. He also emphasizes that such a correspondence is not even desirable. Alternative decomposition techniques extract different types of information from the data, and each can be used to focus on different aspects of economic theory. Which filter is appropriate depends on the question at hand.

The Beveridge–Nelson (1981) (BN) decomposition developed from the observation that economic growth is not predictable, in the sense that it has an intrinsic stochastic component. Deterministic models of trend in economic output, often represented by polynomial functions of time, are

[1] Dr. Chin Nam Low, Aretae Pty Ltd, Singapore.

[2] Professor Heather Anderson, School of Economics, Australian National University, Australia.

then inappropriate indicators of "trend." Related work undertaken by Nelson and Plosser (1982) provided compelling empirical evidence that nearly all macroeconomic time series contain a (single) unit root with drift, and this finding had a profound impact on economists, first because stochastic trends are not predictable, and second because innovations to stochastic trends are not dampened but persist into the future. The innovations could be interpreted as technological advances, consistent with standard models of economic growth (as in Solow 1956). Innovations in this context could also be interpreted as fiscal or monetary policy shocks that had permanent long-run effects on the economy. Of particular importance was the implication that innovations to a stochastic trend had a persistent effect, and this, together with Friedman's (1957) concept of "permanent" and "transitory" components in income, led Beveridge and Nelson to develop their decomposition of macroeconomic time series into "permanent" and "transitory" components.

Economists use BN decompositions to provide measures of persistence in economic output (GDP). They also use the BN decomposition to date and predict various features of the business cycle. An interesting statistical aspect of this decomposition is that it implies perfect correlation between innovations to the permanent and transitory components, and this perfect correlation implies an innovations (single source of error) state-space representation. An interesting aspect of the "trend" and "cycle" interpretation of this decomposition is that the perfect correlation between the innovations to each component will imply that shocks to an economic variable will affect both trend and cycle.

Economists are often interested in the relative contributions of trend and cycle to the total variation in macroeconomic variables, and base their measures of relative contribution on BN components, so as to account for stochastic trends in the data. Empirically, the variation in the BN permanent component is usually very close to the total variation, and the contribution of the transitory component is usually very small. Stock and Watson (1988) discuss this finding, pointing out that economists need to recognize the substantial trend component in output, even if they are primarily interested in short term variation. In practice, the short term variation in the transitory component (cycle) is almost negligible, and its serial dependence properties are typically very weak.

Another interesting property of the BN decomposition arises from the inclusion of (positive) drift in the BN "trend." This drift ensures that economic growth (i.e., the BN trend) will be positive more often than it is negative, automatically creating an asymmetry in the growth process. This is consistent with the classical economic view of the business cycle that associates recessions with lower growth. It runs counter to a more statistical view that there are asymmetries in business cycles, but one can reconcile the apparent anomaly as a simple artifact of the use of different definitions.

Some economists view recessions as periods of low growth, while others view recessions as low points in cycles.

Macroeconomists often use "growth cycle" data to study asymmetries in economic time series, where growth cycles are taken to be the cycles discussed in Zarnowitz and Boschan (1977). Growth cycle data are constructed by taking the first differences (of the logarithms) of the raw time series, and the asymmetries are typically modeled using regime switching autoregressive specifications that allow the intercept and/or autoregressive parameters to change over time. Examples include Markov switching (MS) models (Hamilton 1989), threshold autoregressive (TAR) models (Potter 1995), and smooth transition autoregressive (STAR) models used by Teräsvirta and Anderson (1992). Most of these models allow for two regimes, and the regimes are loosely interpreted as recessionary and expansionary phases of the business cycle. Changes in the intercept correspond to changes in growth rates, while changes in the autoregressive parameters indicate changes in the short-run dynamic characteristics of growth. Low et al. (2006) show that it is possible to extend the BN innovations state space representation to allow for asymmetries in growth cycles, and this is briefly discussed at the end of the chapter.

The aim of this chapter is to familiarize readers with the BN decomposition of economic time series, because it provides an interesting application of the linear innovations state space approach to economics. The treatise provided here focusses on the decomposition of a single variable (in practice, this is usually the logarithm of the gross domestic product (GDP) of a country) into just two components. We do not consider more general linear innovations state space approaches that might explicitly account for seasonality, because economic theory has little to say about seasonal effects, and economists typically think in terms of seasonally adjusted data. We also do not consider local trend models, because there is very little empirical evidence that macroeconomic time series follow processes that are integrated of order two, and economic theory does not distinguish between different types of low frequency data. A state space framework that can be used to undertake the BN decomposition is outlined in Sect. 20.1. This framework is based on Anderson et al. (2006) and uses the perfect correlation between permanent and transitory components to recast the BN decomposition as an innovations state space model. It turns out that this state space approach avoids a computational problem associated with other techniques for estimating the BN permanent component (see Miller 1988; Newbold 1990; Morley 2002), and it also facilitates the direct estimation of various measures of "persistence in output."

In Sect. 20.2 we discuss the application of the state space approach to decomposing data from the USA, the UK and Australia. Some extensions to nonlinear processes are discussed in Sect. 20.3.

20.1 The Beveridge–Nelson Decomposition

The starting point for the Beveridge–Nelson (1981) decomposition is that most economic time series can be approximated by an ARIMA$(p,1,q)$ model. The permanent component of the ARIMA$(p,1,q)$ series is taken to be the limiting forecast of the series as the forecast horizon goes to infinity (given all history up to time t), and the transitory component is then the difference between the present observed value of the series and the permanent component. The definition of the permanent component embodies a clear focus on forecasting. Further, Beveridge and Nelson (1981) show that the BN transitory component consists of the forecastable momentum of the series at each point in time.

The original derivation of the BN decomposition of a time series y_t assumes that y_t is a linear $I(1)$ variable with a stationary Wold representation given

$$\Delta y_t = b + \gamma(L)\varepsilon_t, \tag{20.1}$$

where b is the long run growth or drift, $\gamma(L)$ is a polynomial in the lag operator L with $\gamma(0) = 1$ and $\sum_{i=0}^{\infty} |\gamma_i| < \infty$, and ε_t is the $\mathrm{IID}(0,\sigma^2)$ one-step-ahead forecast error of y_t. Economists are typically interested in $\gamma(1)$, the multiplier which measures the long-run effect of a shock ε_t on y_t. The 1981 implementation of the decomposition defined the permanent component as

$$\tau_t = \lim_{h\to\infty} \mathrm{E}(y_{t+h} - hb \mid y_t, y_{t-1}, \ldots y_1),$$

and then set the transitory component to be $c_t = y_t - \tau_t$. Beveridge and Nelson then showed that $\Delta\tau_t = b + \gamma(1)\varepsilon_t$, which implied that this permanent component followed a random walk with drift. They also showed that the transitory component was stationary.

The approach used by Anderson et al. (2006) assumes an ARIMA$(p,1,q)$ model with drift for y_t, so that

$$\gamma(L) = \frac{\theta_q(L)}{\phi_p(L)} = \frac{1-\theta_1 L - \theta_2 L^2 - \cdots - \theta_q L^q}{1-\phi_1 L - \phi_2 L^2 - \cdots - \phi_p L^p}.$$

The notation α is used to denote the long-run multiplier given by $\gamma(1) = \theta_q(1)/\phi_p(1)$. Equation (20.1) can then be written as

$$(1-L)y_t = b + \left(\frac{\theta_q(L)}{\phi_p(L)} - \alpha\right)\varepsilon_t + \alpha\varepsilon_t,$$

so that y_t can be decomposed into two components:

$$y_t = \left(\frac{b}{1-L} + \frac{\alpha\varepsilon_t}{1-L}\right) + \left(\frac{\theta_q(L) - \alpha\phi_p(L)}{(1-L)\phi_p(L)}\right)\varepsilon_t = \tau_t^* + c_t^*. \tag{20.2}$$

It is clear that $\Delta\tau_t^* = b + \alpha\varepsilon_t$, so that $\tau_t^* = \tau_t$, the BN permanent component. It follows that $c_t^* = c_t$, the BN transitory component. Therefore, we simply use the notation τ_t and c_t for these two components from now on. Note that the numerator of the c_t term in the decomposition (20.2) has a unit root by construction.[3] It is also useful to note that the expression for c_t in (20.2) measures the short-run effects of ε_t on y_t; that is, the difference between the total and long-run effects of ε_t on y_t.

Inspection of the two components in (20.2) shows that they are driven by the same innovation, so that innovations to τ_t and c_t are perfectly correlated. This perfect correlation is a by-product of the BN decomposition rather than an assumption, but its presence allows one to model the decomposition within a linear innovations state space framework.

We can rewrite the expression for τ_t as

$$\tau_t = b + \tau_{t-1} + \alpha\varepsilon_t, \tag{20.3}$$

which shows that the permanent component is a random walk with drift b and an uncorrelated innovation given by $\alpha\varepsilon_t$. Thus, what economists call the *permanent component* or a *stochastic trend* corresponds to a *local level with drift* model (see Sect. 3.5.2). We can also rewrite the expression for c_t as

$$c_t = \left(\frac{\theta_q(L) - \alpha\phi_p(L)}{(1-L)\phi_p(L)}\right)\varepsilon_t = \frac{\psi_r(L)}{\phi_p(L)}\varepsilon_t, \tag{20.4}$$

where $\psi_r(0) = 1 - \alpha$, and the order of $\psi_r(L)$ satisfies the condition $r \leq \max(p-1, q-1)$. Letting $\phi_p^*(L) = \phi_1 L + \phi_2 L^2 + \ldots + \phi_p L^p$ and $\psi_r^*(L) = \psi_1 L + \psi_2 L^2 + \ldots + \psi_r L^r$, the expression for the transitory component becomes

$$c_t = \phi_p^*(L)c_t - \psi_r^*(L)\varepsilon_t + (1-\alpha)\varepsilon_t. \tag{20.5}$$

Substituting (20.3) and (20.5) into $y_t = \tau_t + c_t$ leads to

$$y_t = b + \tau_{t-1} + \phi_p^*(L)c_t - \psi_r^*(L)\varepsilon_t + \varepsilon_t. \tag{20.6}$$

Equations (20.6), (20.3) and (20.5) are a complete representation of a difference-stationary time series. The associated lag polynomials can be of degree greater than one, so it is not a linear innovations state space representation. However, as shown in the next section, equations like this can always be recast into the first-order form associated with linear innovations state space models.

Following the convention of calling the permanent component of the BN decomposition "the trend" and the transitory component "the cycle," the parameter of interest in empirical studies of (the logarithms of) output is typically α, which measures the long run percentage increase in GDP resulting from a 1% shock in GDP in one quarter. This parameter is often used as a

[3] This is confirmed by substituting $L = 1$ into the numerator to give a value of zero.

measure of persistence (see Campbell and Mankiw 1987). It also determines the relative size of (contemporaneous) innovations to each component. In practice, if $\alpha < 1$ then the trend and cycle will have perfect positive correlation and both components will share in the variation of the data. However, if $\alpha > 1$, then the innovations in the trend and cycle will have perfect negative correlation, and the trend τ_t will be more variable than y_t. Some researchers, such as Proietti (2002), have questioned whether one should call τ_t a "trend" when it is more volatile than the output itself, but as was pointed out by Morley et al. (2003) (who observed that $\alpha > 1$ for real US GDP), a shock to output can shift the trend so that the output is behind the trend until it catches up. Thus, it is quite reasonable for "trend innovations" to be negatively correlated with "cycle innovations," and for the former innovations to be more variable than output innovations.

20.2 State Space Form and Applications

We now explore the use of the linear innovations state space approach for computing the BN permanent/transitory decompositions for ARIMA(0,1,1), ARIMA(1,1,0) and ARIMA(2,1,2) models of the logarithms of real output for the United States, the United Kingdom and Australia. The US models coincide with those used by Stock and Watson (1988) in their study of the contribution of the trend component to real US GNP, and the scope is broadened to include decompositions for the UK and Australia to demonstrate the relative contribution of trends in other countries.

In this study, we use quarterly GNP data for the USA (from 1947:1 to 2003:1), quarterly GDP data for the UK (from 1960:1 to 2003:1) and quarterly GDP data for Australia (from 1979:3 to 2003:3).[4] As noted above, the primary parameter of interest is Campbell and Mankiw's (1987) persistence measure α. Because researchers are often interested in the fraction of the variance of the quarterly change in real output that can be attributed to changes in its stochastic trend, the computed BN trends are used to calculate Stock and Watson's (1988) R^2 measure of this ratio. The empirical results are presented in Table 20.1, and the details relating to the innovations state space formulation are outlined below.

20.2.1 ARIMA(0,1,1) Model

The BN components of an ARIMA$(0,1,1)$ model are

$$\begin{aligned}\tau_t &= b + \tau_{t-1} + \alpha\varepsilon_t,\\ c_t &= (1-\alpha)\varepsilon_t,\end{aligned}$$

[4] The US data were obtained from the Federal Reserve Bank of St Louis, the UK data from the Office of National Statistics, and the Australian data from the Australian Bureau of Statistics. The data were transformed so that y_t is 100 times the natural logarithm of the GNP or GDP in each quarter.

Table 20.1. Measures of the importance of trend in real log GNP/GDP.

Univariate statistical model	Long-run change in GNP predicted from a 1% shock change in GNP in one quarter ($\hat{\alpha}$)	Variance ratios R^2
US GNP (1947:1–2003:1)		
ARIMA(0,1,1)	1.2701 (0.0552)	0.9339
ARIMA(1,1,0)	1.5226 (0.1464)	0.8817
ARIMA(2,1,2)	1.2653 (0.1459)	0.8458
UK GDP (1960:1–2003:1)		
ARIMA(0,1,1)	0.9945 (0.0724)	0.9999
ARIMA(1,1,0)	0.9940 (0.0759)	0.9999
ARIMA(2,1,2)	1.2267 (0.1587)	0.9686
Australia GDP (1979:1–2003:3)		
ARIMA(0,1,1)	1.3000 (0.0878)	0.9175
ARIMA(1,1,0)	1.4942 (0.0110)	0.8882
ARIMA(2,1,2)	1.3733 (0.0460)	0.8822

Standard errors are given in parentheses. The R^2 statistic is obtained by regressing the quarterly change in GNP against the change in the BN trend.

where, in terms of the ARMA coefficients for Δy_t, $\alpha = \gamma(1) = 1 - \theta_1$. These equations can be cast into innovations state space form with

$$y_t = b + \begin{bmatrix} 1 & 0 \end{bmatrix} \begin{bmatrix} \tau_{t-1} \\ c_{t-1} \end{bmatrix} + \varepsilon_t$$

as the measurement equation and

$$\begin{bmatrix} \tau_t \\ c_t \end{bmatrix} = \begin{bmatrix} b \\ 0 \end{bmatrix} + \begin{bmatrix} 1 & 0 \\ 0 & 0 \end{bmatrix} \begin{bmatrix} \tau_{t-1} \\ c_{t-1} \end{bmatrix} + \begin{bmatrix} \alpha \\ 1 - \alpha \end{bmatrix} \varepsilon_t$$

as the transition equation. Forecasts for these state space equations can be computed by using a suitable version of the information or Kalman filters, and the maximum likelihood estimates of the parameters (α and b) are obtained using the prediction error decomposition of the likelihood function. Note that it is α rather than the MA(1) parameter that is directly estimated.

The estimated αs and implied variance ratios for the USA, the UK and Australian output are shown in Table 20.1. Here it is interesting to note that

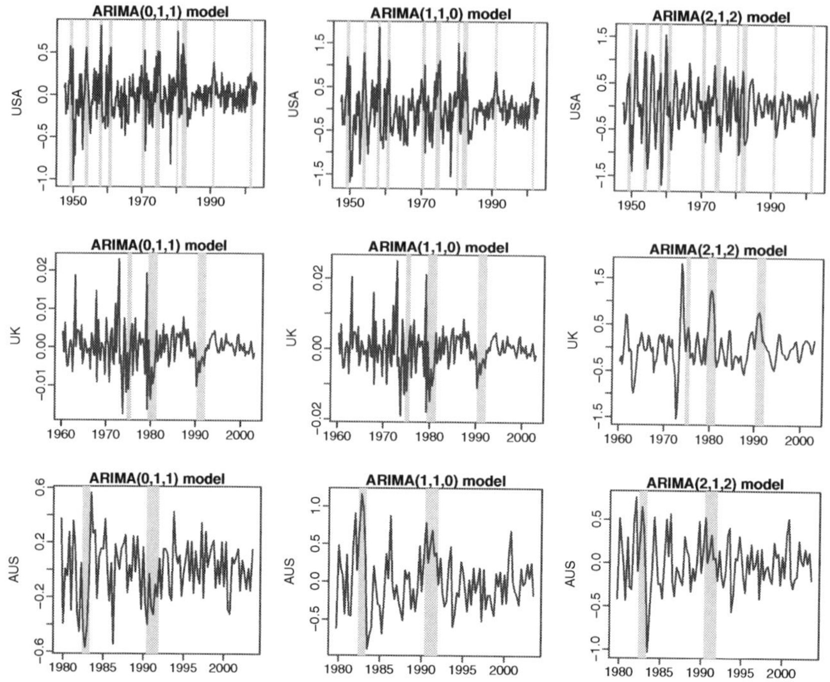

Fig. 20.1. Implied transitory components based on different ARIMA models for the USA, the UK and Australia. The *shaded areas* indicate peak-to-trough episodes (recessions) recorded by the NBER for the USA and by the ECRI for the UK and Australia.

while $\alpha > 1$ for the USA and Australia, implying that innovations to the "trend" and "cycle" are negatively correlated, the same is not true for the UK. In the case of the UK, the standard error associated with the parameter α suggests that the data are consistent with the possibility that $\alpha = 1$. Turning to the R^2 measures of the fraction of the variance in the quarterly change in real output that can be attributed to changes in its stochastic trend, it can be seen that the permanent component makes a relatively lower contribution in the USA and Australia than it does in the UK.

The implied transitory components are illustrated in the left hand side graphs in Fig. 20.1, together with reference recessions published by the NBER[5] and the ECRI.[6] While there are often pronounced declines in the transitory components around the NBER/ECRI peak-to-trough episodes, there are also clear differences between BN-cycles based on ARIMA(0,1,1) models of output and conventional business cycles. This is hardly surprising, given

[5] http://www.nber.org/cycles.html.

[6] http://www.businesscycle.com.

that the two types of cycle have been constructed to serve different purposes, and have been based on quite different information sets.

20.2.2 ARIMA(1,1,0) Model

For an ARIMA(1,1,0) model, the permanent trend component is the same as above, although in this case $\alpha = 1/(1-\phi_1)$ in terms of the ARMA coefficients for Δy_t. The cycle component is given by

$$c_t = \phi_1 c_{t-1} + (1-\alpha)\varepsilon_t.$$

Arranging the model into innovations state space form, the measurement equation is

$$y_t = b + \begin{bmatrix}1 \; \phi_1\end{bmatrix} \begin{bmatrix}\tau_{t-1} \\ c_{t-1}\end{bmatrix} + \varepsilon_t$$

and the transition equation is

$$\begin{bmatrix}\tau_t \\ c_t\end{bmatrix} = \begin{bmatrix}b \\ 0\end{bmatrix} + \begin{bmatrix}1 & 0 \\ 0 & \phi_1\end{bmatrix} \begin{bmatrix}\tau_{t-1} \\ c_{t-1}\end{bmatrix} + \begin{bmatrix}\alpha \\ 1-\alpha\end{bmatrix} \varepsilon_t.$$

Estimation of the innovations state space model imposes the identity that $\phi_1 = (\alpha-1)/\alpha$ (which arises from the observation that $\alpha = 1/(1-\phi_1)$) and provides a direct estimate of α. Results are provided in Table 20.1, and the implied transitory components are illustrated in the center graphs of Fig. 20.1. As for the ARIMA(0,1,1) model, $\alpha > 1$ for the USA and Australia, while $\alpha < 1$ for the UK. Again, because of the standard error of the estimate of α, it is conceivable that $\alpha = 1$, which suggests that UK GDP is largely governed by a random walk with drift. The implied R^2 values for the USA and Australia are much smaller than that for the UK, reflecting a comparatively more noisy transitory component in the former countries.

20.2.3 ARIMA(2,1,2) Model

The ARIMA(2,1,2) model of output has been used by Morley et al. (2003) for US GDP. If the focus is restricted to just ARIMA(0,1,1), ARIMA(1,1,0) and ARIMA(2,1,2) models, it is the model chosen with the AIC for both the USA and the UK. For Australia, however, the ARIMA(1,1,0) model is selected. As usual, the permanent component is given by (20.3), while the transitory component is given by

$$c_t = \phi_1 c_{t-1} + \phi_2 c_{t-2} - \psi_1 \varepsilon_{t-1} + (1-\alpha)\varepsilon_t.$$

In this case $\alpha = (1-\theta_1-\theta_2)/(1-\phi_1-\phi_2)$ in terms of the ARMA coefficients for Δy_t, although this relationship does not affect the following estimation.

The model can be cast into an innovations state space form with

$$y_t = b + \begin{bmatrix} 1 & \phi_1 & 1 \end{bmatrix} \begin{bmatrix} \tau_{t-1} \\ c_{t-1} \\ d_{t-1} \end{bmatrix} + \varepsilon_t$$

being the measurement equation, and

$$\begin{bmatrix} \tau_t \\ c_t \\ d_t \end{bmatrix} = \begin{bmatrix} b \\ 0 \\ 0 \end{bmatrix} + \begin{bmatrix} 1 & 0 & 0 \\ 0 & \phi_1 & 1 \\ 0 & \phi_2 & 0 \end{bmatrix} \begin{bmatrix} \tau_{t-1} \\ c_{t-1} \\ d_{t-1} \end{bmatrix} + \begin{bmatrix} \alpha \\ 1-\alpha \\ -\psi_1 \end{bmatrix} \varepsilon_t$$

being the transition equation.

Table 20.1 reports the estimation results and Fig. 20.1 illustrates the implied transitory components. In the case of the UK, $\hat{\theta}_1$ is statistically insignificant and is set to zero. The reported results are similar to those above. The estimated value of α for the UK is now greater than one, but its standard error suggests the possibility that its value could still be one. Once again, the results suggest that the permanent component in the US and Australian decompositions are relatively less volatile than the corresponding component in the UK decomposition.

20.3 Extensions of the Beveridge–Nelson Decomposition to Nonlinear Processes

The analysis in the previous two sections is based on the assumption that y_t is a linear process, so that all parameters are constant over time and the transitory component is symmetric around zero. This is inconsistent with observed asymmetries in business cycles, and it has led to a small literature on the decomposition of nonlinear time series that has attempted to address this problem.

Clarida and Taylor (2003) have proposed the simulation of the long-run forecast of a variable that follows a nonlinear data generating process. Given an estimated model of the data, the empirical conditional means of simulated long-run forecast densities based on the estimated parameters of this model will deliver the permanent component of the data, provided the data are integrated short-memory in mean. This implies that the long horizon forecast of the first differenced series is a constant (which rules out limit cycles, chaotic behavior and other time variations in the drift). The empirical conditional means of simulated long-horizon forecasts (adjusted for drift) are, of course, analogous to the original definition of the permanent component given by Beveridge and Nelson (1981), in which one had to evaluate the expected conditional mean of the series (adjusted for drift). As before, the transitory component is defined as the difference between the present value of the series and its permanent component.

A few researchers have worked with explicit models, and found direct ways of estimating the permanent-transitory decomposition. For instance, Chen and Tsay (2006) added an exogenous Markov switching (MS) mechanism to the drift in the permanent BN decomposition of GDP, so as to allow for lower growth rates of GDP during recessions. Rather than simulate the implied permanent component, they extended Newbold's (1990) procedure for BN decomposition to allow for regime switches in drift, and found that the transitory component became much less variable as a result.

Low et al. (2006) extended Chen and Tsay's (2006) model by allowing the parameters in both the permanent and transitory components of the decomposition to switch between regimes. As in Chen and Tsay (2006), the regime switches are generated by an exogenously determined MS process, and the regimes are interpreted as periods of recession and expansion. The extended model is given by $y_t = \tau_t + c_t$, with

$$\tau_t = b_{S_t} + \tau_{t-1} + \alpha_{S_t}\varepsilon_t \tag{20.7}$$

and

$$c_t = \phi^*_{p,S_t}(L)c_t - \psi^*_{r,S_t}(L)\varepsilon_t + (1 - \alpha_{S_t})\varepsilon_t, \tag{20.8}$$

in which the subscripted transition variable S_t can take just one of two discrete values at time t, so that the random parameters μ_{S_t}, $\phi^*_{p,S_t}(L)$, $\psi^*_{r,S_t}(L)$, and α_{S_t} all depend on S_t. The transition of S_t between its two regimes is driven by a probability transition matrix P, where

$$P = \begin{bmatrix} p_{11} \; p_{12} \\ p_{21} \; p_{22} \end{bmatrix}, \tag{20.9}$$

$p_{ij} = \Pr(S_t = j|S_{t-1} = i)$ and $p_{i1} + p_{i2} = 1$ for all i.

The innovation in y_t is again $\varepsilon_t \sim \text{IID}(0, \sigma^2)$, and this provides the single source of innovation. The variance σ^2 of ε_t is restricted to be constant in this model, although in principle it could depend on S_t without loss of identification. An example, expressed in innovations form, is the Markov switching version of an ARIMA(2,1,2) specification given by

$$y_t = b_{S_t} + \begin{bmatrix} 1 \; \phi_{1,S_t} \; 1 \end{bmatrix} \begin{bmatrix} \tau_{t-1} \\ c_{t-1} \\ d_{t-1} \end{bmatrix} + \varepsilon_t$$

as the measurement equation, and

$$\begin{bmatrix} \tau_t \\ c_t \\ d_t \end{bmatrix} = \begin{bmatrix} b_{S_t} \\ 0 \\ 0 \end{bmatrix} + \begin{bmatrix} 1 & 0 & 0 \\ 0 & \phi_{1,S_t} & 1 \\ 0 & \phi_{2,S_t} & 0 \end{bmatrix} \begin{bmatrix} \tau_{t-1} \\ c_{t-1} \\ d_{t-1} \end{bmatrix} + \begin{bmatrix} \alpha_{S_t} \\ 1 - \alpha_{S_t} \\ -\psi_{1,S_t} \end{bmatrix} \varepsilon_t$$

as the transition equation. The parameters in this model can be estimated using a maximum likelihood approach, replacing the standard Kalman filter used in Kim's (1994) approximation procedure for estimating the MS innovations state space models with a method similar to the Kalman filter described in Sect. 12.7. Low et al. (2006) estimate the above model, using the same US data as that in Sect. 20.2, and they find evidence of switches in both the permanent and transitory regimes. Growth in the expansionary regime is about 3.5%, compared with 2% in the recessionary regime, and the probability of staying in an expansionary regime is about 85%, compared with a probability of about 63% of staying in the recessionary regime. Perhaps one of the most interesting features of this model is that the long-run multiplier changes if there is switching in the transitory components, and the results presented in Low et al. (2006) suggest that this multiplier is 1.35 during expansions, but only 1.14 during recessions.

20.4 Conclusion

An advantage of the innovations state space approach is that it offers a simple and straightforward formulation of the permanent and transitory components of linear economic time series, and it allows direct study of the long-run multiplier α, a key parameter in macroeconomic analysis. Moreover, because it allows α to be estimated directly, it is then possible to obtain auxiliary statistics such as its standard error and t-statistic. Another benefit of the approach, as shown in Sect. 20.3, is that it can easily be adapted to deal with asymmetries in the data generating process. Furthermore, although this has not been studied here, it could be expanded to undertake multivariate decompositions. The interesting aspect of the latter suggestion is that standard real business cycle theories, as in King et al. (1988), assert that output, consumption and investment are all driven by a common unit root process trend, so that economic theory predicts that the permanent and transitory components of these three variables could all be estimated using a single source of error approach.

20.5 Exercises

The following exercises apply to the monthly copper price series in the data set `mcopper`.

Exercise 20.1. Fit a state space form of the Beveridge–Nelson model with an AR(1) transitory component to the logarithm of all but the last 2 years of the series. To simplify matters fix $\hat{c}_0 = 0$.

You should find the following two local minima:

Solution 1: $\hat{\tau}_0 = 5.529$; $\hat{\alpha} = 0$; $\hat{b} = 0.0038$; $\hat{\phi} = 0.9819$; $\hat{\sigma} = 0.0633$;

Solution 2: $\hat{\tau}_0 = 5.524$; $\hat{\alpha} = 1.37$; $\hat{b} = 0.0034$; $\hat{\phi} = -0.01$; $\hat{\sigma} = 0.0595$.

(Try using different starting values in order to find the two minima.) The first local minimum implies that the series is trend stationary; the second that it is difference stationary. The latter, where the series is dominated by a stochastic trend, has the lowest standard error.

Exercise 20.2. Use both fitted models to predict the series (in the original data space; not the log space) from the the *fixed* forecast origin 1 January 2005. Show that the MAPEs are 31.17% (solution 1) and 34.46% (solution 2). This confirms that a good fit does not guarantee good forecasts.

Exercise 20.3. Repeat Exercise 20.2 using a rolling origin from 1 January 2005. Show that the MAPEs based on the one-step-ahead prediction errors are 5.24 (solution 1) and 5.06 (solution 2). The second solution adapts slightly better to unexpected changes in the market.

Exercise 20.4. Assume that the time series consists of a stochastic trend and an AR(1) cycle according to the multiplicative model:

$$\begin{aligned} y_t &= \tau_{t-1}(1+b)(1+\phi c_{t-1})(1+\varepsilon_t), \\ \tau_t &= \tau_{t-1}(1+b)(1+\alpha\varepsilon_t), \\ c_t &= \phi c_{t-1} + (1-\alpha)\varepsilon_t. \end{aligned}$$

Fit this model to the original series (withholding the final 2 years of data as before) so as to minimize the augmented sum of squared errors (see 5.4). Compare the results with those from the previous exercises.

References

Pages on which each reference is cited are given in square brackets.

Abraham, B. and J. Ledolter (1983) *Statistical methods for forecasting*, Wiley, New York. [282]

Abramowitz, M. and I. A. Stegun (1964) *Handbook of mathematical functions with formulas, graphs, and mathematical tables*, vol. 55 of *NBS Applied Mathematics Series*, National Bureau of Standards, Washington, DC, tenth corrected printing ed. [322]

Agrawal, N. and S. A. Smith (1996) Estimating negative binomial demand for retail inventory management with unobservable lost sales, *Naval Research Logistics*, **43**(6), 839–861. [305]

Akaike, H. (1973) Information theory and an extension of the maximum likelihood principle, in B. N. Petrov and F. Csaki (eds.) *Second International Symposium on Information Theory*, pp. 267–281, Akdemiai Kiado, Budapest. [7]

Akaike, H. (1974) A new look at the statistical model identification, *IEEE Transactions on Automatic Control*, **19**, 716–723. [7, 106, 212]

Akaike, H. (1977) On entropy maximization principle, in P. R. Krishnaiah (ed.) *Applications of Statistics*, pp. 27–41, North Holland. [106]

Akram, M., R. J. Hyndman and J. K. Ord (2007) Non-linear exponential smoothing and positive data, Working paper 14/07, Department of Econometrics & Business Statistics, Monash University. [255]

Allen, P. G. and B. J. Morzuch (2006) Twenty-five years of progress, problems, and conflicting evidence in econometric forecasting. What about the next 25 years?, *International Journal of Forecasting*, **22**, 475–492. [294]

Andersen, T. G., T. Bollerslev and N. Meddahi (2004) Analytical evaluation of volatility forecasts, *International Economic Review*, **45**(4), 1079–1110. [324]

Anderson, T. W. (1958) *An introduction to multivariate statistical analysis*, Wiley, New York. [198]

Anderson, T. W. (1971) *The statistical analysis of time series*, Wiley, New York. [176]

Anderson, T. W. and D. A. Darling (1952) Asymptotic theory of certain "goodness of fit" criteria based on stochastic processes, *The Annals of Mathematical Statistics*, **23**(2), 193–212. [146]

Anderson, B. D. O. and J. B. Moore (1979) *Optimal filtering*, Prentice-Hall, Englewood Cliffs. [6, 7, 214, 219, 288]

Anderson, H. M., C. N. Low and R. D. Snyder (2006) Single source of error state-space approach to the Beveridge–Nelson decomposition, *Economics Letters*, **91**, 104–109. [327, 328]

Ansley, C. F. and R. Kohn (1985) A structured state space approach to computing the likelihood of an ARIMA process and its derivatives, *Journal of Statistical Computation & Simulation*, **21**, 135–169. [200, 214]

Aoki, M. (1987) *State space modeling of time series*, Springer, Berlin Heidelberg New York. [6, 212]

Aoki, M. and A. Havenner (1991) State space modelling of multiple time series, *Econometric Reviews*, **10**, 1–59. [7]

Archibald, B. C. (1984) Seasonal exponential smoothing models, Working paper 30/1984, School of Business Administration, Dalhousie University, Halifax, Canada. [158]

Archibald, B. C. (1990) Parameter space of the Holt-Winters' model, *International Journal of Forecasting*, **6**, 199–209. [16, 158]

Archibald, B. C. (1991) Invertible region of damped trend, seasonal, exponential smoothing model, Working paper 10/1991, School of Business Administration, Dalhousie University, Halifax, NS, Canada. [156, 158]

Archibald, B. C. and A. B. Koehler (2003) Normalization of seasonal factors in Winters' methods, *International Journal of Forecasting*, **19**, 143–148. [128]

Assimakopoulos, V. and K. Nikolopoulos (2000) The theta model: a decomposition approach to forecasting, *International Journal of Forecasting*, **16**, 521–530. [15, 48]

Athanasopoulos, G. and F. Vahid (2008a) A complete VARMA modelling methodology based on scalar components, *Journal of Time Series Analysis*, **29**(3), 533–554. [293]

Athanasopoulos, G. and F. Vahid (2008b) VARMA versus VAR for macroeconomic forecasting, *Journal of Business & Economic Statistics*, **26**(2), 237–252. [293]

Bachelier, L. (1900) Théorie de la spéculation, *Annales Scientifiques de l'École Normale Supérieure*, **3**(17), 21–86. [317]

Bell, W. R. (1984) Signal extraction for nonstationary time series, *The Annals of Statistics*, **12**(2), 646–664. [224]

Belsley, D. A., E. Kuh and R. E. Welsch (1980) *Regression diagnostics: identifying influential data and sources of collinearity*, Wiley, New York. [144]

Beveridge, S. and C. R. Nelson (1981) A new approach to decomposition of economic time series into permanent and transitory components with particular attention to measurement of the business cycle, *Journal of Monetary Economics*, **7**, 151–174. [325, 328, 334]

Billah, B., R. J. Hyndman and A. B. Koehler (2003) Empirical information criteria for time series forecasting model selection, Working paper 02/03, Department of Econometrics & Business Statistics, Monash University. [106, 107, 118]

Billah, B., R. J. Hyndman and A. B. Koehler (2005) Empirical information criteria for time series forecasting model selection, *Journal of Statistical Computation & Simulation*, **75**(10), 831–840. [107, 118]

Black, F. and M. Scholes (1973) The pricing of options and corporate liabilities, *Journal of Political Economy*, **81**(3), 637–654. [317]

Bollerslev, T. (1986) Generalized autoregressive conditional heteroscedasticity, *Journal of Econometrics*, **31**, 307–327. [319]

Bowerman, B. L., R. T. O'Connell and A. B. Koehler (2005) *Forecasting, time series and regression: an applied approach*, Thomson Brooks/Cole, Belmont CA. [16]

Bowman, K. O. and L. R. Shenton (1975) Omnibus test contours for departures from normality based on $\sqrt{b_1}$ and b_2, *Biometrika*, **62**(2), 243–250. [146]

Box, G. E. P. and D. R. Cox (1964) An analysis of transformations, *Journal of the Royal Statistical Society, Series B*, **26**(2), 211–252. [66]

Box, G. E. P. and G. M. Jenkins (1970) *Time series analysis: forecasting and control*, Holden-Day, San Francisco. [163]

Box, G. E. P., G. M. Jenkins and G. C. Reinsel (1994) *Time series analysis: forecasting and control*, Prentice-Hall, Englewood Cliffs, New Jersey, 3rd ed. [6, 36, 69, 141, 142, 167, 168, 215, 231]

Brockwell, P. J. and R. A. Davis (1991) *Time series: theory and methods*, Springer, Berlin Heidelberg New York, 2nd ed. [37]

Brown, R. G. (1959) *Statistical forecasting for inventory control*, McGraw-Hill, New York. [5, 13, 14, 41, 232, 277, 284, 303, 307]

Brown, R. G. (1963) *Smoothing, forecasting and prediction of discrete time series*, Prentice Hall, Englewood Cliffs, New Jersey. [5]

Burridge, P. and K. F. Wallis (1988) Prediction theory for autoregressive-moving average processes, *Econometric Reviews*, **7**(1), 65–95. [224]

Caines, P. and D. Mayne (1970) On the discrete time matrix Riccati equation of optimal control, *International Journal of Control*, **12**, 785–794. [215]

Caines, P. and D. Mayne (1971) On the discrete time matrix Riccati equation of optimal control: a correction, *International Journal of Control*, **14**, 205–207. [215]

Campbell, J. and N. G. Mankiw (1987) Permanent and transitory components in macroeconomic fluctuations, *The American Economic Review*, **77**, 111–117. [330]

Canova, F. (1998) Detrending and business cycle facts, *Journal of Monetary Economics*, **41**, 475–512. [325]

Chan, K. S. and J. Ledolter (1995) Monte Carlo EM estimation for time series models involving counts, *Journal of the American Statistical Association*, **90**(429), 242–252. [278]

Charnes, A. and W. W. Cooper (1962) Programming with linear fractional functionals, *Naval Research Logistics Quarterly*, **9**, 181–186. [227]

Chatfield, C. (1993) Calculating interval forecasts, *Journal of Business & Economic Statistics*, **11**, 121–135. [75]

Chatfield, C. and M. Yar (1991) Prediction intervals for multiplicative Holt-Winters, *International Journal of Forecasting*, **7**, 31–37. [84]

Chen, C. C. and W. J. Tsay (2006) The Beveridge–Nelson decomposition of Markov-switching processes, *Economics Letters*, **91**, 83–89. [335]

Clarida, R. H. and M. P. Taylor (2003) Nonlinear permanent-temporary decompositions in macroeconometrics and finance, *Economic Journal*, **113**, C125–C139. [334]

Cook, R. D. and S. Weisberg (1999) *Applied regression including computing and graphics*, Wiley, New York. [144]

Cottet, R. and M. Smith (2003) Bayesian modeling and forecasting of intraday electricity load, *Journal of the American Statistical Association*, **98**(464), 839–849. [240]

Croston, J. D. (1972) Forecasting and stock control for intermittent demands, *Operational Research Quarterly*, **23**(3), 289–304. [278, 281, 282, 284]

de Jong, P. (1991a) The diffuse Kalman filter, *The Annals of Statistics*, **19**, 1073–1083. [200, 214]

de Jong, P. (1991b) Stable algorithms for the state space model, *Journal of Time Series Analysis*, **12**, 143–157. [214]

de Silva, A., R. J. Hyndman and R. D. Snyder (2007) The vector innovation structural time series framework: a simple approach to multivariate forecasting, Working paper 3/07, Department of Econometrics & Business Statistics, Monash University. [287, 292, 296, 298, 299]

Doornik, J. A. and H. Hansen (1994) An omnibus test for univariate and multivariate normality, Working paper W4&91, Nuffield College, Oxford University. [146]

Duncan, D. B. and S. D. Horn (1972) Linear dynamic recursive estimation from the viewpoint of regression analysis, *Journal of the American Statistical Association*, **67**, 815–821. [7]

Durbin, J. and S. J. Koopman (2001) *Time series analysis by state space methods*, Oxford University Press, Oxford. [7]

Engle, R. F. (1982) Autoregressive conditional heteroscedasticity with estimates of the variance of United Kingdom inflation, *Econometrica*, **50**, 987–1007. [319]

Engle, R. F. and C. W. J. Granger (1987) Cointegration and error-correction: representation, estimation and testing, *Econometrica*, **55**, 251–276. [293]

Fildes, R. (1992) The evaluation of extrapolative forecasting methods, *International Journal of Forecasting*, **8**, 81–98. [49]

Friedman, M. (1957) *The theory of the consumption function*, Princeton University Press, Princeton, NJ. [326]

Gallant, A. R. (1987) *Nonlinear statistical methods*, Wiley, New York. [67]

Gardner, Jr, E. S. (1985) Exponential smoothing: the state of the art, *Journal of Forecasting*, **4**, 1–28. [12]

Gardner, Jr, E. S. (2006) Exponential smoothing: the state of the art – part II, *International Journal of Forecasting*, **22**(4), 637–666. [5]

Gardner, Jr, E. S. and E. McKenzie (1985) Forecasting trends in time series, *Management Science*, **31**(10), 1237–1246. [15, 290]

Gentleman, W. M. (1973) Least squares computations by Givens transformations without square roots, *IMA Journal of Applied Mathematics*, **12**(3), 329–336. [207]

Geweke, J. F. and R. A. Meese (1981) Estimating regression models of finite but unknown order, *International Economic Review*, **22**, 55–70. [107]

Gijbels, I., A. Pope and M. P. Wand (1999) Understanding exponential smoothing via kernel regression, *Journal of the Royal Statistical Society, Series B*, **61**(1), 39–50. [224]

Gilchrist, W. G. (1967) Methods of estimation involving discounting, *Journal of the Royal Statistical Society, Series B*, **29**(2), 355–369. [279]

Golub, G. H. and C. F. Van Loan (1996) *Matrix computations*, Johns Hopkins University Press, 3rd ed. [186, 205, 207]

Gould, P., A. B. Koehler, F. Vahid-Araghi, R. D. Snyder, J. K. Ord and R. J. Hyndman (2008) Forecasting time series with multiple seasonal patterns, *European Journal of Operational Research*, **191**(1), 205–220. [229]

Granger, C. W. J. and P. Newbold (1986) *Forecasting economic time series*, Academic Press, New York, 2nd ed. [215, 295]

Graves, S. C. (1999) A single-item inventory model for a nonstationary demand process, *Manufacturing & Service Operations Management*, **1**(1), 50–61. [90, 303, 311]

Grunwald, G. K., K. Hamza and R. J. Hyndman (1997) Some properties and generalizations of non-negative Bayesian time series models, *Journal of the Royal Statistical Society, Series B*, **59**, 615–626. [280]

Hamilton, J. D. (1989) A new approach to the economic analysis of nonstationary time series and the business cycle, *Econometrica*, **57**(2), 357–384. [327]

Hamilton, J. D. (1994) *Time series analysis*, Princeton University Press, Princeton, NJ. [67, 70]

Hannan, E. J. (1980) The estimation of the order of an ARMA process, *The Annals of Statistics*, **8**, 1071–1081. [107]

Hannan, E. J. and M. Deistler (1988) *The statistical theory of linear systems*, Wiley, New York. [6, 7, 150, 152, 219]

Hannan, E. J. and B. Quinn (1979) The determination of the order of an autoregression, *Journal of the Royal Statistical Society, Series B*, **41**(2), 190–195. [106, 107]

Harrison, P. J. (1967) Exponential smoothing and short-term sales forecasting, *Management Science*, **13**(11), 821–842. [82, 90]

Harrison, P. J. (1997) Convergence and the constant dynamic linear model, *Journal of Forecasting*, **16**(5), 287–292. [214]

Harrison, P. J. and C. F. Stevens (1976) Bayesian forecasting, *Journal of the Royal Statistical Society, Series B*, **38**(3), 205–247. [7]

Harvey, A. C. (1985) Trends and cycles in macroeconomic time series, *Journal of Business & Economic Statistics*, **3**, 216–237. [325]

Harvey, A. C. (1986) Analysis and generalisation of a multivariate exponential smoothing model, *Management Science*, **32**(3), 374–380. [289]

Harvey, A. C. (1989) *Forecasting, structural time series models and the Kalman filter*, Cambridge University Press, Cambridge. [7, 137, 144, 145, 147, 176, 197, 211, 212, 214, 231, 287, 289, 295]

Harvey, A. C. and C. Fernandes (1989) Time series models for count or qualitative observations, (with discussion) *Journal of Business & Economic Statistics*, **7**(4), 407–422. [278, 279]

Harvey, A. C. and S. J. Koopman (2000) Signal extraction and the formulation of unobserved components models, *Econometrics Journal*, **3**(1), 84–107. [223]

Harvey, A. C., E. Ruiz and N. Shephard (1994) Multivariate stochastic variance models, *Review of Economic Studies*, **61**, 247–264. [324]

Heinen, A. (2003) Modeling time series count data: an autoregressive conditional Poisson model, CORE Discussion Paper 2003/62, Center of Operations Research and Econometrics, Université catholique de Louvain. [278]

Heligman, L. and J. H. Pollard (1980) The age pattern of mortality, *Journal of the Institute of Actuaries*, **107**, 49–80. [325]

Hendry, D. F. (1995) *Dynamic econometrics*, Oxford University Press, Oxford. [294]

Hillmer, S. C. and G. C. Tiao (1982) An ARIMA-model-based approach to seasonal adjustment, *Journal of the American Statistical Association*, **77**, 63–70. [224]

Hodrick, R. and E. Prescott (1980) Post-war U.S. business cycles: An empirical investigation, Working paper, Carnegie Mellon University. [325]

Holt, C. C. (1957) Forecasting trends and seasonals by exponentially weighted averages, O.N.R. Memorandum 52/1957, Carnegie Institute of Technology. [5, 14, 15, 44]

Holt, C. C. (2004) Forecasting seasonals and trends by exponentially weighted moving averages, *International Journal of Forecasting*, **20**, 5–10. [5, 14]

Hull, J. and A. White (1987) Hedging the risks from writing foreign currency options, *Journal of International Money and Finance*, **6**, 131–152. [318]

Hurvich, C. M. and C. Tsai (1989) Regression and time series model selection in small samples, *Biometrika*, **76**, 297–307. [107]

Hyndman, R. J. (2001) It's time to move from 'what' to 'why' – comments on the M3-competition, *International Journal of Forecasting*, **17**(4), 567–570. [75]

Hyndman, R. J. (2004) The interaction between trend and seasonality, *International Journal of Forecasting*, **20**(4), 561–563. [11]

Hyndman, R. J. and M. Akram (2006) Some nonlinear exponential smoothing models are unstable, Working paper 3/06, Department of Econometrics & Business Statistics, Monash University. [255, 258, 262]

Hyndman, R. J., M. Akram and B. C. Archibald (2008) The admissible parameter space for exponential smoothing models, *Annals of the Institute of Statistical Mathematics*, **60**(2), 407–426. [149, 153, 156, 158]

Hyndman, R. J. and B. Billah (2003) Unmasking the Theta method, *International Journal of Forecasting*, **19**(2), 287–290. [15, 48]

Hyndman, R. J. and A. B. Koehler (2006) Another look at measures of forecast accuracy, *International Journal of Forecasting*, **22**, 679–688. [25, 26, 108]

Hyndman, R. J., A. B. Koehler, J. K. Ord and R. D. Snyder (2005) Prediction intervals for exponential smoothing using two new classes of state space models, *Journal of Forecasting*, **24**, 17–37. [77]

Hyndman, R. J., A. B. Koehler, R. D. Snyder and S. Grose (2002) A state space framework for automatic forecasting using exponential smoothing methods, *International Journal of Forecasting*, **18**(3), 439–454. [9, 12, 15, 23, 28, 72, 73, 78, 114, 256, 261]

Jarque, C. M. and A. K. Bera (1980) Efficient tests for normality, homoscedasticity and serial independence of regression residuals, *Economics Letters*, **6**(3), 255–259. [146]

Jazwinski, A. H. (1970) *Stochastic processes and filtering theory*, Academic Press, New York. [7]

Johansen, S. (1988) Statistical analysis of cointegration vectors, *Journal of Economic Dynamics & Control*, **12**, 231–254. [293]

Johnston, F. R. and J. E. Boylan (1996) Forecasting for items with intermittent demand, *Journal of the Operational Research Society*, **47**, 113–121. [281]

Johnston, F. R. and P. J. Harrison (1986) The variance of lead-time demand, *Journal of the Operational Research Society*, **37**(3), 303–308. [82, 90, 303]

Jung, R. C., M. Kukuk and R. Liesenfeld (2006) Time series of count data: modeling, estimation and diagnostics, *Computational Statistics & Data Analysis*, **51**, 2350–2364. [278]

Kalman, R. E. (1960) A new approach to linear filtering and prediction problem, *Journal of Basic Engineering*, **82**(1), 35–45. [7, 179]

Kalman, R. E. and R. S. Bucy (1961) New results in linear filtering and prediction theory, *Journal of Basic Engineering*, **83**(3), 95–108. [7]

Kendall, M. G. (1953) The analysis of economic time-series – part I: prices, *Journal of the Royal Statistical Society, Series A*, **116**(1), 11–34. [317]

Kim, C. J. (1994) Dynamic linear models with Markov-switching, *Journal of Econometrics*, **60**, 1–22. [336]

King, R. G., C. I. Plosser and S. Rebelo (1988) Production, growth and business cycles II: new directions, *Journal of Monetary Economics*, **21**, 309–341. [336]

Koehler, A. B., R. D. Snyder and J. K. Ord (2001) Forecasting models and prediction intervals for the multiplicative Holt-Winters' method, *International Journal of Forecasting*, **17**, 269–286. [84]

Koning, A. J., P. H. Franses, M. Hibon and H. O. Stekler (2005) The M3 competition: statistical tests of the results, *International Journal of Forecasting*, **21**(3), 397–409. [111]

Lawton, R. (1998) How should additive Holt-Winters' estimates be corrected?, *International Journal of Forecasting*, **14**, 393–403. [153]

Leeds, M. (2000) *Error structures for dynamic linear models: single source versus multiple source*, Ph.D. thesis, Department of Statistics, The Pennsylvania State University. [215, 221]

Lilliefors, H. W. (1967) On the Kolmogorov–Smirnov test for normality with mean and variance unknown, *Journal of the American Statistical Association*, **62**(318), 399–402. [146]

Ljung, G. M. and G. E. P. Box (1978) On a measure of lack of fit in time series models, *Biometrika*, **65**(2), 297–303. [145]

Low, C. N., H. M. Anderson and R. D. Snyder (2006) Beveridge–Nelson decomposition with Markov switching, Melbourne Institute Working Paper 14/06, The University of Melbourne. [327, 335, 336]

Lütkepohl, H. (2005) *New introduction to multiple time series analysis*, Springer, Berlin Heidelberg New York. [287, 298]

Makridakis, S. (1993) Accuracy measures: theoretical and practical concerns, *International Journal of Forecasting*, **9**, 527–529. [26]

Makridakis, S., A. Anderson, R. Carbone, R. Fildes, M. Hibon, R. Lewandowski, J. Newton, E. Parzen and R. Winkler (1982) The accuracy of extrapolation (time series) methods: results of a forecasting competition, *Journal of Forecasting*, **1**, 111–153. [28]

Makridakis, S. and M. Hibon (2000) The M3-competition: results, conclusions and implications, *International Journal of Forecasting*, **16**, 451–476. [26, 28, 105, 108, 109, 246]

Makridakis, S., S. C. Wheelwright and R. J. Hyndman (1998) *Forecasting: methods and applications*, Wiley, New York, 3rd ed. [11, 16, 71, 72, 78]

McClain, J. O. and L. J. Thomas (1973) Response-variance tradeoffs in adaptive forecasting, *Operations Research*, **21**, 554–568. [155]

McKenzie, E. (1976) A comparison of some standard seasonal forecasting systems, *The Statistician*, **25**(1), 3–14. [223]

McKenzie, E. (1986) Error analysis for Winters' additive seasonal forecasting system, *International Journal of Forecasting*, **2**, 373–382. [125]

Merton, R. C. (1973) Theory of rational option pricing, *Bell Journal of Economics & Management Science*, **4**(1), 141–183. [317]

Miller, M. (1988) The Beveridge–Nelson decomposition of economic time series: another economical computational method, *Journal of Monetary Economics*, **21**, 141–142. [327]

Morgan, F. (2005) *Real analysis and applications: including Fourier series and the calculus of variations*, American Mathematical Society, Providence, R.I. [97]

Morley, J. C. (2002) A state-space approach to calculating the Beveridge–Nelson decomposition, *Economics Letters*, **75**, 123–127. [327]

Morley, J. C., C. R. Nelson and E. Zivot (2003) Why are the Beveridge–Nelson and unobserved-components decompositions of GDP so different?, *The Review of Economics & Statistics*, **85**, 235–243. [330, 333]

Muth, J. F. (1960) Optimal properties of exponentially weighted forecasts, *Journal of the American Statistical Association*, **55**(290), 299–306. [6]

Nahmias, S. (1994) Demand estimation in lost sales inventory systems, *Naval Research Logistics*, **41**(6), 739–757. [305]

Nelson, C. R. and C. I. Plosser (1982) Trends and random walks in macroeconomic time series, *Journal of Monetary Economics*, **10**, 132–162. [326]

Nelson, D. B. (1991) Conditional heteroskedasticity in asset returns: a new approach, *Econometrica*, **59**(2), 347–370. [320]

Newbold, P. (1990) Precise and efficient computation of the Beveridge–Nelson decomposition of economic time series, *Journal of Monetary Economics*, **26**, 453–457. [327, 335]

Ord, J. K., A. B. Koehler and R. D. Snyder (1997) Estimation and prediction for a class of dynamic nonlinear statistical models, *Journal of the American Statistical Association*, **92**, 1621–1629. [7, 9, 16, 78, 79, 240]

Ord, J. K. and P. Young (2004) Estimating the impact of recent interventions on transportation indicators, *Journal of Transportation & Statistics*, **7**, 69–85. [148]

Ouwehand, P., R. J. Hyndman, T. G. de Kok and K. H. van Donselaar (2007) A state space model for exponential smoothing with group seasonality, Working paper 07/07, Department of Econometrics & Business Statistics, Monash University. [289]

Paige, C. C. and M. A. Saunders (1977) Least squares estimation of discrete linear dynamic systems using orthogonal transformations, *SIAM Journal on Numerical Analysis*, **14**(2), 180–193. [179]

Park, J. W., M. G. Genton and S. K. Ghosh (2007) Censored time series analysis with autoregressive moving average models, *Canadian Journal of Statistics*, **35**(1), 151–168. [305]

Pearlman, J. G. (1980) An algorithm for the exact likelihood of a high-order autoregressive-moving average process, *Biometrika*, **67**(1), 232–233. [174]

Pegels, C. C. (1969) Exponential forecasting: some new variations, *Management Science*, **15**(5), 311–315. [11]

Potter, S. (1995) A nonlinear approach to US GNP, *Journal of Applied Econometrics*, **10**(2), 109–125. [327]

Press, W. H., S. A. Teukolsky, W. T. Vetterling and B. P. Flannery (2002) *Numerical recipes in C++: the art of scientific computing*, Cambridge University Press, Cambridge, 2nd ed. [314]

Proietti, T. (2002) Forecasting with structural time series models, in Clements and Hendry (eds.) *A Companion to Economic Forecasting*, pp. 105–132, Prentice-Hall. [330]

Proietti, T. and A. C. Harvey (2000) A Beveridge–Nelson smoother, *Economics Letters*, **67**, 139–146. [224]

Ramanathan, R., R. F. Engle, C. W. J. Granger, F. Vahid-Araghi and C. Brace (1997) Short-run forecasts of electricity loads and peaks, *International Journal of Forecasting*, **13**, 161–174. [240]

Roberts, S. A. (1982) A general class of Holt-Winters type forecasting models, *Management Science*, **28**(8), 808–820. [125, 158]

Robinson, P. M. (1980) Estimation and forecasting for time series containing censored or missing observations, in O. D. Anderson (ed.) *Time Series*, pp. 167–182, North-Holland, Amsterdam. [305]

Schott, J. R. (2005) *Matrix analysis for statistics*, Wiley, Hoboken, NJ, 2nd ed. [86]

Schwarz, G. (1978) Estimating the dimension of a model, *The Annals of Statistics*, **6**, 461–464. [106, 107]

Schweppe, F. (1965) Evaluation of likelihood functions for Gaussian signals, *IEEE Transactions on Information Theory*, **11**, 61–70. [184, 213]

Shenstone, L. and R. J. Hyndman (2005) Stochastic models underlying Croston's method for intermittent demand forecasting, *Journal of Forecasting*, **24**, 389–402. [282]

Shibata, R. (1976) Selection of the order of an autoregressive model by Akaike's information criterion, *Biometrika*, **63**, 117–126. [107]

Shiryaev, A. N. (1984) *Probability*, Springer, Berlin Heidelberg New York. [262]

Silver, E. A., D. F. Pyke and R. Peterson (1998) *Inventory management and production planning and scheduling*, Wiley, New York, 3rd ed. [303]

Silverman, B. W. (1986) *Density estimation for statistics and data analysis*, Chapman and Hall, London. [78]

Sims, C. A. (1980) Macroeconomics and reality, *Econometrica*, **48**(1), 1–48. [293]

Snyder, R. D. (1980) The safety stock syndrome, *Journal of the Operational Research Society*, **31**, 833–837. [309]

Snyder, R. D. (2002) Forecasting sales of slow and fast moving inventories, *European Journal of Operational Research*, **140**, 684–699. [282]

Snyder, R. D. and C. S. Forbes (2003) Reconstructing the Kalman filter for stationary and non stationary time series, *Studies in Nonlinear Dynamics and Econometrics*, **7**(2), 1–18. [153, 200]

Snyder, R. D., A. B. Koehler, R. J. Hyndman and J. K. Ord (2004) Exponential smoothing models: means and variances for lead-time demand, *European Journal of Operational Research*, **158**(2), 444–455. [77, 90]

Snyder, R. D., A. B. Koehler and J. K. Ord (1999) Lead time demand for simple exponential smoothing: an adjustment factor for the standard deviation, *Journal of the Operational Research Society*, **50**, 1079–1082. [90, 303]

Snyder, R. D., G. M. Martin, P. Gould and P. D. Feigin (2008) An assessment of alternative state space models for count time series, Statistics and Probability Letters, forthcoming. [278]

Snyder, R. D., J. K. Ord and A. B. Koehler (2001) Prediction intervals for ARIMA models, *Journal of Business & Economic Statistics*, **19**(2), 217–225. [79]

Solow, R. M. (1956) A contribution to the theory of economic growth, *Quarterly Journal of Economics*, **70**, 65–94. [326]

Stirling, W. D. (1981) Least squares subject to linear constraints, *Journal of Applied Statistics*, **30**, 204–212. [186, 207]

Stock, J. H. and M. W. Watson (1988) Variable trends in economic time series, *Journal of Economic Perspectives*, **2**(3), 147–174. [326, 330]

Stuart, A. and J. K. Ord (1994) *Kendall's advanced theory of statistics. Vol. 1: distribution theory*, Hodder Arnold, London, 6th ed. [258, 263, 265, 267, 277]

Sugiura, N. (1978) Further analysis of the data by Akaike's information criterion and the finite corrections, *Communications in Statistics*, **A7**, 13–26. [106, 107]

Sweet, A. L. (1985) Computing the variance of the forecast error for the Holt-Winters seasonal models, *Journal of Forecasting*, **4**, 235–243. [153]

Syntetos, A. A. and J. E. Boylan (2001) On the bias of intermittent demand estimates, *International Journal of Production Economics*, **71**, 457–466. [281, 282]

Syntetos, A. A. and J. E. Boylan (2005) The accuracy of intermittent demand estimates, *International Journal of Forecasting*, **21**, 303–314. [281, 282]

Taylor, J. W. (2003a) Exponential smoothing with a damped multiplicative trend, *International Journal of Forecasting*, **19**, 715–725. [12, 64]

Taylor, J. W. (2003b) Short-term electricity demand forecasting using double seasonal exponential smoothing, *Journal of the Operational Research Society*, **54**, 799–805. [231, 232, 246]

Teräsvirta, T. and H. M. Anderson (1992) Characterizing nonlinearities in business cycles using smooth transition autoregressive models, *Journal of Applied Econometrics*, **7**, S119–S136. [327]

Tsay, R. S. (2005) *Analysis of financial time series*, Wiley, New York, 2nd ed. [142, 320, 324]

Watson, M. W. (2003) Macroeconomic forecasting using many predictors, in *Advances in economics and econometrics, theory and applications*, vol. 3, pp. 87–115, Eighth World Congress of the Econometric Society. [299]

West, M. and P. J. Harrison (1997) *Bayesian forecasting and dynamic models*, Springer, Berlin Heidelberg New York, 2nd ed. [7, 55, 212]

Willemain, T. R., C. N. Smart, J. H. Shockor and P. A. DeSautels (1994) Forecasting intermittent demand in manufacturing: a comparative evaluation of Croston's method, *International Journal of Forecasting*, **10**, 529–538. [282]

Williams, D. (1991) *Probability with martingales*, Cambridge University Press, Cambridge. [262]

Winters, P. R. (1960) Forecasting sales by exponentially weighted moving averages, *Management Science*, **6**, 324–342. [5, 15, 46, 230, 231]

Working, H. (1960) Note on the correlation of first differences of averages in a random chain, *Econometrica*, **28**, 916–918. [317, 324]

Yar, M. and C. Chatfield (1990) Prediction intervals for the Holt-Winters forecasting procedure, *International Journal of Forecasting*, **6**, 127–137. [83, 90]

Zarnowitz, V. and C. Boschan (1977) Cyclical indicators, in *57th annual report*, pp. 34–38, National Bureau of Economic Research. [327]

Zellner, A. (1962) An efficient method of estimating seemingly unrelated regressions and tests for aggregation bias, *Journal of the American Statistical Association*, **57**, 348–368. [289]

Author Index

Data Index

Subject Index

Springer Series in Statistics

Alho/Spencer: Statistical Demography and Forecasting
Andersen/Borgan/Gill/Keiding: Statistical Models Based on Counting Processes
Atkinson/Riani: Robust Diagnostic Regression Analysis
Atkinson/Riani/Ceriloi: Exploring Multivariate Data with the Forward Search
Berger: Statistical Decision Theory and Bayesian Analysis, 2nd edition
Borg/Groenen: Modern Multidimensional Scaling: Theory and Applications, 2nd edition
Brockwell/Davis: Time Series: Theory and Methods, 2nd edition
Bucklew: Introduction to Rare Event Simulation
Cappé/Moulines/Rydén: Inference in Hidden Markov Models
Chan/Tong: Chaos: A Statistical Perspective
Chen/Shao/Ibrahim: Monte Carlo Methods in Bayesian Computation
Coles: An Introduction to Statistical Modeling of Extreme Values
Devroye/Lugosi: Combinatorial Methods in Density Estimation
Diggle/Ribeiro: Model-based Geostatistics
Dudoit/Van der Laan: Multiple Testing Procedures with Applications to Genomics
Efromovich: Nonparametric Curve Estimation: Methods, Theory, and Applications
Eggermont/LaRiccia: Maximum Penalized Likelihood Estimation, Volume I: Density Estimation
Fahrmeir/Tutz: Multivariate Statistical Modeling Based on Generalized Linear Models, 2nd edition
Fan/Yao: Nonlinear Time Series: Nonparametric and Parametric Methods
Ferraty/Vieu: Nonparametric Functional Data Analysis: Theory and Practice
Ferreira/Lee: Multiscale Modeling: A Bayesian Perspective
Fienberg/Hoaglin: Selected Papers of Frederick Mosteller
Frühwirth-Schnatter: Finite Mixture and Markov Switching Models
Ghosh/Ramamoorthi: Bayesian Nonparametrics
Glaz/Naus/Wallenstein: Scan Statistics
Good: Permutation Tests: Parametric and Bootstrap Tests of Hypotheses, 3rd edition
Gu: Smoothing Spline ANOVA Models
Györfi/Kohler/Krzyżak/Walk: A Distribution-Free Theory of Nonparametric Regression
Hamada/Wilson/Reese/Martz: Bayesian Reliability
Harrell: Regression Modeling Strategies: With Applications to Linear Models, Logistic Regression, and Survival Analysis
Hart: Nonparametric Smoothing and Lack-of-Fit Tests
Hastie/Tibshirani/Friedman: The Elements of Statistical Learning: Data Mining, Inference, and Prediction
Hedayat/Sloane/Stufken: Orthogonal Arrays: Theory and Applications
Heyde: Quasi-Likelihood and its Application: A General Approach to Optimal Parameter Estimation
Huet/Bouvier/Poursat/Jolivet: Statistical Tools for Nonlinear Regression: A Practical Guide with S-PLUS and R Examples, 2nd edition
Hyndman/Koehler/Ord/Snyder: Forecasting with Exponential Smoothing. The State Space Approach
Iacus: Simulation and Inference for Stochastic Differential Equations: With R Examples

Ibrahim/Chen/Sinha: Bayesian Survival Analysis
Jiang: Linear and Generalized Linear Mixed Models and Their Applications
Jolliffe: Principal Component Analysis, 2nd edition
Konishi/Kitagawa: Information Criteria and Statistical Modeling
Knottnerus: Sample Survey Theory: Some Pythagorean Perspectives
Kosorok: Introduction to Empirical Processes and Semiparametric Inference
Küchler/Sørensen: Exponential Families of Stochastic Processes
Kutoyants: Statistical Inference for Ergodic Diffusion Processes
Lahiri: Resampling Methods for Dependent Data
Lavallée: Indirect Sampling
Le/Zidek: Statistical Analysis of Environmental Space-Time Processes
Liese/Miescke: Statistical Decision Theory: Estimation, Testing, Selection
Liu: Monte Carlo Strategies in Scientific Computing
Manski: Partial Identification of Probability Distributions
Mielke/Berry: Permutation Methods: A Distance Function Approach, 2nd edition
Molenberghs/Verbeke: Models for Discrete Longitudinal Data
Morris/Tibshirani: The Science of Bradley Efron, Selected Papers
Mukerjee/Wu: A Modern Theory of Factorial Designs
Nelsen: An Introduction to Copulas, 2nd edition
Pan/Fang: Growth Curve Models and Statistical Diagnostics
Politis/Romano/Wolf: Subsampling
Ramsay/Silverman: Applied Functional Data Analysis: Methods and Case Studies
Ramsay/Silverman: Functional Data Analysis, 2nd edition
Reinsel: Elements of Multivariate Time Series Analysis, 2nd edition
Rosenbaum: Observational Studies, 2nd edition
Rosenblatt: Gaussian and Non-Gaussian Linear Time Series and Random Fields
Särndal/Swensson/Wretman: Model Assisted Survey Sampling
Santner/Williams/Notz: The Design and Analysis of Computer Experiments
Schervish: Theory of Statistics
Shaked/Shanthikumar: Stochastic Orders
Simonoff: Smoothing Methods in Statistics
Song: Correlated Data Analysis: Modeling, Analytics, and Applications
Sprott: Statistical Inference in Science
Stein: Interpolation of Spatial Data: Some Theory for Kriging
Taniguchi/Kakizawa: Asymptotic Theory for Statistical Inference for Time Series
Tanner: Tools for Statistical Inference: Methods for the Exploration of Posterior Distributions and Likelihood Functions, 3rd edition
Tillé: Sampling Algorithms
Tsaitis: Semiparametric Theory and Missing Data
van der Laan/Robins: Unified Methods for Censored Longitudinal Data and Causality
van der Vaart/Wellner: Weak Convergence and Empirical Processes: With Applications to Statistics
Verbeke/Molenberghs: Linear Mixed Models for Longitudinal Data
Weerahandi: Exact Statistical Methods for Data Analysis

15503934R00202

Made in the USA
Lexington, KY
31 May 2012